RIVERSIDE COMMUNITY COLLEGE
1916

O9-CFT-202

Categories of Commutative Algebras

Categories of Commutative Algebras

YVES DIERS
Department of Mathematics
University of Valenciennes

CLARENDON PRESS · OXFORD
1992

s, Walton Street, Oxford OX2 6DP
New York Toronto
Calcutta Madras Karachi
Petaling Jaya Singapore Hong Kong Tokyo
Nairobi Dar es Salaam Cape Town
Melbourne Auckland
and associated companies in
Berlin Ibadan

Oxford is a trade mark of Oxford University Press

Published in the United States
by Oxford University Press, New York

A catalogue record for this book is available from the British Library

Library of Congress Cataloging in Publication Data
Diers, Yves, 1937–
Categories of commutative algebras / Yves Diers.
Includes index.
ISBN 0-19-853586-4
1. Commutative algebra. 2. Categories (Mathematics) I. Title.
QA251.3.D54 1992 512′.24—dc20 91-37359

Typeset by Colset Pte. Ltd., Singapore
Printed in Great Britain by
Biddles Ltd, Guildford & King's Lynn

PREFACE

This book is a study of the properties and structures of categories of commutative algebras. Its purpose is to bring out the universal properties of these categories and to provide a universal framework for commutative algebra and algebraic geometry. Its aim is to extend the domain of validity and application of the large number of results obtained in commutative algebra and algebraic geometry. The author takes the viewpoint that the concepts and constructions arising in commutative algebras are not bound so tightly to the universe of rings, but possess a universality that is independent of them and can be interpreted in various categories of discourse. The axiomatic method is used to deal with categories equipped with structure instead of with categories defined concretely by their objects and morphisms. This work contributes to the solution of the classification problem for categories, which is to characterize by means of universal properties any important class of categories and eventually any important category up to equivalence. In this respect it follows the same path as the work of A. Grothendieck on Abelian categories and categories of sheaves, F.W. Lawvere on algebraic categories and categories of sets, C. Ehresmann on categories of models, P. Gabriel and F. Ulmer on locally finitely presentable categories, and A. Kock on categories of differential manifolds.

I am grateful to the international community of categoricians for stimulation, co-operation, and friendship.

Valenciennes Y.D.
March 1992

CONTENTS

INTRODUCTION

The motivation for a study of the properties of categories of commutative algebras was the natural outcome of the publication of several papers developing techniques which mimic those of classical commutative algebral and algebraic geometry in areas such as the study of ordered algebra, lattice algebra, real algebra, differential algebra, and C^∞-algebra. For instance, G. W. Brumfiel's *Partially ordered rings and semi-algebraic geometry* [4] develops a kind of ordered commutative algebra, the paper by K. Keimel entitled 'The representation of lattice ordered groups and rings by sections in sheaves' [28] develops a kind of lattice-ordered commutative algebra, the paper by M. E. Alonso and M. F. Roy entitled 'Real strict localizations' [1] develops real commutative algebra, the paper by W. K. Keigher entitled 'Prime differential ideals in differential rings' [27] develops differential commutative algebra and the paper by I. Moerdijk and G. E. Reyes 'Rings of smooth functions and their localizations' [32] develops C^∞ commutative algebra. There are many other similar papers intent on extending the domain of interpretation of commutative algebra and transcending the particularity of ring structure. Thus the idea arose that the essence of commutative algebra is to be sought not in its calculus in commutative rings, but rather in the universal calculus in the category of commutative rings. It became natural to look for the universal properties in categories of commutative algebras which make commutative algebra work.

The so-called Zariski categories are precisely the categories equipped with such a structure on which a basic commutative algebra and algebraic geometry can be developed. This structure is completely defined by means of universal properties, so that the concepts, techniques, tools, and constructions of category theory can be used extensively. For example, all the usual constructions involving different kinds of rings are performed in any Zariski category as universal constructions, and properties of categories of affine schemes are obtained by applying the categorical duality principle, which simply consists in inverting the domain and codomain of morphisms and reversing composition.

The possibility of varying the domain of interpretation of commutative algebra afforded by category theory, by replacing the absolute category **CRng** of commutative rings by a plurality of Zariski categories, brings new flexibility to many concepts of classical commutative algebra. For example, the notion of schemes over an affine scheme **Spec**(A) is just the notion of schemes on the Zariski category **CAlg** (A) of commutative A-algebras, and reduced schemes are precisely schemes on the Zariski category **RedCRng** of reduced commutative rings, quasi-coherent modules are exactly schemes on the Zariski category **Mod** of all modules on variable commutative rings, quasi-coherent algebras are precisely schemes on the Zariski category **CAlg** of all commutative algebras on variable commutative rings, and locally boolean spaces are exactly schemes on the Zariski category **Bool** of boolean algebras. Similarly, algebraic varieties on an algebraically closed field k are precisely finitely presentable integral ultraschemes on the Zariski category **RedCAlg** (k) of commutative reduced k-algebras, while real algebraic varieties on a real closed field R are precisely finitely presentable integral ultraschemes on the Zariski category **RedRlAlg** (R) of reduced real R-algebras. Homogenous coordinates for projective algebraic varieties on k just make sense in the Zariski category **RedGradCAlg** (k) of reduced graded commutative k-algebras. Galois theory of extensions of a field k takes naturally its place in the Zariski category **AlgCAlg** (k) of algebraic commutative k-algebras.

The pluralism of Zariski categories is counterbalanced by the existence of a network of morphisms of Zariski categories which bring coherence to the system. Consider for example the three Zariski categories **CRng** of commutative rings, **RegCRng** of von Neumann regular commutative rings, and **NtshCRng** of neatish, i.e. ind-neat, commutative rings. The last two categories are Zariski categories of a special kind, called locally simple categories, in which simple objects such as fields play a central role. The category **RegCRng** is a reflective subcategory of **CRng**, i.e. the inclusion functor has a left adjoint. This inclusion functor is indeed a morphism of Zariski categories and **RegCRng** is the co-universal locally simple category associated to **CRng**. The category **NtshCRng** is a coreflective subcategory of **CRng**, i.e. the inclusion functor has a right adjoint called the coreflector. This coreflector is also a morphism of Zariski categories, and **NtshCRng** is the universal locally simple category associated to **CRng**. Morphisms of Zariski categories also provide a means of shifting from one Zariski category of discourse to another in order, for instance, to simplify the formulation of a problem. Consider as an example the notion of the real prime spectrum of an arbitrary commutative ring A. Let us consider the Zariski category **RlRng** of real rings. It is a reflective subcategory of **CRng** and the inclusion functor is a morphism of Zariski categories. If B is the real ring associated to A by the reflector, then the real prime spectrum of A is just the prime spectrum of B computed in the Zariski category **RlRng**. Similarly, the patch spectrum of A is the prime spectrum computed in the Zariski category **RegCRng** of the universal von Neumann regular ring associated to A, and the indecomposable spectrum of A is the prime spectrum computed in the Zariski category **NtshCRng** of the co-universal neatish ring associated to A.

What are the particular properties of categories of commutative algebras? Their algebraic nature is clear and well understood. For example, **CRng** and

CAlg (k) are algebraic categories in the sense of Lawvere, i.e. categories of sets equipped with an algebraic structure given by operations and identities. **RedGradCAlg** (k) and **AlgCAlg** (k) are algebraic categories in the Gabriel–Ulmer sense, i.e. categories of families of sets equipped with algebraic structure. They are complete and cocomplete categories i.e. any small diagram has a limit and a colimit. Indeed, all of them are locally finitely presentable categories in the Gabriel–Ulmer sense; i.e. cocomplete categories having a set of finitely presentable objects that generates the whole category. They are also regular categories in the Barr sense i.e. categories in which any morphism factors naturally into an epimorphism that is a co-equalizer of some pair of morphisms and is called a regular epimorphism, followed by a monomorphism. The initial object in categories of commutative algebras is never trivial. If, following the definition of an initial object, there is just one morphism starting from it, there are, in contrast, many morphisms arriving in it. After all, it is the ring of integers in **CRng**. The terminal object in categories of commutative algebras is strict, i.e. not only is there just one morphism arriving in it, following the definition of a terminal object, but also any morphism starting from it is an isomorphism. It is the zero ring in any category of commutative algebras. Categories of commutative algebras share with duals of categories of topological spaces and toposes the property that finite products of objects are codisjoint and co-universal, i.e. the pushout of two different projections is terminal and the pushout along any morphism is still a product.

The difficulty with categories concerns the calculus of objects of fractions: it is impossible to perform objects of fractions with the usual categorical constructions such as equalizers, products, pullbacks, co-equalizers, coproducts, pushouts, adjoint functors, etc. If it is true that a ring of fractions is a universal ring making a set of elements invertible, it is necessary to interpret the notion of invertible elements in an arbitrary category. What is the categorical nature of rings of fractions? It is indeed a universal construction, but it is different from the usual constructions, with independent existence and behaviour. The notion of codisjunctor has succeeded in providing a good description of objects of fractions and localization processes in categories of commutative algebras. This notion is definable in an arbitrary category and takes its place with the notions of equalizers, co-equalizers, pullbacks, etc. It has not been available previously, but is used extensively here. The dual notion is used in categories of schemes. For example, any pair $(f, g) : \mathrm{Spec}(A) \rightrightarrows \mathrm{Spec}(B)$ of parallel morphisms between affine schemes has a disjunctor: it is an open subscheme of $\mathrm{Spec}(A)$, which is not necessarily affine, and its underlying space is the complement of the underlying space of the equalizer of (f, g).

Codisjunctors are defined in an arbitrary category as follows. A pair of morphisms $(g, h) : C \rightrightarrows A$ is said to be codisjointed if any morphism $u : A \to X$ which co-equalizes (g, h) necessarily has a terminal object for its codomain. A morphism $f: A \to B$ is said to codisjoint a pair of morphisms $(g, h) : C \rightrightarrows A$ if the pair $(fg, fh) : C \rightrightarrows B$ is codisjointed. A codisjunctor of a pair of morphisms $(g, h) : C \rightrightarrows A$ is a universal morphism $d: A \to D$ which codisjoints (g, h). Whenever such a codisjunctor exists, the pair (g, h) is said to be codisjunctable. These notions apply to any relation on an object, and in particular to any equivalence relation. The basic example is the codisjunctor of the equivalence relation modulo a principal ideal a of a commutative ring A; it is the ring of fractions $A \to A[a^{-1}]$. A complete description of the calculus of codisjunctors in the category of commutative rings has been given in [12], and its extension to some other categories of algebras in [11]. It is immediately obvious that the behaviour of the forgetful functors and their adjoints with respect to codisjunctors is not the same as their behaviour with respect to limits or colimits.

Together with the notion of codisjunctor appear the notions of singular epimorphism and singular quotient object. A singular epimorphism is just a morphism which is the codisjunctor of some pair of morphisms or, equivalently, of some equivalence relation. The role of these epimorphisms will balance the well-known role of regular epimorphisms in algebraic categories. Epimorphisms which are simultaneously regular an singular are precisely the direct factor morphisms, i.e. projections of products of objects. The singular quotients of an object A are the quotient objects of A represented by singular epimorphisms. For commutative algebras they are in one-to-one correspondence with affine open sets of the prime spectrum of A, and the restriction morphisms of the structure sheaf of A along affine open subsets are precisely the codisjunctors of the codisjunctable equivalence relations on A. The presingular epimorphisms, a presingular factorization of any morphism, and a presingular hull for any object are derived from singular epimorphisms. Presingular epimorphisms include singular and semisingular epimorphisms, and they all belong to the important class of flat epimorphisms.

In an arbitrary Zariski category the traditional role of ideals in categories of commutative algebras is played by congruences. A congruence, also called an effective equivalence relation, is an internal equivalence relation in the category which is a kernel pair of some morphism. Different kinds of congruences appear: proper, maximal, prime, radical, irreducible, direct factor, rational, codisjunctable, etc. They satisfy the usual properties of ideals: for example, proper congruences are included in maximal congruences, maximal congruences are prime, etc.

The importance of flatness properties in Zariski categories is undeniable, but it turns out that a weak notion of flatness is sufficient for all purposes. An object A in a category $\mathbf{A}$ is said to be flat if the endofunctor $A \amalg (-)$ preserves monomorphisms, and a morphism $f: A \to B$ is flat if the pushout along f functor $A/\mathbf{A} \to B/\mathbf{A}$ preserves monomorphisms. This notion is weak in the sense that the previous functors need not preserve finite limits, i.e. are not exact. In any Zariski category codisjunctors are flat epimorphisms, but in some Zariski categories any object is flat, or any simple object is flat, or any morphism is flat.

The proof of the Chevalley theorem asserting that any finitely presentable morphism of schemes preserves constructible sets led to the discovery of another previously unknown universal construction in categories of commutative algebras: the calculus of terminators. The idea is similar to that of codisjunctors, but pushouts are used instead of co-equalizers. In an arbitrary category **A** with a strict terminal object denoted by 1, let $f: A \to B$ be a given morphism. A morphism $g: A \to C$ is said to terminate f if the pushout of (f, g) is terminal, i.e. the pushout object is 1. In categories of commutative algebras, a universal von Neumann regular object $n_c: C \to N(C)$ is associated to any object C, and it turns out that a morphism $g: A \to C$ terminates f if and only if the morphism $n_c g: A \to N(C)$ does. Thus, the study of morphisms terminating f can, to some extent, be restricted to those morphisms $g: A \to C$ which have a von Neumann regular codomain and are called von Neumann morphisms. A terminator of f is then defined as being a universal von Neumann morphism terminating f, i.e. a von Neumann morphism terminating f and such that any von Neumann morphism terminating f factors through it in a unique way. The existence of a terminator is not assured for any morphism, but it can be checked by using a 'Freyd theorem' with a finite terminating solution set condition. In the category **CRng**, any finitely presentable morphism has a terminator. This result is not obvious and is proved by noetherian induction. It is still valid in any category **CAlg** (k) of commutative algebras over some ring k, but not in an arbitrary Zariski category. In a locally noetherian Zariski category this property is equivalent to the fact that any finitely presentable integral extension has a terminator.

In any Zariski category **A** satisfying the amalgamation property for simple extensions, terminating morphisms and terminators are meaningful from the prime spectra point of view. A morphism $t: A \to T$ terminates a morphism $f: A \to B$ if and only if the images ImSpec(f) and ImSpec (t) of the maps Spec (f) : Spec $(B) \to$ Spec (A) and Spec (t) : Spect $(T) \to$ Spec (A) respectively are disjoint subsets in Spec (A). If t is the terminator of f, then ImSpec (f) and ImSpec (t) are complemented subsets in Spec (A). A morphism $f: A \to B$ has a terminator if and only if ImSpec (f) is a constructible subset of Spec (A). Furthermore, any constructible subset of Spec (A) is of the form ImSpec (f) for some morphism $f: A \to B$ having a terminator, and is also of the form ImSpec (t) for some terminator $t: A \to T$ of some morphism $f: A \to B$. Finally, the existence of a terminator for any finitely presentable morphism in **A** implies the Chevalley theorem. Two special kinds of morphism are also described: interminable morphisms and preterminal morphisms. From a spectra point of view, the former are such that the map Spec (f) is surjective, while the latter are such that the interior of ImSpec (f) is empty.

In any Zariski category we can define a whole range of objects which correspond to the usual types of rings. For example, simple rings, local rings, and etale rings can all be realized in this way, as can many other classes of ring which will be met in the text. An appropriate kind of morphism which allows the usual constructions of commutative algebras to be performed as universal constructions is associated to any kind of object. For example, are obtains the universal simple object associated to an integral object, the universal reduced object associated to an object, the universal simple object associated to a pseudosimple object, etc. An exception to this is that the construction of an algebraically closed extension of a simple object is not a universal construction.

The set of prime congruences on an object equipped with the Zariski topology, whose open sets are defined by congruences and whose affine open sets are defined by codisjunctable congruences, is the prime spectrum of the object. It is a spectral space in the sense of Hochster. The subspace of maximal congruences is the maximal spectrum of the object. For Jacobson objects, the maximal spectrum is very dense in the prime spectrum and thus it can replace it for many purpose. The patch spectrum of an object which carries the constructible sets is the co-universal boolean space associated to the prime spectrum. It is homeomorphic to the spectrum of the universal regular object associated to the object.

The localization process associates to any object a sheaf of local objects on its prime spectrum, while the globalization process rebuilds any object from the continuous family of its localized objects. By these processes, any object turns into a geometrical object—its affine scheme. By making these processes functorial, one obtains an equivalence between the category of affine schemes on a Zariski category **A** and the dual of **A**. As in commutative algebra, numerous properties of objects and morphisms in **A** have nice geometrical interpretations in the associated category of affine schemes. For example, an object in **A** is pseudosimple (quasi-primary, indecomposable, Jacobson, local) if and only if its affine scheme is a singleton (is irreducible, is non-empty connected, is Jacobson, has a unique closed point), and a morphism in **A** is a regular epimorphism (singular epimorphism, semisingular epimorphism, direct factor, premonomorphism) if and only if its affine spectrum is a closed immersion (open immersion, embedding, clopen immersion, dominant). Other properties can be localized: for example, an object is reduced (regular) if and only if its local objects are reduced (simple), and a morphism is flat if and only if its local morphisms are flat. These properties are essentially due to the fact that in a Zariski category the class of local objects is a strong cogenerating class.

Schemes on a Zariski category are built up by the classical method of gluing affine schemes together using modelled spaces and locally modelled spaces instead of ringed spaces and locally ringed spaces. Numerous classical properties and constructions of schemes are valid, finite limits and arbitrary coproducts of schemes exist, and the general gluing construction can be performed. Moreover, separated schemes are disjunctable; hence all disjunctors exist in the category of separated schemes, and they are open immersions. In particular, affine schemes are disjunctable and the open immersions into an affine scheme are precisely its singular subobjects. Jacobson schemes are obtained by gluing together affine schemes of Jacobson objects. For Jacobson objects and Jacobson schemes, the notion of ultrascheme is introduced in order to remove the generic points and deal only with the closed points.

The rational Zariski categories form a special class of Zariski categories, whose main feature is that the initial object K is simple and algebraically closed. The finitely presentable objects and locally finitely presentable schemes on such a category are Jacobson and so it is enough to deal with closed points. But these points turn out to be rational, i.e. their residue simple object is isomorphic to K: they are the usual points. Then the structure sheaf of a locally finitely presentable ultrascheme X can be canonically represented as a sheaf of K-valued functions on X, and this representation is faithful if and only if X is reduced.

The reduced rational Zariski categories are precisely the rational Zariski categories in which any object is reduced. For such a category, the category of locally finitely presentable ultraschemes becomes a concrete category, i.e. a category in which any object is an actual set equipped with some structure and any morphism is an actual map preserving the structure. Then one can develop the analogue of the classical sketch of algebraic geometry on an algebraic closed field. Algebraic spaces are defined as being separated finitely presentable ultraschemes. They are the objects of a concrete category and can be defined as sets equipped with a noetherian topology and a sheaf of regular functions on it. Any algebraic space is a union of finitely many irreducible algebraic spaces called algebraic varieties. Any algebraic variety X has its simple object $K(X)$ of rational functions on it in such a way that the values of its sheaf of regular functions are subobjects of $K(X)$. In addition to the morphisms and isomorphisms of algebraic varieties, rational morphisms and birational equivalences of algebraic varieties also exist. One can prove that two algebraic varieties are birationally equivalent if and only if their objects of rational functions are isomorphic. The category of affine algebraic spaces is equivalent to the dual of the category of finitely presentable objects, while the category of affine algebraic varieties is equivalent to the dual of the category of finitely presentable integral objects.

The category of schemes on a Zariski category can be fully and left-exactly embedded in a large topos—its large Zariski topos—while the category of locally finitely presentable schemes can be fully and left-exactly embedded in a topos—its Zariski topos. Modelled toposes and locally modelled toposes can be defined relative to a Zariski category. They efficiently play the role of ringed toposes and locally ringed toposes. The Zariski topos is the classifying topos for locally modelled toposes, and to any modelled topos is associated a universal locally modelled topos—its prime spectrum. In the case of spatial toposes, this last construction does not necessarily give the same result as the construction of the prime spectrum of a modelled space that we also perform.

Zariski categories are linked together by morphisms, geometrical morphisms, or cogeometrical morphisms. These morphisms preserve simple, integral, reduced, regular, and local objects, and their left-adjoints preserve finite products, codisjunctors, and preneat, neat, and neatish objects. Geometrical and cogeometrical morphisms extend naturally to the categories of schemes, modelled spaces, and modelled toposes, and so they provide relations between the different kinds of schemes, modelled spaces, and modelled toposes associated to different Zariski categories.

Zariski categories can be built up using general constructions of categories. Finite products of Zariski categories are Zariski categories. If **A** is a Zariski category, then the category $A/\mathbf{A}$ of objects of **A** under a fixed object A, the category $\mathbf{A}^{\mathbf{C}}$ of functors with values in **A** defined on a small category **C** with a strict initial object, the category $\mathbf{A}^{\rightarrow}$ of morphisms of **A**, the category **Sh** $[X, \mathbf{A}]$ of sheaves with values in **A** defined on a boolean space X, and the full subcategory of reduced (or regular, neatish, or etalish) objects of **A** are Zariski categories.

1
ZARISKI CATEGORIES

We use the axiomatic method, which deals with categories equipped with a specified structure rather than categories described concretely by their objects and their morphisms. The so-called Zariski categories have such a structure, and we claim here that it is the basic structure of categories of commutative algebras. We shall prove the validity of this claim by developing, in an arbitrary Zariski category, elementary commutative algebra and algebraic geometry.

The structure of Zariski categories relies on the notion of codisjunctors. This universal construction is introduced in order to describe the calculus of fractions and is used extensively. In Zariski categories, the congruences, also called effective equivalence relations, play the important role of ideals. A kind of De Morgan law for codisjunctors of congruences is assumed. Among the basic properties of Zariski categories are the flatness properties. We use the following definition: a morphism $f: A \to B$ is flat in $\mathbf{A}$ if the pushout functor $A/\mathbf{A} \to B/\mathbf{A}$ along f preserves monomorphisms. Codisjunctors are assumed to be flat morphisms.

1.1 CODISJUNCTORS

Let us be in an arbitrary category $\mathbf{A}$. A *terminal* object in $\mathbf{A}$ will be denoted by 1, i.e. for any object A, there is a unique morphism $A \to 1$. Such a terminal object is said to be *strict* [21] if any morphism $f: 1 \to A$ is an isomorphism.

Definition (1.1.1). *A pair of parallel morphisms* $(g, h) : C \rightrightarrows A$ *is said to be* codisjointed *if any morphism* $u : A \to X$ *which satisfies* $ug = uh$ *necessarily has, as its codomain, a terminal object.*

If the category $\mathbf{A}$ has no terminal object, a pair $(g, h) : C \rightrightarrows A$ is codisjointed if and only if there exists no morphism $u : A \to X$ satisfying $ug = uh$. If $\mathbf{A}$ has a terminal object 1, the existence of a pair of codisjointed morphisms implies that 1 is a strict terminal object. In this case, a pair $(g, h) : C \rightrightarrows A$ is codisjointed if and only if the morphism $A \to 1$ is the coequalizer of (g, h). In a category with a non-strict terminal object, for instance a null object, there exists no pair of codisjointed morphisms.

Definition (1.1.2).

(i) *A morphism* $f: A \to B$ codisjoints *a pair of morphisms* $(g, h): C \rightrightarrows A$, *if the pair* (fg, fh) *is codisjointed.*

(ii) A codisjunctor *of a pair of morphisms* $(g, h): C \rightrightarrows A$ *is a morphism* $f: A \to B$ *which codisjoints the pair* (g, h) *and is such that any morphism* $u: A \to X$ *which codisjoints* (g, h) *factors in a unique way through* f.

A codisjunctor of (g, h) is an epimorphism which is defined up to unique isomorphism.

Definition (1.1.3) *A pair of parallel morphisms is said to be* codisjunctable *if it admits a codisjunctor.*

Proposition (1.1.4). *A pair* $(g, h): C \rightrightarrows A$ *is codisjointed if and only if it has as a codisjunctor the identity* 1_A.

Proof. Immediate. ■

Proposition (1.1.5). *If* $(g, h): C \rightrightarrows A$ *and* $(u, v): E \rightrightarrows F$ *are pairs of morphisms and* $m: C \to E$, $n: A \to F$ *are morphisms such that* $um = ng$ *and* $vm = nh$, *then*

(i) (g, h) *is codisjointed* $\Rightarrow$ (u, v) *is codisjointed, and*

(ii) *If* m *is epimorphic and pushouts along* n *exist,*
(g, h) *is codisjunctable with codisjunctor* $f \Rightarrow (u, v)$ *is codisjunctable with codisjunctor the pushout of* f *along* n.

Proof.
(i) If (g, h) is codisjointed and $w: F \to W$ is a morphism such that $wu = wv$, then $wng = wum = wvm = wnh$. Thus $W \simeq 1$, and hence (u, v) is codisjointed.

(ii) Let $f: A \to B$ be the codisjunctor of (g, h) and let $(\bar{f}: F \to \bar{B}, \bar{n}: B \to \bar{B})$ be the pushout of (n, f). According to (i), $(\bar{f}u, \bar{f}v)$ is codisjointed since (fg, fh) is codisjointed. Thus $\bar{f}$ codisjoints (u, v). Let $w: F \to W$ be a morphism which codisjoints (u, v). We shall now show that wn codisjoints (g, h). Let $t: W \to T$ be a morphism satisfying $twng = twnh$. Then $twum = twng = twnh = twvm$, and since m is epimorphic, $twu = twv$. Also, since (wu, wv) is codisjointed, $T \simeq 1$. Thus (wng, wnh) is codisjointed, i.e. wn codisjoints (g, h). Consequently, there exists a unique morphism $k: B \to W$ such that $kf = wn$. If follows that there exists a unique morphism $r: \bar{B} \to W$ such that $r\bar{n} = k$ and $r\bar{f} = w$. As a result, $\bar{f}$ is the codisjunctor of (u, v). ■

Definition (1.1.6). *An object* A *is said to be* codisjunctable *if the coproduct of* A *with itself exists and the pair of inductions* $A \rightrightarrows A \amalg A$ *is codisjunctable.*

Proposition (1.1.7). *In a finitely cocomplete category, an object A is codisjunctable if and only if any pair of morphisms $(g, h) : A \rightrightarrows C$ with domain A is codisjunctable.*

Proof. Let A be a codisjunctable object and $d : A \amalg A \to D$ be the codisjunctor of the pair of inductions $(q, r) : A \rightrightarrows A \amalg A$. For a pair of morphisms $(g, h) : A \rightrightarrows C$, we shall denote by $k : A \amalg A \to C$ the morphism defined by $kq = g$ and $kr = h$. Let $(f : C \to B, e : D \to B)$ be the pushout of $(k : A \amalg A \to C,\ d : A \amalg A \to D)$. According to proposition (1.1.5), the morphism f is a codisjunctor of (g, h), and consequently the pair (g, h) is codisjunctable. Conversely, if any pair $(g, h) : A \rightrightarrows C$ is codisjunctable, in particular the pair $(q, r) : A \rightrightarrows A \amalg A$ is codisjunctable, then the object A is codisjunctable. ■

Corollary (1.1.8). *In a finitely cocomplete category any quotient object of a codisjunctable object is codisjunctable.*

Proof. Let A be a codisjunctable object and let $f : A \to B$ be an epimorphism. For any pair of morphisms $(g, h) : B \rightrightarrows C$, the codisjunctor $d : C \to D$ of (gf, hf) is also a codisjunctor of (g, h). The object B is therefore codisjunctable. ■

Codisjunctors of relations and congruences

Given an object A of $\mathbf{A}$, let us write $(p_1, p_2) : A \times A \rightrightarrows A$ for the product of A with itself. A relation on A is a subobject $r : R \to A \times A$ of $A \times A$ which can be identified with the pair of morphisms $(r_1, r_2) : R \rightrightarrows A$ defined by $r_1 = p_1 r$ and $r_2 = p_2 r$. Thus, Definitions (1.1.1), (1.1.2), and (1.1.3) can be applied to relations on A. A congruence on A is a relation on A which is the kernel pair of a morphism $f : A \to B$ ([19], § 5). Thus Definitions (1.1.1), (1.1.2), and (1.1.3) can be applied to congruences on A.

The notion of a codisjunctor of one pair of parallel morphisms extends immediately to a family of pairs of parallel morphisms with the same codomain in the following way.

Definition (1.1.9). *A* simultaneous codisjunctor *of a family of pairs of morphisms $((g_i, h_i) : C_i \rightrightarrows A)_{i \in I}$ is a morphism $f : A \to B$ which simultaneously codisjoints each pair of the family and is such that any morphism $u : A \to X$ which simultaneously codisjoints each pair of the family factors in a unique way through f.*

A simultaneous codisjunctor of $((g_i, h_i))_{i \in I}$ is an epimorphism which is defined up to a unique isomorphism. When it exists, the family $((g_i, h_i))_{i \in I}$ is said to be *simultaneously codisjunctable*. These notions also apply to families of relations on an object A and in particular to families of congruences on A.

Examples

Examples in the category **CRng** *of unitary commutative rings*

The zero ring $\{0\}$ is a strict terminal object. Let $(g,h):\mathbf{Z}[X]\rightrightarrows A$ be a pair of parallel morphisms. The coequalizer of (g,h) is the canonical morphism $A\to A/(a)$ with $a=g(X)-h(X)$. The pair (g,h) is codisjointed if and only if $A/(a)=\{0\}$, i.e. if and only if a is invertible. Thus a morphism $f:A\to B$ codisjoints (g,h) if and only if $f(a)$ is invertible. Therefore a codisjunctor of (g,h) is a universal morphism $f:A\to B$ which makes the element a invertible, i.e. it is the canonical morphism $A\to A[a^{-1}]$. Then any pair of morphisms $(g,h):\mathbf{Z}[X]\rightrightarrows A$ is codisjunctable, and so the object $\mathbf{Z}[X]$ is codisjunctable. An exhaustive study of codisjunctors in **CRng** can be found in [12]. It is easy to see that the canonical morphism $A\to A[S^{-1}]$, where S is a multiplicative subset in A, appears as a simultaneous codisjunctor.

Examples in the category **GradCRng** *of graded commutative rings*

The category **GradCRng** has, as objects, **Z**-graded commutative rings and, as morphisms, **Z**-graded homomorphisms of rings, i.e. homomorphisms preserving the graduation. Let $\mathbf{Z}[X]$ be the canonically graded ring of polynomials, and let $(g,h):\mathbf{Z}[X]\rightrightarrows A$ be a pair of parallel morphisms. Let $a=g(X)-h(X)$. Then a is a homogeneous element of degree 1, (a) is a homogeneous ideal of A, and $A/(a)$ is canonically graded. The coequalizer of (g,h) is the canonical morphism $A\to A/(a)$. Therefore (g,h) is codisjointed if and only if $A/(a)=\{0\}$, i.e. a is invertible. Thus a morphism $f:A\to B$ codisjoints (g,h) if and only if it makes the element a invertible. It follows that the codisjunctor of (g,h) is the canonical morphism of graded rings $A\to A[a^{-1}]$ where the graduation on $A[a^{-1}]$ has, as homogeneous elements of degree n, elements of the form b/a^p where b is a homogeneous element of degree $n+p$ in A. As a result, the graded ring $\mathbf{Z}[X]$ is codisjunctable in **GradCRng**.

Examples in the category **Mod** *of modules on variable commutative rings*

Mod is the category whose objects are ordered pairs (A,E) of a commutative ring A and an A-module E, and whose morphisms are pairs $(\varphi,f):(A,E)\to(B,F)$ of a ring morphism $\varphi:A\to B$ and a group morphism $f:E\to F$ which satisfies $f(ax)=\varphi(a)f(x)$ for any $(a,x)\in A\times E$. The zero module $(\{0\},\ \{0\})$ is a strict terminal object. Let us consider a pair of morphisms $((\psi, g), (\theta, h)) : (\mathbf{Z}[X], \mathbf{Z}[X]) \rightrightarrows (A, E)$. Let $a = \psi(X) - \theta(X)$. The coequalizer of this pair of morphisms is the canonical morphism $(A, E) \to (A/(a), A/(a) \otimes_A E)$. Thus this pair is codisjointed if and only if $A/(a) = \{0\}$, i.e. a is invertible. Hence a morphism $(\varphi, f) : (A, E) \to (B, F)$ codisjoints this pair if and only if the element $\varphi(a)$ is invertible. Consequently, if we denote the $A[a^{-1}]$ module $A[a^{-1}] \otimes_A E$ by $E[a^{-1}]$, and the canonical morphisms by $\varphi_a : A \to A[a^{-1}]$ and $f_a : E \to E[a^{-1}]$, the pair $(\varphi_a, f_a) : (A, E) \to (A[a^{-1}], E[a^{-1}])$ is a codisjunctor of the pair $((\psi, g), (\theta, h))$. As a result, the $\mathbf{Z}[X]$-module $\mathbf{Z}[X]$ is a codisjunctable object in **Mod**.

*Examples in the category **RlRng** of real rings*

RlRng is the full subcategory of **CRng** whose objects are the rings satisfying the following properties:

$$(x_1^2 + \dots + x_n^2)y = 0 \Rightarrow x_1^2 y = 0 \tag{1.1.10}$$

$$1 + x_1^2 + \dots + x_n^2 \text{ is invertible.} \tag{1.1.11}$$

Here, these rings are called real rings. The real field $\mathbf{Q}$ is the initial object while the real zero ring $\{0\}$ is the terminal object. Let A be a real ring. We call an ideal I of A real if it is such that $(x_1^2 + \dots + x_n^2)y \in I \Rightarrow x_1^2 y \in I$. These are precisely the ideals of A such that the quotient ring A/I is real. Any maximal ideal I of A is real since the quotient field A/I obviously satisfies the second property of real rings and thus also satisfies the first property. It follows that for an ideal I of A, whose real hull is denoted by ${}^R I$, we have $I = A \Leftrightarrow {}^R I = A$. Let us prove by induction on n that for an ideal I of A generated by n elements $a_1, \dots, a_n$, we have that $I = A \Leftrightarrow a_1^2 + \dots + a_n^2$ is invertible. The condition is sufficient since $a_1^2 + \dots + a_n^2 \in I$. Let us suppose that $I = A$. There exist $u_1, \dots, u_n \in A$ such that $u_1 a_1 + \dots + u_n a_n = 1$. Let $b = u_{n-1} a_{n-1} + u_n a_n$. Then $u_1 a_1 + \dots + u_{n-2} a_{n-2} + b = 1$. Thus $(a_1, \dots, a_{n-2}, b) = A$. By the induction hypothesis $a_1^2 + \dots + a_{n-2}^2 + b^2$ is invertible. It follows from the second axiom of real rings that any element of A of the form $a_1^2 + \dots + a_{n-2}^2 + b^2 + \text{(sum of squares)}$ is invertible. Consequently, the element

$$\begin{aligned}
&(a_1^2 + \dots + a_n^2)(1 + 2u_{n-1}^2 + 2u_n^2) \\
&= a_1^2 + \dots + a_{n-2}^2 + 2a_{n-1}^2 u_{n-1}^2 + 2a_n^2 u_n^2 + \text{(sum of squares)} \\
&= a_1^2 + \dots + a_{n-2}^2 + (a_{n-1}u_{n-1} + a_n u_n)^2 + (a_{n-1}u_{n-1} - a_n u_n)^2 \\
&+ \text{(sum of squares)} = a_1^2 + \dots + a_{n-2}^2 + b^2 + \text{(sum of squares)}
\end{aligned}$$

is invertible in A. Hence $a_1^2 + \dots + a_n^2$ is invertible in A.

Let us consider a finitely generated real ring R and a pair of parallel morphisms $(g, h) : R \rightrightarrows A$. Let $(x_i)_{i \in [1,n]}$ be a set of generators in R, let

$a_i = g(x_i) - h(x_i)$ for any $i \in [1, n]$, let $I = (a_1, \ldots, a_n)$, and let $^R I$ be the real hull of I. The coequalizer of (g, h) is the canonical morphism $A \to A/^R I$. Consequently, we have that (g, h) is codisjointed $\Leftrightarrow A/^R I = \{0\}$ $\Leftrightarrow {}^R I = A \Leftrightarrow I = A \Leftrightarrow a_1^2 + \ldots + a_n^2$ is invertible. It follows that a morphism $f: A \to B$ codisjoints (g, h) if and only if it makes the element $a_1^2 + \ldots + a_n^2$ of A invertible in B, and that a codisjunctor of (g, h) is a universal morphism of this type. Let us consider the multiplicative subset of A defined by

$$S = \{ (a_1^2 + \ldots + a_n^2)^p + x_1^2 + \ldots + x_m^2 ; p, m \in \mathbb{N} \quad \text{and} \quad x_1, \ldots, x_m \in A \}.$$

The ring of fractions $A[S^{-1}]$ is real and the canonical morphism $A \to A[S^{-1}]$ is the codisjunctor of (g, h). As a result any finitely generated real ring is a codisjunctable object in **RlRng**.

*Examples in the category **RlOrdRng** of real ordered rings*

RlOrdRng is the category whose objects are the unitary commutative ordered rings satisfying the axioms

$$x^2 \geqslant 0 \tag{1.1.12}$$

$$(0 \leqslant x \leqslant y \quad \text{and} \quad yz = 0) \Rightarrow xz = 0 \tag{1.1.13}$$

$$1 \leqslant x \Rightarrow x \text{ is invertible,} \tag{1.1.14}$$

and whose morphisms are order-preserving morphisms of unitary rings. Here, these rings are called real ordered rings. The ordered field $\mathbf{Q}$ is the initial object, and the ordered zero ring is the terminal object. Let A be a real ordered ring. According to [3([4], § 2.5), 2.5] an ideal I of A is absolutely convex if it is such that $(0 \leqslant x \leqslant y$ and $zy \in I) \Rightarrow zx \in I$. These ideals are precisely the ideals I such that the quotient ring A/I is real ordered by the order defined by $\bar{a} \geqslant 0 \Leftrightarrow (\exists x \geqslant 0, \bar{x} = \bar{a})$. Any ideal I of A has an absolute hull [4, § 2.5] denoted by ^{AC}I. Let us suppose that $^{AC}I = A$. Then $\mathrm{rad}(^{AC}I) = A$ and, according to ([4], Propositions 2.2.3, 2.5.2), there are elements $x_1, \ldots, x_n \in I$ and elements $y_1, \ldots, y_n \in A$ such that $1 \leqslant x_1 y_1 + \ldots + x_n y_n$, and hence such that $x_1 y_1 + \ldots + x_n y_n$ is invertible. Then $I = A$. As a result, $^{AC}I = A \Leftrightarrow I = A$.

Let us consider a finitely generated real ordered ring R and a pair of parallel morphisms $(g, h) : R \rightrightarrows A$. Let $(x_i)_{i \in [1,n]}$ be a set of generators in R, $a_i = g(x_i) - h(x_i)$ for each $i \in [1, n]$ and $I = (a_1, \ldots, a_n)$. The coequalizer of (g, h) is the canonical morphism $A \to A/_{\mathrm{AC}_I}$. Therefore (g, h) is codisjointed $\Leftrightarrow A/_{\mathrm{AC}_I} = \{0\} \Leftrightarrow {}^{\mathrm{AC}}I = A \Leftrightarrow I = A \Leftrightarrow a_1^2 + \ldots + a_n^2$ is invertible (according to the results of the previous example). It follows that a morphism $f: A \to B$ codisjoints (g, h) if and only if it inverts the element $a_1^2 + \ldots + a_n^2$. Let S be the multiplicative subset of A defined by $S = \{x \in A : \exists p \in \mathbb{N},\ (a_1^2 + \ldots + a_n^2)^p \leqslant x\}$. The ring of fractions $A[S^{-1}]$ with the order defined by $a/s \geqslant 0 \Leftrightarrow \exists t \in S,\ ast^2 \geqslant 0$ is a real ordered ring ([4], Propositions 3.1.1, 3.8.1, 3.8.3). Then the canonical morphism $A \to A[S^{-1}]$ is the codisjunctor of (g, h). As a result, any finitely generated real ordered ring is a codisjunctable object in **RlOrdRng**.

1.2 DEFINITION OF ZARISKI CATEGORIES

Notation

We shall use the results of Gabriel and Ulmer [16] on locally finitely presentable categories, the results of Barr [2] and Grillet [19] on regular categories, the notion of codisjunctors introduced by Diers [12] and reviewed in 1.1. and the following notation. Let $\mathbf{A}$ be a locally finitely presentable category and A an object in $\mathbf{A}$. The set $\mathrm{Rel}(A)$ of relations on A is a complete lattice whose meets $\wedge$ are called intersections and whose joins $\vee$ are called unions [17; 31, p. 122]. The set $\mathrm{Cong}(A)$ of congruences on A is a subset of $\mathrm{Rel}(A)$ closed under intersections. Thus it is a complete lattice whose meets $\wedge$ are intersections, but whose joins $\overset{c}{\vee}$ need not be unions. On the other hand, the set $\mathrm{Epi}(A)$ of epimorphisms with domain A is a complete prelattice, and the set $\mathrm{Quot}(A)$ of quotient objects of A is a complete lattice whose meets $\wedge$ are called co-intersections and whose joins $\vee$ are called co-unions ([35], § 14.2). Following Barr [2], a *regular epimorphism* is an epimorphism which is the coequalizer of some pair of morphisms, a regular epimorphism is *universal* if its pullback along any morphism is still a regular epimorphism and a category is said to be *fully supported* if any morphism whose codomain is the terminal object 1 is a regular epimorphism. For any pair of morphisms $(f: A \to B,\ g: A \to C)$, whose pushout is the pair $(h: B \to D,\ k: C \to D)$, the morphism h will be called *the pushout of g along f*. By pushing out along a fixed morphism $f: A \to B$, we obtain the *pushout functor $A/\mathbf{A} \to B/\mathbf{A}$ along f*. The product $(p_1: A \times B \to A,\ p_2: A \times B \to B)$ of two objects is said to be *co-universal* ([10], § 1.1.0) if, for any morphism $f: A \times B \to C$, the pushouts q_1, q_2 of p_1, p_2 along f give rise to a product (q_1, q_2). A morphism $f: A \to B$ is *flat* (cf. Definition (1.6.1) below) if the pushout functor $A/\mathbf{A} \to B/\mathbf{A}$ along f preserves monomorphisms. An object A is *flatly codisjunctable* if it is codisjunctable and the codisjunctor of the inductions $A \rightrightarrows A \amalg A$ is a flat morphism. Let us recall that a category is said to be *complete* (cocomplete) if any small diagram has a limit (colimit).

Definition

A category $\mathbf{A}$ *is called a Zariski category* if it satisfies the following axioms.

Axiom (1.2.1). *A is cocomplete.*

Axiom (1.2.2). *A has a proper, i.e. strong, generating set whose objects are finitely presentable and flatly codisjunctable.*

Axiom (1.2.3). *Regular epimorphisms are universal.*

Axiom (1.2.4). *The terminal object of A is finitely presentable and has no proper subobject.*

Axiom (1.2.5). *Products of two objects of A are co-universal.*

Axiom (1.2.6). *For any finite sequence of codisjunctable congruences $r_1, \ldots, r_n$ on an object A, with respective codisjunctors $d_1, \ldots, d_n$: $r_1 \overset{c}{\vee} \ldots \overset{c}{\vee} r_n = 1_{AXA} \Rightarrow d_1 \vee \ldots \vee d_n = 1_A$.*

According to [16], a Zariski category is a locally finitely presentable category. Thus it is complete and in particular has a terminal object. According to [2], a Zariski category is also a regular category. In what follows, we considers a Zariski category **A** with initial object Z and terminal object 1.

1.3 EXAMPLES OF ZARISKI CATEGORIES

All the rings and algebras considered are commutative and unitary, and all the homomorphisms of rings or algebras preserve units.

Example (1.3.1) *CRng is the category of commutative rings.*

As **CRng** is an algebraic category ([35], Chapter 18) and a locally indecomposable category ([10], § 1.2.1), it satisfies Axioms (1.2.1), (1.2.3), (1.2.4), and (1.2.5). The polynomial ring **Z[X]** is a finitely presentable proper generator in **CRng**. It is a codisjunctable object, as any parallel pair of morphisms $(g, h) : \mathbf{Z}[X] \rightrightarrows A$ has for codisjunctor the canonical morphism $l_a : A \to A[a^{-1}]$ with $a = g(X) - h(X)$ [12]. The morphism l_a is flat in **Rngc** as, in fact, is any codisjunctor in **CRng** ([12], Theorem 5.2). Thus Axiom (1.2.2) is satisfied.

Let A be an object in **CRng** and let $r_1, \ldots, r_n$ be a sequence of codisjunctable congruences on A such that $r_1 \overset{c}{\vee} \ldots \overset{c}{\vee} r_n = 1_{A \times A}$. Each congruence r_k is the congruence modulo an ideal I_k of A. Then $I_1 + \ldots + I_n = A$. Thus there exist $a_1 \in I_1, \ldots, a_n \in I_n$ such that $a_1 + \ldots + a_n = 1$ in A. It is a classical result that the sequence of morphisms $(l_{a_1} : A \to A[a_1^{-1}], \ldots, l_{a_n} : A \to A[a_n^{-1}])$ is an effective co-union of quotient objects of A(cf. [8]), and, hence it is a co-union. For any $k \in [1, n]$, let d_k be the codisjunctor of r_k. As the morphism l_{a_k} codisjoints r_k, it factors through d_k. It follows that the sequence $(d_k)_{k\in[1,n]}$ is a co-union, i.e. $d_1 \vee \ldots \vee d_n = 1_A$. Thus Axiom (1.2.6) is satisfied.

Example (1.3.2). *RedCRng is the category of reduced commutative rings.*

RedCRng is an algebraic category full reflective subcategory of **CRng** and a locally indecomposable category ([10], § 1.9.2). Thus it satisfies Axioms (1.2.1), (1.2.3), (1.2.4), and (1.2.5). The object $\mathbf{Z}[X]$ is a flatly codisjunctable finitely presentable proper generator since the morphism l_a: $A \to A[a^{-1}]$ is still a codisjunctor for a parallel pair $(g, h): \mathbf{Z}[X] \rightrightarrows A$ in **RedCRng**. Thus Axiom (1.2.2) holds. If $r_1, \ldots, r_n$ is a sequence of congruences on an object A in **RedCRng** such that $r_1 \overset{c}{\vee} \ldots \overset{c}{\vee} r_n = 1_{A\times A}$, then each r_k is the congruence modulo a radical ideal I_k of A, and we have $(I_1 + \ldots + I_n)^{1/2} = A$ which implies $I_1 + \ldots + I_n = A$, and hence the arguments developed for Example (1.3.1) are still valid. Thus Axiom (1.2.6) holds.

Example (1.3.3). *RegCRng is the category of commutative regular rings* [33].

Example (1.3.4). *CRng* (p) *are commutative rings of characteristic* p, *where* p *is a prime, and including the zero ring.*

Example (1.3.5). *CAlg* (k) *are commutative algebras over a field* k.

Example (1.3.6). *AlgCAlg* (k) *are commutative algebraic algebras over a field* k.

AlgCAlg (k) is the full subcategory of **CAlg** (k) whose objects are the k-algebras whose elements are algebraic over k. It is a locally indecomposable category ([10], § 1.12.3) whose objects of the form $k[X]/(P(X))$ with $P(X) \neq 0$ are finitely presentable and flatly codisjunctable and make up a proper generating set.

Example (1.3.7). *RedAlgCAlg* (k) *are reduced commutative algebraic algebras on a field* k.

Example (1.3.8). *RedCAlg* (k) *are reduced commutative algebras over a field* k.

Example (1.3.9). *RegCAlg* (k) *are regular commutative algebras over a field* k.

Example (1.3.10). *Mod is the category of modules on variable commutative rings.*

Mod is the category whose objects are ordered pairs (A, E) of a commutative ring A and an A-module E, and whose morphisms are pairs $(\varphi, f)\colon (A, E) \to (B, F)$ of a ring homomorphism $\varphi\colon A \to B$ and a group homomorphism $f\colon E \to F$ which satisfies $f(ax) = \varphi(a)\, f(x)$ for any $(a, x) \in A \times E$. It is an algebraic locally indecomposable category ([9], § 9.1), and thus it satisfies Axioms (1.2.1), (1.2.3), (1.2.4), and (1.2.5). The object $(\mathbf{Z}[X], \mathbf{Z}[X])$ is a finitely presentable proper generator in **Mod**.

Let $((\psi, g), (\theta, h)) : (\mathbf{Z}[X], \mathbf{Z}[X]) \rightrightarrows (A, E)$ be a pair of parallel morphisms in **Mod**. Let $a = \psi(X) - \theta(X)$, let $A[a^{-1}]$ be the ring of fractions, and let $E[a^{-1}] = E \otimes_A A[a^{-1}]$ be the A-module of fractions. The canonical morphisms $\varphi_a\colon A \to A[a^{-1}]$ and $f_a\colon E \to E[a^{-1}]$ yield a morphism (φ_a, f_a) in **Mod** which is the codisjunctor of $((\psi, g), (\theta, h))$. Let $(\alpha, u) : (A, E) \to (B, F)$ be a monomorphism in **Mod**, i.e. α and u are injective. The pushout of (α, u) along (φ_a, f_a) is the natural extension $(\bar{\alpha}, \bar{u}) : (A[a^{-1}], E[a^{-1}]) \to (B[b^{-1}], F[b^{-1}])$ of (α, u) with $b = \alpha(a)$. The morphism $\bar{\alpha}$ is monomorphic in **CRng** because φ_a is flat in **CRng**. Let $x/a^n \in \mathrm{Ker}(\bar{u})$. Then $u(x)/\alpha(a)^n = 0/1$ in $F[b^{-1}]$. There exists $p \in \mathbb{N}$ such that $u(x)b^p = 0$; thus $u(xa^p) = u(x)u(a^p) = u(x)b^p = 0$. As u is monomorphic, $xa^p = 0$ and thus $x/a^n = 0/1$ in $E[a^{-1}]$. It follows that $\bar{u}$ is monomorphic. Consequently $(\bar{\alpha}, \bar{u})$ is monic and (φ_a, f_a) is flat in **Mod**. Therefore $(\mathbf{Z}[X], \mathbf{X}[X])$ is flatly codisjunctable and Axiom (1.2.2) holds.

Let $(A, E) \in \mathbf{Mod}$. Any congruence on (A, E) is an ordered pair $(\mathrm{mod}(I), \mathrm{mod}(X))$ of a congruence modulo an ideal I on A and a congruence modulo an A-submodule X of E containing $I \cdot E$. Let $(\mathrm{mod}(I_1), \mathrm{mod}(X_1)), \ldots, (\mathrm{mod}(I_n), \mathrm{mod}(X_n))$ be a sequence of congruences on (A, E) such that $(\mathrm{mod}(I_1), \mathrm{mod}(X_1)) \overset{c}{\vee} \ldots \overset{c}{\vee} (\mathrm{mod}(I_n), \mathrm{mod}(X_n)) = 1_{(A,E)\times(A,E)}$. Then $\mathrm{mod}(I_1) \overset{c}{\vee} \ldots \overset{c}{\vee} \mathrm{mod}(I_n) = 1_{A \times A}$ in **CRng** so that $I_1 + \ldots + I_n = A$ and there exists $a_1 \in I_1, \ldots, a_n \in I_n$ such that $a_1 + \ldots + a_n = 1$. Then the sequence $\varphi_{a_1}\colon A \to A[a_1^{-1}], \ldots, \varphi_{a_n}\colon A \to A[a_n^{-1}]$ is an effective co-union in **CRng**, while the sequence $f_{a_1}\colon E \to E[a_1^{-1}], \ldots, f_{a_n}\colon E \to E[a_n^{-1}]$ is an effective co-union in A-**Mod**. It follows that the sequence $(\varphi_{a_1}, f_{a_1}), \ldots, (\varphi_{a_n}, f_{a_n})$ is a co-union in **Mod**. Let (δ_k, d_k) be the codisjunctor of the congruence $(\mathrm{mod}(I_k), \mathrm{mod}(X_k))$. As (φ_{a_k}, f_{a_k}) factors through (δ_k, d_k), it follows that $(\delta_1, d_1) \vee \ldots \vee (\delta_n, d_n) = 1_{(A,E)}$. Consequently Axiom (1.2.6) holds.

Example (1.3.11). *Mod(Red) are modules on variable reduced commutative rings.*

Example (1.3.12). *Mod(Reg) are modules on variable regular commutative rings.*

Example 1.13.13. *Mod(Bool) are modules on variable boolean rings.*

Example (1.3.14). *CAlg are commutative algebras on variable rings.*

Example (1.3.15). *RedCAlg are reduced commutative algebras on variable reduced rings.*

Example (1.3.16). *RlRng is the category of real rings.*

RlRng is the full subcategory **A** of **CRng** whose objects are the rings fulfilling the following axioms ([4], §1, 2; [29], p. 228, §5.3):

$$(x_2^1 + \dots + x_n^2)y = 0 \Rightarrow x_1^2 y = 0 \tag{1.3.17}$$

$$1 + x_1^2 + \dots + x_n^2 \text{ is invertible.} \tag{1.3.18}$$

As **A** is a full reflective subcategory of **CRng** closed under filtered colimits and direct factors, it is a locally indecompasable category ([9], Proposition 10.0). If $f: A \to B$ is a morphism in **A**, the factorization $f = mg$, where $g: A \to C$ is a regular epimorphism and $m: C \to B$ is a monomorphism in the category **CRng**, is such that the object C belongs to **A**. It follows that the category **A** is regular. Thus **A** satisfies Axioms (1.2.1), (1.2.3), (1.2.4), and (1.2.5).

We now show that **A** satisfies Axiom (1.2.2). Let us denote by $\mathbf{Z}\langle X\rangle$ the ring of fractions $\mathbf{Z}[X]\,[S_1^{-1}]$ where $S_1 = \{1 + P_1^2 + \dots + P_n^2 : n \in N$ and $P_1, \dots, P_n \in \mathbf{Z}[X]\}$. $\mathbf{Z}\langle X\rangle$ is the universal object of **A** associated to the object $\mathbf{Z}[X]$ of **CRng**, and thus it is a finitely presentable proper generator in **A**. Let $(g, h) : \mathbf{Z}\langle X\rangle \rightrightarrows A$ be a parallel pair of morphisms in **A**. Let $a = g(X) - h(X)$, $S_a = \{a^{2n} + x_1^2 + \dots + x_n^2 : n \in \mathbb{N}$ and $x_1, \dots, x_n \in A\}$. The ring of fractions $A[S_a^{-1}]$ belongs to **A**. The canonical morphism $l_a : A \to A[S_a^{-1}]$ codisjoints the pair (g, h) because any morphism $k : A[S_a^{-1}] \to H$ in **A** such that $kl_a g = kl_a h$ is such that $k(a)$ must be both invertible and null, and so $H = \{0\}$. Let $f: A \to B$ be a morphism which codisjoints (g, h) in **A**. As the inclusion functor $\mathbf{A} \to \mathbf{CRng}$ preserves codisjoint pairs ([11], §2.4.3), the morphism f codisjoints (g, h) in **CRng**. If $l: \mathbf{Z}[X] \to \mathbf{Z}\langle X\rangle$ is the canonical morphism, then the morphism f codisjoints (gl, hl) in **CRng**, and hence the element $f(g(l(X))) - f(h(l(X))) = f(g(X) - h(X)) = f(a)$ is invertible in B. Therefore any element $a^{2n} + x_1^2 + \dots + x_n^2$ of S_a has an image $f(a^{2n} + x_1^2 + \dots + x_n^2) = (f(a))^{2n} + (f(x_1))^2 + \dots + (f(x_n))^2$ which is invertible in B by Axiom (1.3.18). It follows that f factors uniquely through l_a. As a result, l_a is a codisjunctor of (g, h) in **A**, and the object $\mathbf{Z}\langle X\rangle$ is codisjunctable in **A**.

Let $f: A \to B$ be a monomorphism in **A**, i.e. an injective morphism in **A**. The pushout of f along l_a *is the canonical extension* $\bar{f}: A[S_a^{-1}] \to B[S_b^{-1}]$ of the morphism f, where $b = f(a)$ and S_b is defined similarly to S_a. Let $x/s \in A[S_a^{-1}]$ such that $\bar{f}(x/s) = f(x)/f(s) = 0/1$ in $B[S_b^{-1}]$. There exists an element $b^{2n} + y_1^2 + \ldots + y_n^2$ of S_b such, that $f(x)(b^{2n} + y_1^2 + \ldots + y_n^2) = 0$. By Axiom (1.3.17), it follows that $f(x)b^{2n} = 0$. Thus $f(xa^{2n}) = f(x)f(a^{2n}) = f(x)b^{2n} = 0$. As f is injective, $xa^{2n} = 0$ and thus $x/s = 0/1$ in $A[S_a^{-1}]$. If follows that $\bar{f}$ is injective and that l_a is a flat morphism in **A**. Consequently, $\mathbf{Z}\langle X\rangle$ is flatly codisjunctable and Axiom (1.2.2) holds.

Finally, we show that **A** satisfies Axiom (1.2.6). Let $A \in \mathbf{A}$. Any congruence on A is the congruence modulo a real ideal of A (cf. § 1.1, p. 13). Let I, J be two real ideals of A such that the related congruences $\text{mod}(I)$, $\text{mod}(J)$ satisfy $\text{mod}(I) \overset{c}{\vee} \text{mod}(J) = 1_{A \times A}$. Then the real hull ${}^{R}(I + J)$ of $I + J$ is the unit ideal of A. It implies that $I + J = A$. Thus there exist $a \in I$ and $b \in J$ such that $a + b = 1$. Let us suppose that congruences $\text{mod}(I)$ and $\text{mod}(J)$ are codisjunctable and have for codisjunctors $d: A \to D$ and $\delta: A \to \Delta$ respectively. The morphism $l_a: A \to A[S_a^{-1}]$ codisjoints $\text{mod}(I)$ and thus it factors through d. Similarly, $l_b: A \to A[S_b^{-1}]$ factors through δ. In order to prove that $d \vee \delta = 1_A$, it is sufficient to prove that $l_a \vee l_b = 1_A$. Let us prove this last statement. Let $\alpha = a^{2n} + x_1^2 + \ldots + x_n^2 \in S_a$ and let $\beta = b^{2n} + y_1^2 + \ldots + y_n^2 \in S_b$. The equality $(a + b)^{2n} = 1$ is of the form $ua^n + vb^n = 1$. Then

$$
\begin{aligned}
&u^2\alpha + v^2\beta = u^2a^{2n} + v^2b^{2n} + (\text{sum of squares}) = \\
&\tfrac{1}{2}\left[(ua^n + vb^n)^2 + (ua^n - vb^n)^2\right] + (\text{sum of squares}) = \\
&\tfrac{1}{2} + (\text{sum of squares}).
\end{aligned}
$$

By Axiom (1.3.18), the element $u^2\alpha + v^2\beta$ is invertible in A. Then

$$
\frac{u^2}{u^2\alpha + v^2\beta}\alpha + \frac{v^2}{u^2\alpha + v^2\beta}\beta = 1
$$

This relation implies that the pair of morphisms $(A \to A[\alpha^{-1}], A \to A[\beta^{-1}])$ is an effective co-union in **CRng**. It follows that the pair $(l_a: A \to A[S_a^{-1}],\ l_b: A \to A[S_b^{-1}])$ is an effective co-union in **CRng** because it is the filtered colimit of the pairs $(A \to A[\alpha^{-1}], A \to A[\beta^{-1}]$ when (α, β) runs through $S_a \times S_b$. The co-intersection of (l_a, l_b) is the morphism $A \to A[(S_a \cdot S_b)^{-1}]$ in **CRng**, while it is the morphism l_{ab}: $A \to A[S_{ab}^{-1}]$ in **A**. Let us now show that the canonical morphism $A[(S_a, S_b)^{-1}] \to A[S_{ab}^{-1}]$ is a monomorphism in **CRng**. Let $x/s \in A[(S_a \cdot S_b)^{-1}]$ such that $x/s = 0/1$ in $A[S_{ab}^{-1}]$. There exists $(ab)^{2n} + x_1^2 + \ldots + x_n^2 \in S_{ab}$ such that $x((ab)^{2n} + x_1^2 + \ldots + x_n^2) = 0$. By Axiom (1.3.17) we have $xa^{2n}b^{2n} = 0$. Thus $x/s = 0/1$ in $A[S_a \cdot S_b)^{-1}]$. It follows that the pair (l_a, l_b) is an effective co-union in **A**, as was the case in

CRng. Thus it is a co-union in **A**, i.e. $l_a \vee l_b = 1_A$.

Let us consider a sequence $I_1, \ldots, I_n$ of real ideals of A such that the related congruences $\mathrm{mod}(I_1), \ldots, \mathrm{mod}(I_n)$ are codisjunctable and satisfy $\mathrm{mod}(I_1) \overset{c}{\vee} \ldots \overset{c}{\vee} \mathrm{mod}(I_n) = 1_{A\times A}$. We prove by induction on n that the codisjunctors $d_1, \ldots, d_n$ of $\mathrm{mod}(I_1), \ldots, \mathrm{mod}(I_n)$ satisfy $d_1 \vee \ldots \vee d_n = 1_A$. The real hull of $I_1 + \ldots + I_n$ is the unit ideal and consequently $I_1 + \ldots + I_n = A$. Therefore there exists $a_1 \in I_1, \ldots, a_n \in I_n$ such that $a_1 + \ldots + a_n = 1$. Let us put $a_2 + \ldots + a_n = b$. Then $a_1 + b = 1$. By the previous result the pair $(l_{a_1}: A \to A[S_{a_1}^{-1}],\ l_b: A \to A[S_b^{-1}])$ is a co-union in **A**. By the induction hypothesis the sequence $(A[S_b^{-1}] \to A[S_{a_2b}^{-1}], \ldots, A[S_b^{-1}] \to A[S^{-1}_{a_nb}])$ is a co-union in **A**. It follows that the sequence $(A \to A[S_{a_1}^{-1}],\ A \to A[S^{-1}_{a_2b}], \ldots, A \to A[S^{-1}_{a_nb}])$ is a co-union in **A**. But for $k \in [2, n]$ the morphism $A \to A[S^{-1}_{a_kb}]$ factors through the morphism $A \to A[S_{a_k}^{-1}]$. Consequently, the sequence $(A \to A[S_{a_1}^{-1}], \ldots, A \to A[S_{a_n}^{-1}])$ is a co-union in **A**. It follows that $d_1 \vee \ldots \vee d_p = 1_A$ because the morphism $l_{a_k}: A \to A[S_{a_k}^{-1}]$ factors through d_k. Consequently Axiom (1.2.6) holds.

Example (1.3.19). ***RedRlRng** is the category of reduced real rings.*

RedRlRng is the full subcategory of **CRng** whose objects are the rings fulfilling the following axioms ([4], §1.2):

$$x_1^2 + \ldots + x_n^2 = 0 \Rightarrow x_1 = 0 \tag{1.3.20}$$

$$1 + x_1^2 + \ldots + x_n^2 \text{ is invertible.} \tag{1.3.21}$$

Example (1.3.22). ***RlOrdRng** is the category of real ordered rings.*

RlOrdRng is the category **A** whose objects are commutative ordered rings satisfying the axioms

$$x^2 \geqslant 0 \tag{1.3.23}$$

$$(0 \leqslant x \leqslant y \quad \text{and} \quad yz = 0) \Rightarrow xz = 0 \tag{1.3.24}$$

$$1 \leqslant x \Rightarrow x \text{ is invertible,} \tag{1.3.25}$$

and whose morphisms are order-preserving homomorphisms of rings. It is a locally finitely presentable category whose terminal object $\{0\}$ is finitely presentable and has no proper subobject. Thus Axioms (1.2.1) and (1.2.4) holds. Products of pairs of objects are defined as in **CRng**. Thus Axiom (1.2.5) holds. Following [4], § 2.5, regular epimorphisms in **A** are the canonical morphisms $q_I: A \to A/I$ where I is an absolutely convex ideal of A and the order on A/I is defined by $\bar{a} \geqslant 0 \Leftrightarrow \exists x \geqslant 0, \bar{x} = \bar{a}$. Let $g: C \to A/I$ be any morphism of **A** and let $(f: B \to A,\ q: B \to C)$ be the pullback of (q_I, g)

in **A**. As the forgetful functor $\mathbf{A} \to \mathbf{CRng}$ preserves $\xleftarrow{lim}$, this pullback is also a pullback in **CRng**. Therefore the morphism q is of the form $q = q_j: B \to B/J$ where J is an ideal of B, and the morphism g is the quotient morphism $\bar{f}: B/J \to A/I$ of f. The ideal J is absolutely convex as it is the kernel of the morphism q_J belonging to **A**. Let x be an element of B/J such that $x \geqslant 0$. Then $\bar{f}(x) \geqslant 0$ in A/I. Thus there exists $a \geqslant 0$ in A such that $q_I(a) = \bar{f}(x)$. By the construction of pullbacks in **A**, there exists an element b in B such that $b \geqslant 0$, $f(b) = a$, and $q_J(b) = x$. It follows that q_J is a regular epimorphism in **A**. Consequently regular epimorphisms are universal in **A** and so Axiom (1.2.3) holds.

We now show that Axiom (1.2.2) holds. Let us denote by $\mathbf{Z}\{X\}$ the subring of $\mathbf{Q}(X)$ whose elements are of the form

$$\frac{P}{1 + N_1 Q_1^2 + \ldots + N_q Q_q^2}$$

with $P, Q_1, \ldots, Q_q \in \mathbf{Z}[X]$, and $N_1, \ldots, N_q \in \mathbf{N}[X]$. Equipped with the order whose positive or null elements are those of the form

$$\frac{M_1 P_1^2 + \ldots + M_p P_p^2}{1 + N_1 Q_1^2 + \ldots + N_q Q_q^2}$$

with $P_i, Q_j \in \mathbf{Z}[X]$ and $M_i, N_j \in \mathbf{N}[X]$, the ring $\mathbf{Z}\{X\}$ is an object of **A**. Let $f: \mathbf{Z}\{X\} \to A$ be a morphism in **A**. As $X \geqslant 0$ in $\mathbf{Z}\{X\}$, the element $a = f(X)$ is $\geqslant 0$. Conversely, if a is an element of A such that $a \geqslant 0$, we can define a morphism $f: \mathbf{Z}\{X\} \to A$ by

$$f\left(\frac{P}{1 + N_1 Q_1^2 + \ldots + N_q Q_q^2}\right) = \frac{P(a)}{1 + N_1(a) Q_1^2(a) + \ldots + N_q^2(a) Q_q(a)}$$

because the element $1 + N_1(a)Q_1^2(a) + \ldots + N_q(a)Q_q^2 \geqslant 1$ is invertible in A. It follows that there is a one-to-one correspondence between $\mathrm{Hom}_{\mathbf{A}}(\mathbf{Z}\{X\}, A)$ and the set of positive or null elements of A. Let $g: A \to B$ be a morphism of **A** such that the map $\mathrm{Hom}_{\mathbf{A}}(\mathbf{Z}\{X\}, g)$ is a bijection. Let $a \in \ker(g)$. We have $(a + \frac{1}{2})^2 = a^2 + a + (\frac{1}{2})^2$ in A. As $a^2 \geqslant 0$, the equality $g(a^2) = (g(a))^2 = 0$ implies $a^2 = 0$. As $(a + \frac{1}{2})^2 \geqslant 0$ and $(\frac{1}{2})^2 \geqslant 0$, the relation

$$g((a + \tfrac{1}{2})^2) = (g(a) + g(\tfrac{1}{2}))^2 = (g(\tfrac{1}{2}))^2 = g((\tfrac{1}{2})^2)$$

implies $(a + \frac{1}{2})^2 = (\frac{1}{2})^2$. Therefore

$$a = (a + \tfrac{1}{2})^2 - a^2 - (\tfrac{1}{2})^2 = 0.$$

It follows that g is monomorphic. Let $b \in B$. There exists $a_1 \geqslant 0$, $a_2 \geqslant 0$, and $a_3 \geqslant 0$ in A such that $g(a_1) = (b + \frac{1}{2})^2$, $g(a_2) = b^2$, and $g(a_3) = (\frac{1}{2})^2$. Then the element $a_1 - a_2 - a_3 \in A$ is such that

$$g(a_1 - a_2 - a_3) = g(a_1) - g(a_2) - g(a_3) = (b + \tfrac{1}{2})^2 - b^2 - (\tfrac{1}{2})^2 = b.$$

It follows that g is a bijection. It is in fact an isomorphism in **A** because, for any element $a \in A$, $a \geqslant 0 \Leftrightarrow g(a) \geqslant 0$. As a result, the object $\mathbf{Z}\{X\}$ is a proper generator in **A**. It is easy to see, by the construction of filtered colimits in **A**, that the functor $\mathrm{Hom}_{\mathbf{A}}(\mathbf{Z}\{X\}, -) : \mathbf{A} \to \mathbf{Set}$ preserves filtered colimits, and so the object $\mathbf{Z}\{X\}$ is finitely presentable in **A**. Let $(g, h) : \mathbf{Z}\{X\} \rightrightarrows A$ be a pair of parallel morphisms in **A**. Let $g(X) - h(X) = a$ and $S_a = \{x \in A : \exists n \in \mathbf{N}, a^{2n} \leqslant x\}$. Then the ring of fractions $A[S_a^{-1}]$ is an object of **A** ([4], Properties 3.1.1, 3.8.1, 3.8.3) and the canonical morphism $l_a : A \to A[S_a^{-1}]$ is a morphism of **A** which codisjoints the pair (g, h). Let $f: A \to B$ be a morphism of **A** which codisjoints (g, h). As the forgetful functor $\mathbf{A} \to \mathbf{CRng}$ preserves codisjoint pairs ([11], § 2.4.5), f codisjoints (g, h) in **CRng**. Then f codisjoints (gk, hk) in **CRng** where $k : \mathbf{Z}[X] \to \mathbf{Z}\{X\}$ is the canonical morphism in **CRng**; hence the element $f(g(k(X))) - f(h(k(X))) = f(g(X) - h(X)) = f(a)$ is invertible in B. By Axiom (1.3.25), for $x \in S_a$ the element $f(x) \geqslant f(a^{2n})$ is invertible in B. It follows that f factors uniquely through l_a in a morphism which belongs to **A**. As a result, l_a is a codisjunctor of (g, f) in **A**. Thus $\mathbf{Z}\{X\}$ is codisjunctable. Let $f: A \to B$ be a monomorphism in **A**, i.e. an injective morphism. The pushout of f along l_a is the natural extension $\bar{f}: A[S_a^{-1}] \to B[S_b^{-1}]$ of f where $b = f(a)$ and S_b is built up in the same way as S_a. Let $\alpha/s \in \mathrm{Ker}(\bar{f})$. Then $f(\alpha)/f(s) = 0/1$ in $B[S_b^{-1}]$. Thus there exists $y \in B$ and $n \in \mathbf{N}$ such that $b^{2n} \leqslant y$ and $f(\alpha)y = 0$. By Axiom (1.3.24) it follows that $f(\alpha)b^{2n} = 0$; thus $f(\alpha a^{2n}) = f(\alpha)b^{2n} = 0$. As f is monomorphic, $\alpha a^{2n} = 0$ and so $\alpha/s = 0/1$ in $A[S_a^{-1}]$. Therefore $\bar{f}$ is monomorphic. As a result, the morphism l_a is flat in **A** and the object $\mathbf{Z}\{X\}$ is flatly codisjunctable in **A**. Consequently, Axiom (1.2.2) holds.

Finally, we show that Axiom (1.2.6) holds. Let $A \in \mathbf{A}$. Any congruence on A is the congruence modulo an absolutely convex ideal I of A ([4], § 2.5). Let I and J be two absolutely convex ideals of A such that $\mathrm{mod}(I) \overset{c}{\vee} \mathrm{mod}(J) = 1_{A \times A}$. Then the absolutely convex hull of $I + J$ is the unit ideal A, and consequently $I + J = A$ ([4], § 2.5). Thus there exist $a \in I$ and $b \in J$ such that $a + b = 1$. Let us suppose that the congruences $\mathrm{mod}(I)$ and $\mathrm{mod}(J)$ are codisjunctable and have codisjunctors $d : A \to D$ and $\delta : A \to \Delta$ respectively. The morphism $l_a : A \to A[S_a^{-1}]$ codisjoints $\mathrm{mod}(I)$ and thus it factors through d. Similarly, l_b factors through δ. In order to prove that $d \vee \delta = 1_A$, it is enough to prove that $l_a \vee l_b = 1_A$. Let us prove this last formula. Let $\alpha \in S_a$ and $\beta \in S_b$. There exists $n \in \mathbf{N}$ such that $a^{2n} \leqslant \alpha$ and $b^{2n} \leqslant \beta$. The relation $(a + b)^{2n} = 1$ is of the form $ua^n + vb^n = 1$. Then

$$u^2\alpha + v^2\beta \geqslant u^2a^{2n} + v^2b^{2n} = \frac{1}{2}\left((ua^n + vb^n)^2 + (ua^n - vb^n)^2\right) \geqslant \frac{1}{2}.$$

By Axiom (1.3.25) $u^2\alpha + v^2\beta$ is invertible in A. Thus

$$\frac{u^2}{u^2\alpha + v^2\beta}\alpha + \frac{v^2}{u^2\alpha + v^2\beta}\beta = 1.$$

It follows that the pair $(A \to A[\alpha^{-1}], A \to A[\beta^{-1}])$ is an effective co-union in **CRng**. As the pair $(l_a: A \to A[S_a^{-1}], l_b: A \to A[S_b^{-1}])$ is a filtered colimit of pairs of the form $(A \to A[\alpha^{-1}], A \to A[\beta^{-1}])$, where (α, β) runs through $S_a \times S_b$, the pair (l_a, l_b) is also an effective co-union in **CRng**. The co-intersection of (l_a, l_b) is the morphism $A \to A[(S_aS_b)^{-1}]$ in **CRng**, while it is the morphism $l_{ab}: A \to A[S_{ab}^{-1}]$ in **A**. Let us show that the canonical morphism $A[(S_aS_b)^{-1}] \to A[S_{ab}^{-1}]$ is monomorphic. Let $x/s \in A[(S_aS_b)^{-1}]$ such that $x/s = 0/1$ in $A[S_{ab}^{-1}]$. There exist $\gamma \in A$ and $n \in \mathbf{N}$ such that $(ab)^{2n} \leqslant \gamma$ and $x\gamma = 0$. By Axiom (1.3.24), $(ab)^{2n}x = 0$. Thus $x/s = 0/1$ in $A[(S_aS_b)^{-1}]$. It follows that the pair (l_a, l_b) is an effective co-union in **A** as it was so in **CRng**. Then $l_a \vee l_b = 1_A$ and $d \vee \delta = 1_A$. The same result follows for a sequence of congruences on A by induction, as seen previously in Example (1.3.16). Thus Axiom (1.2.6) holds.

Example (1.3.26). *RedRlOrdRng is the category of reduced real ordered rings* [4].

RedRlOrdRng is the category whose objects are commutative ordered rings satisfying the axioms

$$x \neq 0 \Rightarrow x^2 > 0 \tag{1.3.27}$$

$$1 \leqslant x \Rightarrow x \text{ is invertible,} \tag{1.3.28}$$

and whose morphisms are order-preserving homomorphisms.

Example (1.3.29). *RlLatRng is the category of real lattice rings* [26].

RlLatRng is the category whose objects are commutative rings equipped with a lattice-ordered structure satisfying the axioms

$$x \leqslant y \Rightarrow x + z \leqslant y + z \tag{1.3.30}$$

$$(x \wedge y = 0 \quad \text{and} \quad z \geqslant 0) \Rightarrow x \wedge yz = 0 \tag{1.3.31}$$

$$1 \leqslant x \Rightarrow x \text{ is invertible,} \tag{1.3.32}$$

and whose morphisms are homomorphisms preserving the joins and meets of two elements. The proof that **A** is a Zariski category is similar to that given in Example (1.3.22), after the following two changes. Regular morphisms are the canonical morphisms $q_I: A \to A/I$ where I is an absolutely convex ideal of A and where the lattice structure on A/I is given by $\bar{a} \wedge \bar{b} = \overline{a \wedge b}$ and $\bar{a} \vee \bar{b} = \overline{a \vee b}$. The proper generator is the object $\mathbf{Z}\{\{X\}\}$ which is the ring of fractions $F[S_1^{-1}]$, where F is the free strongly lattice-ordered ring generated by $\{X\}$ and $S_1 = \{f \in F : 1 \leqslant f\}$, equipped with the lattice structure defined by

$$\frac{x}{s} \wedge \frac{y}{t} = \frac{xt \wedge ys}{st} \qquad \text{and} \qquad \frac{x}{s} \vee \frac{y}{t} = \frac{xt \vee ys}{st}.$$

Example (1.3.33). ***RedRlLatRng*** *is the category of reduced real lattice rings.*

Example (1.3.34). ***GradCRng*** *is the category of graded commutative rings and graded homomorphisms, (i.e. Z*-gradation).

Example (1.3.35). ***RedGradCRng*** *is the category of reduced graded commutative rings.*

Example (1.3.36). ***RegGradCRng*** *is the category of regular graded commutative rings.*

Example (1.3.37). ***GradMod*** *is the category of graded modules on variable commutative graded rings.*

Example (1.3.38). ***GradMod(Red)*** *is the category of graded modules on variable commutative graded reduced rings.*

Example (1.3.39). ***GradMod(Reg)*** *is the category of graded modules on variable commutative graded regular rings.*

Example (1.3.40). ***RegDifCRng*** *is the category of regular differential rings* (cf. [10]).

Example (1.3.41). ***RegDifPerfCRng(p)*** *is the category of regular differential rings differentially perfect of prime characteristic p* (cf. [10]).

Example (1.3.42). ***Bool*** *is the category of boolean algebras.*

Example (1.3.43). *Any locally simple category* ([10], Chapter 2).

1.4 PRODUCTS OF OBJECTS AND MORPHISMS

Proposition (1.4.1). *The terminal object is strict and the category is fully supported.*

Proof. If the proper generating set mentioned in Axiom (1.2.2) of Zariski categories is empty, then any morphism is an isomorphism and 1 is strict. Otherwise, if A is an object belonging to this generating set, then the pair

$(1_A, 1_A) : A \rightrightarrows A$ has a codisjunctor whose codomain must be a strict terminal object. Let $f: A \to 1$ be any morphism. It factors in the form $f = mg$ where g is a regular epimorphism and m is a monomorphism. According to Axiom (1.2.4), m is an isomorphism. Thus f is a regular epimorphism. As a result, the category is fully supported. ■

Proposition (1.4.2). *Any projection of a product is a regular epimorphism.*

Proof. The morphism $p_i : \Pi_{i \in I} A_i \to A_i$ is the pullback of the regular epimorphism $\Pi_{j \in I - \{i\}} A_j \to 1$ along the morphism $A_i \to 1$.

Proposition (1.4.3). *Products are codisjoint* ([21], II, § 4.5).

Proof. Projections of products are epimorphisms by Proposition (1.4.2). Let us consider first the product $(u : Z \times Z \to Z, v : Z \times Z \to Z)$ of Z by Z. The pair $(u, v) : Z \times Z \rightrightarrows Z$ is the congruence on Z kernel pair of the regular epimorphism $f: Z \to 1$. Thus f is the coequalizer of (u, v). Consequently, the pair (u, v) is codisjointed. Let $(m : Z \to M, n : Z \to M)$ be the pushout of (u, v). Then $m = n$. It follows that $mu = mv$ and $M \simeq 1$. As a result, the product $Z \times Z$ is codisjoint. Let $(p : A \times B \to A, q : A \times B \to B)$ be the product of A and B, and let $(t : A \to T, s : B \to T)$ be the pushout of (p, q). Let $i_A : Z \to A$ and $i_B : Z \to B$ be the unique morphisms, and let $i_A \times i_B : Z \times Z \to A \times B$ be their products. Then we have $t i_A u = tp(i_A \times i_B) = sq(i_A \times i_B) = s i_B v$. Since the product $Z \times Z$ is codisjoint, it follows that $T \simeq 1$. As a result, the product $A \times B$ is codisjoint. Let $(p_i : \Pi_{i \in I} A_i \to A_i)_{i \in I}$ be a product and $i, j \in I$ and $i \neq j$. Let $p : \Pi_{i \in I} A_i \to \Pi_{k \in I - \{i\}} A_k$ and $q_j : \Pi_{k \in I - \{i\}} A_k \to A_j$ be the canonical projections. Then $p_j = q_j p$. If $(t : A_i \to T, s : A_j \to T)$ is the pushout of (p_i, p_j), then $tp_i = sp_j = sq_j p$. Since the product $\Pi_{i \in I} A_i = A_i \times \Pi_{k \in I - \{i\}} A_k$ is codisjoint, $T \simeq 1$. As a result, the product $\Pi_{i \in I} A_i$ is codisjoint. ■

Proposition (1.4.4). *Let $\Pi_{i \in I} f_i : \Pi_{i \in I} A_i \to \Pi_{i \in I} B_i$ be the product of a family of morphisms $(f_i : A_i \to B_i)_{i \in I}$.* If $i \in I$ and $p_i : \Pi_{i \in I} A_i \to A_i$ and $q_i : \Pi_{i \in I} B_i \to B_i$ *are the projections indexed by i, then (f_i, q_i) is the pushout of $(p_i, \Pi_{i \in I} f_i)$.*

Proof. Let $f: A \to B, g : C \to D$ be a pair of morphisms and let $f \times g : A \times C \to B \times D$ be their product with projections $p_1 : A \times C \to A$, $p_2 : A \times C \to C, q_1 : B \times D \to B$, and $q_2 : B \times D \to D$. Since the product $A \times C$ is co-universal, the pushout of $(f \times g, p_1)$ is of the form $(m \times n : B \times D \to \bar{B} \times \bar{D}, (h, k) : A \to \bar{B} \times \bar{D})$. If $\bar{q}_2 : \bar{B} \times \bar{D} \to \bar{D}$ is the projection, then $ngp_2 = nq_2(f \times g) = \bar{q}_2(m \times n)(f \times g) = \bar{q}_2(h, k)p_1 = kp_1$. Since the product $A \times C$ is codisjoint, $\bar{D} \simeq 1$. It follows that $(mq_1 : B \times D \to \bar{B}, h : A \to \bar{B})$ is the pushout of $(f \times g, p_1)$. The relation $q_1(f \times g) = fp_1$

implies the existence of a morphism $t: \bar{B} \to B$ such that $tmq_1 = q_1$. Then the morphisms m and t must be isomorphisms. Consequently (f, q_1) is the pushout of $(p_1, f \times g)$. The proof of the proposition follows, for $\Pi_{i \in I} f_i = f_i \times \Pi_{j \in I - \{i\}} f_j$. ■

Proposition (1.4.5). *For two families of morphisms $(f_i : A_i \to B_i)_{i \in I}$, $(g_i : A_i \to B_i)_{i \in I}$ indexed by I, $\Pi_{i \in I} f_i = \Pi_{i \in I} g_i \Leftrightarrow \forall\, i \in I, f_i = g_i$.*

Proof. The proof follows immediately from Proposition (1.4.4) or from the fact that the projections are epimorphisms. ■

Proposition (1.4.6). *For a family of morphisms $(f_i : A_i \to B_i)_{i \in I}$, we have that $\Pi_{i \in I} f_i$ is monomorphic $\Leftrightarrow \forall\, i \in I, f_i$ is monomorphic.*

Proof. It is well known that products of monomorphisms are themselves monomorphisms. Let us consider a pair of morphisms $f: A \to B$, $g: C \to D$ such that the product $f \times g: A \times C \to B \times D$ is monomorphic. Let $(m, n): M \rightrightarrows A$ such that $fm = fn$. Then the pair of morphisms $(m \times 1_C, n \times 1_C): M \times C \rightrightarrows A \times C$ is such that $(f \times g)(m \times 1_C) = (f \times g)(n \times 1_C)$. Thus $m \times 1_C = n \times 1_C$ and, by Proposition (1.4.5), $m = n$. If follows that f is monomorphic. The proof of the proposition follows since $\Pi_{i \in I} f_i = f_i \times \Pi_{j \in I - \{i\}} f_j$. ■

Proposition (1.4.7). *Finite products commute with coequalizers and with codisjunctors.*

Proof. Let $(g_i, h_i): C_i \rightrightarrows A_i$ for $i \in \{1, 2\}$ be pairs of morphisms with coequalizers $q_i: A_i \to Q_i$. Since the product $A_1 \times A_2$ is co-universal, any morphism $m: A_1 \times A_2 \to M$ is of the form $m = m_1 \times m_2$ with $m_1: A_1 \to M_1$, $m_2: A_2 \to M_2$, and $M = M_1 \times M_2$. Then m coequalizes $(g_1 \times g_2, h_1 \times h_2)$ if and only if m_i coequalizes (g_i, h_i) for $i \in \{1, 2\}$ (Proposition (1.4.5)). It follows easily that $q_1 \times q_2$ is the coequalizer of $(g_1 \times g_2, h_1 \times h_2)$. Therefore finite products commute with coequalizers. It follows that m codisjoints $(g_1 \times g_2, h_1 \times h_2) \Leftrightarrow (m_1 g_1 \times m_2 g_2, m_1 h_1 \times m_2 h_2)$ is codisjointed $\Leftrightarrow (m_1 g_1, m_1 h_1)$ and $(m_2 g_2, m_2 h_2)$ are codisjointed $\Leftrightarrow m_1$ codisjoints (g_1, h_1) and m_2 codisjoints (g_2, h_2). Consequently, if $f_i: A_i \to B_i$ is the codisjunctor of (g_i, h_i) for $i \in \{1, 2\}$, then $f_1 \times f_2$ is the codisjunctor of $(g_1 \times g_2, h_1 \times h_2)$. Therefore finite products commute with codisjunctors. ■

Proposition (1.4.8). *For a finite product of morphisms $\Pi_{i \in I} f_i$, we have that $\Pi_{i \in I} f_i$ is epimorphic (regular epimorphic) $\Leftrightarrow \forall\, i \in I, f_i$ is epimorphic (regular epimorphic).*

Proof. If $\Pi_{i \in I} f_i$ is epimorphic (regular epimorphic), then any f_i is epimorphic (regular epimorphic) by Proposition (1.4.4). Conversely, if any f_i is epimorphic, then the pushout of (f_i, f_i) is of the form $(1_{B_i}, 1_{B_i})$ for any $i \in I$ and, according to Proposition (1.4.7), the pushout of $(\Pi_{i \in I} f_i, \Pi_{i \in I} f_i)$ is of the form $(\Pi_{i \in I} 1_{B_i}, \Pi_{i \in I} 1_{B_i})$. Therefore $\Pi_{i \in I} f_i$ is epimorphic. Moreover, in any regular category finite products of regular epimorphisms are regular epimorphisms. ∎

Proposition (1.4.9). *For a finite family of morphisms $(f_i : A \to B_i)_{i \in I}$ defining the morphism $f = (f_i)_{i \in I} : A \to \Pi_{i \in I} B_i$, the following statements are equivalent.*

(i) *f is an epimorphism.*

(ii) (a) *For each $i \in I$, f_i is an epimorphism.*

(b) *For each $(i, j) \in I \times I$ such that $i \neq j$, $f_i \wedge f_j = O_A$.*

Then $f = \vee_{i \in I} f_i$.

Proof. It is sufficient to prove the proposition for a pair of morphisms $f_1 : A \to B_1$, $f_2 : A \to B_2$. Let $(p_1 : B_1 \times B_2 \to B_1, p_2 : B_1 \times B_2 \to B_2)$ be the product of B_1 and B_2.

(i) ⇒ (ii) According to Proposition (1.4.2), p_1 and p_2 are epimorphisms. Thus $f_1 = p_1 f$ and $f_2 = p_2 f$ are epimorphisms. Let $(m : B_1 \to M, n : B_2 \to M)$ be the pushout of (f_1, f_2). Then $mp_1 f = mf_1 = nf_2 = np_2 f$. Thus $mp_1 = np_2$. Since products are codisjoint (Proposition (1.4.3)), $M \simeq 1$. Consequently, $f_1 \wedge f_2 = O_A$.

(ii) ⇒ (i) Since f_1 is epimorphic, the pushout of f_1 along f_1 is 1_{B_1}. Since $f_1 \wedge f_2 = O_A$, the pushout of f_2 along f_1 is O_{B_1}. Since finite products are co-universal, the pushout of f along f_1 is 1_{B_1}. Similarly, the pushout of f along f_2 is 1_{B_2}. Therefore the pushout of f along f is $1_{B_1} \times 1_{B_2} = 1_{B_1 \times B_2}$. It follows that f is epimorphic.

Moreover, if $g : A \to C$ is an epimorphism such that $f_1 \leqslant g$ and $f_2 \leqslant g$, there exist morphisms $u_1 : C \to B_1$ and $u_2 : C \to B_2$ such that $u_1 g = f_1$ and $u_2 g = f_2$, and the morphism $u = (u_1, u_2) : C \to B \times B_2$ satisfies $ug = f$. Hence $f \leqslant g$. Consequently, $f = f_1 \vee f_2$. ∎

Corollary (1.4.10). *Finite disjoint co-unions of epimorphisms are effective.*

1.5 CONGRUENCES

Let us recall that, following Grillet [19], § 5, a *congruence* on an object A is a relation on A which is the kernel pair of some morphism $f : A \to B$. It is an equivalence relation on A ([20], § 10.9), which is called an effective equivalence relation in ([21], § 10.10). A congruence on A will be denoted by $r = (r_1, r_2) : R \rightrightarrows A$, or $r = (r_1, r_2) : R \to A \times A$ or (r_1, r_2) or r or R. The least congruence on A is the diagonal $\Delta_A : A \to A \times A$, which is also called the equality $(1_A, 1_A) : A \rightrightarrows A$, while the greatest is the unit congruence $1_{A\times A} : A \times A \to A \times A$ or $(p_1, p_2) : A \times A \rightrightarrows A$. The quotient of a congruence $r = (r_1, r_2) : R \rightrightarrows A$ is the coequalizer of (r_1, r_2) and will be denoted by $q_r : A \to A/r$.

Proposition (1.5.1). *A congruence is unit if and only if it is codisjointed.*

Proof. A congruence (r_1, r_2) on A is codisjointed if and only if its coequalizer is the morphism $f : A \to 1$. Therefore any codisjointed congruence on A is the kernel pair of f, and thus is the unit congruence on A. Since f is a regular epimorphism (Proposition (1.4.1)), its kernel pair $(1_A, 1_A)$ is codisjointed. ∎

Proposition (1.5.2). *The set* Cong(A) *of congruences on an object A is a compact algebraic lattice whose compact elements are the finitely generated congruences. It is isomorphic to the opposite of the lattice* QuotReg(A) *of regular quotient objects of A.*

Proof. The set Sub(A) of subobjects of A is a complete lattice whose meets and joins are the intersections and unions respectively ([31], p. 122). Let $m : X \to A$ be a subobject of A whose domain X is a finitely generated object in **A** [16]. Let us show that it is a compact element in Sub(A) [17]. Let $m = \vee_{i\in I} m_i$. Denote by $p_0(I)$ the set of finite subsets of I and, for each $I_0 \in P_0(I)$, let $m_{I_0} = \vee_{i\in I_0} m_i$. Then $m = \vee_{I_0 \in P_0(I)} m_{I_0}$ is a directed union of subobjects of A. Thus the morphism m is the filtered colimit of $(m_{I_0})_{I_0 \in P_0(I)}$, because filtered colimits of monomorphisms are monomorphisms in any locally finitely presentable category. Since the object X is finitely generated, there exists $I_0 \in P_0(I)$ such that $m = m_{I_0}$. Thus $m = \vee_{i\in I_0} m_i$ where I_0 is a finite subset of I. Therefore m is a compact element in Sub(A). As any object of **A** is a monomorphic filtered colimit of finitely generated objects [16], any subobject of A is a union of compact subobjects of A, so that the lattice Sub(A) is compactly generated, i.e. is algebraic [17]. The set Rel(A) of relations on A is precisely the set Sub$(A \times A)$ and so it is an algebraic lattice. The set Cong(A) of congruences on A is a subset of Rel(A), closed under intersections, and thus it is a complete lattice. It is also closed in Rel(A) under directed unions, because directed unions in Rel(A) are computed as filtered colimits and directed colimits commute with kernel pairs [16]. We shall denote by $\overset{c}{\vee}$ arbitrary joins in Cong(A). Any relation r on A generates a congruence on A: it is the intersection of all congruences on A greater then r. It defines an order-preserving map Rel$(A) \to$ Cong(A) which is left-adjoint to the inclusion map Cong$(A) \to$ Rel(A). It is then easy to see that finitely generated congruences, i.e. congruences generated by compact relations $r \in$ Rel(A), are compact elements of Cong(A) and that any element in Cong(A) is the union of compact elements of Cong(A), so that Cong(A) is an algebraic lattice.

Let us show that the unit congruence $1_{A\times A}$ on A is a compact element in Cong(A). Let $(r_i)_{i\in I}$ be a family of congruences on A such that $\overset{c}{\vee}_{i\in I} r_i = 1_{A\times A}$. If $P_0(I)$ denotes the set of finite subsets of I and $r_{I_0} = \overset{c}{\vee}_{i\in I_0} r_i$ for $I_0 \in P_0(I)$, then we have a directed union

$$\overset{c}{\bigvee_{I_0\in P_0(I)}} r_{I_0} = \bigvee_{I_0\in P_0(I)} r_{I_0} = r.$$

Thus r is a filtered colimit of $(r_{I_0})_{I_0\in P_0(I)}$. Let us denote by q_{I_0} the coequalizer of r_{I_0}. We obtain a filtered colimit $\xrightarrow{lim}_{I_0\in P_0(I)} q_{I_0}$ which is the morphism $O_A : A \to 1$. As the terminal object 1 is finitely presentable, there exists $I_0 \in P_0(I)$ such that $q_{I_0} = O_A$ and thus $r_{I_0} = 1_{A\times A}$. It follows that $1_{A\times A}$ is compact in Cong(A). As a result, Cong(A) is compact.

Let us denote by QuotReg(A) the set of regular quotient objects of A, i.e. the set of quotient objects of A represented by regular epimorphisms, i.e. equalizers [2]. It is easy to check that we obtain an isomorphism between Cong(A) and QuotReg$(A)^{\text{op}}$ by assigning the coequalizer to a congruence in one direction, and by assigning the kernel pair to a regular epimorphism in the other direction. ■

Proposition (1.5.3). *Any congruence on A is the join of finitely generated codisjunctable congruences.*

Proof. Let $r, s \in$ Cong(A) be such that $r \not\leqslant s$. Let q_r and q_s be the quotients of A by r and s respectively. Then $q_s \not\leqslant q_r$, i.e. q_s does not factor through q_r. According to Axiom (1.2.2) of Zariski categories, there is a finitely generated codisjunctable object X of A and a pair of morphisms $(m, n) : X \rightrightarrows A$ such that $q_r m = q_r n$ and $q_s m \neq q_s n$. Let q be the coequalizer of (m, n) and let t be the kernel pair of q. Then q is the quotient of A by t, and the codisjunctor of (m, n) is the codisjunctor of t. The morphism q_r factors through q while q_s does not, i.e. $q_r \leqslant q$ and $q_s \not\leqslant q$. Thus $t \leqslant r$ and $t \not\leqslant s$. As a result, the relation $r \not\leqslant s$ implies the existence of a finitely generated codisjunctable congruence t such that $t \leqslant r$ but $t \not\leqslant s$. Alternatively, if, for any finitely generated codisjunctable congruence t, we have $t \leqslant r \Rightarrow t \leqslant s$, then $r \leqslant s$. Consequently, r is the join of all finitely generated codisjunctable congruences which are less than r. ■

Proposition (1.5.4). *Codisjunctors commute with finite intersections of congruences, i.e. for any pair of codisjunctable congruences r, s on A, the intersection $r \wedge s$ is codisjunctable and* codis $(r \wedge s) =$ codis$(r) \wedge$ codis(s).

Proof. Let $p : A \to P$, $q : A \to Q$, and $m : A \to M$ be the coequalizers of r, s, and $r \wedge s$ respectively. Let us denote by $v : M \to P$ and $w : M \to Q$ the morphisms such that $vm = p$ and $wm = q$. Let $en = (v,w)$ be the regular factorization of the morphism $(v, w) : M \to P \times Q$, where n is a regular epimorphism and e is monomorphic. Let u be the kernel pair of the regular epimorphism nm. We have $r \wedge s \leqslant u$ because $r \wedge s$ is the kernel pair of m. But we also have $u \leqslant r$ because p factors through nm, and $u \leqslant s$ because q factors through nm. Thus $u \leqslant r \wedge s$ so that $r \wedge s = u$, n is isomorphic, and $(v, w) : M \to P \times Q$ is monic. Let us denote by $d : A \to D$ and $\delta : A \to \Delta$ the codisjunctors of r and s respectively. Let $(g : D \to B, h : \Delta \to B)$ be the pushout of (d, δ). We now show that $f = gd = h\delta$ is the codisjunctor of $r \wedge s$. Let $(\bar{f} : M \to \bar{M}, \bar{m} : B \to \bar{M})$ be the pushout of (m, f). Let $\bar{v} : \bar{M} \to \bar{P}$ be the pushout of v along $\bar{f}$ and let $\bar{w} : \bar{M} \to \bar{Q}$ be the pushout of w along $\bar{f}$. As products of two objects are co-universal, the pushout of $(v, w) : M \to P \times Q$ along $\bar{f}$ is $(\bar{v}, \bar{w}) : \bar{M} \to \bar{P} \times \bar{Q}$. As the morphism f is flat (Propositions (1.6.2) and (1.8.3) below), $(\bar{v}, \bar{w}) : \bar{M} \to \bar{P} \times \bar{Q}$ is monomorphic. But $\bar{P} \simeq 1$ because the pushout of p along d is the morphism $O_D : D \to 1$ and because f factors through d. Similarly, $\bar{Q} \simeq 1$. Thus $\bar{P} \times \bar{Q} \simeq 1$, the morphism $(\bar{v}, \bar{w}) : \bar{M} \to \bar{P} \times \bar{Q}$ is isomorphic, and $\bar{M} = 1$. It follows that the morphism f codisjoints the relation $r \wedge s$. Moreover, let $h : A \to H$ be a morphism which codisjoints $r \wedge s$. Then it codisjoints r and so it factors through d. Similarly, it codisjoints s and so it factors through δ. Consequently, the morphism h factors through f in a unique way as f is an epimorphism. As a result, f is the codisjunctor of $r \wedge s$. ■

Direct and inverse images of congruences

Let $f : A \to B$ be any morphism in $\mathbf{A}$. It induces a *direct image map* $f_* : \mathrm{Sub}(A) \to \mathrm{Sub}(B)$ which assigns to a subobject $m : X \to A$ the subobject $f_*(m) : Y \to B$ which appears in the factorization $fm = f_*(m)e$ where e is a regular epimorphism. It also induces an *inverse image map* $f^* : \mathrm{Sub}(B) \to \mathrm{Sub}(A)$ which assigns to a subobject $n : Y \to B$ the subobject $f^*(n) : X \to A$ appearing in the pullback $(f^*(n), g)$ of (f, n). The maps f_*, f^* are order-preserving and f_* is left-adjoint to f^*. The morphism f also induces the morphism $f \times f : A \times A \to B \times B$ which induces the direct image map $(f \times f)_* : \mathrm{Sub}(A \times A) \to \mathrm{Sub}(B \times B)$ and the inverse image map $(f \times f)^* : \mathrm{Sub}(B \times B) \to \mathrm{Sub}(A \times A)$. These two maps are simply denoted by $f_* : \mathrm{Rel}(A) \to \mathrm{Rel}(B)$ and $f^* : \mathrm{Rel}(B) \to \mathrm{Rel}(A)$. Let $r = (r_1, r_2) : R \rightrightarrows A$ be a congruence on A. Let us denote by $q : A \to Q$ the quotient of A by r, and by $(p : B \to P, g : Q \to P)$ the pushout of (f, q). The kernel pair of p is a *congruence* on B called the *direct image of r by f*, and is denoted by $f_{*c}(r)$. It defines the *direct image congruence map* $f_{*c} : \mathrm{Cong}(A) \to \mathrm{Cong}(B)$ induced by f. Notice that f_{*c} is not the restriction of the map $f_* : \mathrm{Rel}(A) \to \mathrm{Rel}(B)$ and that the quotient by $f_{*c}(r)$ is obtained from the quotient by r by pushing out along f. If $s = (s_1, s_2) : S \rightrightarrows B$ is a congruence on B, which is the kernel pair of a morphism

$g : B \to C$, the inverse image $f^*(s)$ of the relation s by the morphism f is the kernel pair of gf. Thus it is a congruence on A. It defines the *inverse image congruence map* $f^{*c} : \mathrm{Cong}(B) \to \mathrm{Cong}(A)$ which is, indeed, the restriction of the map $f^* : \mathrm{Rel}(B) \to \mathrm{Rel}(A)$. Notice that the quotient by $f^{*c}(s)$ is the regular epimorphism $q : A \to Q$ which appears in the factorization $pf = mq$, where p is the quotient of B by s and m is a monomorphism. It is easy to see that f^{*c} is right-adjoint to f_{*c}. In the special case where f is a regular epimorphism, $f_{*c} \circ f^{*c} = Id$.

Congruence functors

Let $f : A \to B$ be a morphism in $\mathbf{A}$. Let $r = (r_1, r_2) : R \rightrightarrows A$ be a congruence on A, let $q_r : A \to A/r$ be the quotient of A by r, let $s = f_{*c}(r)$, and let $q_s : B \to B/s$ be the quotient of B by s. Then q_s is the pushout of q_r along f. If r is finitely generated, q_r is a finitely generated regular quotient and thus q_s is also; hence s is finitely generated. Let us assume that r is codisjunctable. Let d be the codisjunctor of r and let δ be the pushout of d along f. According to Proposition (1.1.5), δ is the codisjunctor of the pair (fr_1, fr_2). But the pair (fr_1, fr_2) has the same coequalizer q_s as the congruence s. Therefore the pair (fr_1, fr_2) has the same codisjunctor as s. Thus δ is the codisjunctor of s and hence s is codisjunctable. As a result, if **Ord** is the category of ordered sets, $\vee$-**SemLat** is the category of bounded join-semilattices, and V-**SemLat** is the category of complete bounded join-semilattices, then we can define the following congruence functors:

(i) the *congruence functor* $\mathrm{Cong} : \mathbf{A} \to V$-**SemLat** which assigns $\mathrm{Cong}(A)$ *to an object* A and f_{*c} to a morphism f;

(ii) the *finitely generated congruence functor* $\mathrm{Cong}_0 : \mathbf{A} \to \vee$-**SemLat** which assigns the set $\mathrm{Cong}_0(A)$ of finitely generated congruences on A to an object A and the restriction of f_{*c} to a morphism f;

(iii) the *codisjunctable congruence functor* $\mathrm{CongCodis} : \mathbf{A} \to \mathbf{Ord}$ which assigns the set $\mathrm{CongCodis}(A)$ *of codisjunctable congruences on* A to an object A and the restriction of f_{*c} to a morphism f;

(iv) the *finitely generated codisjunctable congruence functor* $\mathrm{CongCodis}_0 : \mathbf{A} \to \mathbf{Ord}$ which assigns the set $\mathrm{CongCodis}_0(A)$ of finitely generated codisjunctable congruences on A to an object A and the restriction of f_{*c} to a morphism f.

Proposition (1.5.5). *The finitely generated congruence functor* Cong_0 : $\mathbf{A} \to \vee$-**SemLat** *preserves filtered colimits.*

Proof. Let $(\alpha_i : A_i \to A)_{i \in I}$ be a filtered colimit in $\mathbf{A}$.

(i) Let $r = (r_1, r_2) : R \rightrightarrows A$ in $\mathrm{Cong}_0(A)$. There exists a finitely presentable object X and a morphism $f : X \to R$ such that $(r_1 f, r_2 f)$ has the same coequalizer as (r_1, r_2). There exists $i \in \mathbf{1}$ and $(m_i, n_i) : X \rightrightarrows A_i$ such that

$(\alpha_i m_i, \alpha_i n_i) = (r_1 f, r_2 f)$. The congruence r_i on A_i, the kernel pair of the coequalizer of (m_i, n_i), is finitely generated and its direct image by α_i is r, i.e. $(\alpha_i)_{*c}(r_i) = r$.

(ii) Let $i \in \mathbf{I}$ and $s_i = (s_{i1}, s_{i2}) : S_i \rightrightarrows A_i$ in $\mathrm{Cong_o}(A_i)$ such that $(\alpha_i)_{*c}(s_i) = \Delta_A$. Then α_i coequalizes (s_{i1}, s_{i2}). There exists a finitely presentable object Y and a morphism $g : Y \to S_i$ such that $(s_{i1}g, s_{i2}g)$ has the same coequalizer as (s_{i1}, s_{i2}). There exists a morphism $u : i \to i'$ in $\mathbf{I}$ such that A_u coequalizes $(s_{i1}g, s_{i2}g)$, and then A_u coequalizes (s_{i1}, s_{i2}). Thence $(A_u)_{*c}(s_i) = \Delta_{A_{i'}}$.

(iii) Let us assume that i_0 is an initial object in $\mathbf{I}$. Let $r_0 \leqslant s_0$ in $\mathrm{Cong_o}(A_{i_0})$ such that $(\alpha_{i_0})_{*c}(r_0) = (\alpha_{i_0})_{*c}(s_0)$. Let $r = (\alpha_{i_0})_{*c}(r_0)$. For any $i \in \mathbf{I}$, let us denote the unique morphism in $\mathbf{I}$ by $u_i : i_0 \to i$, the congruence $(A_{u_i})_{*c}(r_0)$ by r_i, and the coequalizer of r_i by $q_i : A_i \to B_i$. For any morphism $u : i \to i'$ in $\mathbf{I}$, let us denote by $B_u : B_i \to B_{i'}$ the morphism such that $B_u q_i = q_{i'} A_u$. In this way we obtain a filtered diagram $(B_i)_{i \in \mathbf{I}}$. Let us denote the coequalizer of r by $q : A \to B$ and, for any $i \in \mathbf{I}$, the morphism such that $\beta_i q_i = q\alpha_i$ by $\beta_i : B_i \to B$. In this way we obtain an inductive cone $(\beta_i : B_i \to B)_{i \in \mathbf{I}}$ which is indeed a filtered colimit in $\mathbf{A}$. The direct image $(q_{i_0})_{*c}(s_0)$ of s_0 by q_{i_0} is such that

$$(\beta_{i_0})_{*c}((q_{i_0})_{*c}(s_0)) = (\beta_{i_0} q_{i_0})_{*c}(s_0) = (q\alpha_{i_0})_{*c}(s_0) =$$
$$q_{*c}((\alpha_{i_0})_{*c}(s_0)) = q_{*c}(r) = \Delta_B.$$

By applying (ii) to the filtered colimit $(\beta_i : B_i \to B)_{i \in \mathbf{I}}$, we find that there exists $i \in \mathbf{I}$ such that the morphism $u_i : i_0 \to i$ in $\mathbf{I}$ is such that $(B_{u_i})_{*c}((q_{i_0})_{*c}(s_0)) = \Delta_{B_i}$. Then $(q_i A_{u_i})_{*c}(s_0) = (B_{u_i} q_{i_0})_{*c}(s_0) = \Delta_{B_i}$. It implies that $q_i A_{u_i}$ coequalizes s_0 and hence that $(A_{u_i})_{*c}(r_0) = (A_{u_i})_{*c}(s_0)$.

(iv) Let $i \in \mathbf{I}$ and $r_i \leqslant s_i$ in $\mathrm{Cong_o}(A_i)$ such that $(\alpha_i)_{*c}(r_i) = (\alpha_i)_{*c}(s_i)$. Let us consider the comma category i/$\mathbf{I}$, the diagram $(A_j)_{(j,u) \in i/\mathbf{I}}$ induced by the diagram $(A_i)_{i \in \mathbf{I}}$, and the colimit $(\alpha_j : A_j \to A)_{(j,u) \in i/\mathbf{I}}$ induced by the colimit $(\alpha_i : A_i \to A)_{i \in \mathbf{I}}$. By applying (iii) to this filtered colimit, we deduce the existence of a morphism $u : i \to j$ in $\mathbf{I}$ such that $(A_u)_{*c}(r_i) = (A_u)_{*c}(s_i)$.

(v) Let $i \in \mathbf{I}$ and let r_i, s_i in $\mathrm{Cong_o}(A_i)$ be such that $(\alpha_i)_{*c}(r_i) = (\alpha_i)_{*c}(s_i)$. Then $r_i \overset{c}{\vee} s_i$ is in $\mathrm{Cong_o}(A_i)$ and is such that

$$(\alpha_i)_{*c}(r_i \overset{c}{\vee} s_i) = (\alpha_i)_{*c}(r_i) \overset{c}{\vee} (\alpha_i)_{*c}(s_i) = (\alpha_i)_{*c}(r_i) = (\alpha_i)_{*c}(s_i).$$

According to (iv), there exists a morphism $u : i \to j$ in $\mathbf{I}$ such that

$$(A_u)_{*c}(r_i) = (A_u)_{*c}(r_i \overset{c}{\vee} s_i) = (A_u)_{*c}(s_i).$$

(vi) As a result of the above, the functor $\mathrm{Cong_o}$ preserves filtered colimits. ∎

Examples

Congruences in **CRng**, **CAlg**(k), **AlgCAlg**(k), **RegCRng**, and **RegCAlg**(k) are congruences modulo ideals. Congruences in **RedCRng** and **RedCAlg**(k) are congruences modulo radical ideals. Congruences in **GradCRng** are congruences modulo graded ideals, i.e. generated by homogeneous elements. Congruences in **RlRng** are congruences modulo real ideals, i.e. ideals I such that

$$(x_1^2 + \ldots + x_n^2)y \in I \Rightarrow x_1^2 y \in I.$$

Congruences in **RedRlRng** are congruences modulo radical real ideals, i.e. ideals I such that

$$x_1^2 + \ldots + x_n^2 \in I \Rightarrow x_1 \in I.$$

Congruences in **RlOrdRng** and **RlLatRng** are congruences modulo absolutely convex ideals ([4], Corollary 2.5.5; [28], Definition 1.8). Congruences in **Mod** on an A-module E are pairs $(\mathrm{mod}(I), \mathrm{mod}(X))$ of a congruence modulo an ideal I of A, and a congruence modulo an A-submodule X of E containing IE.

1.6 FLAT MORPHISMS

Definition (1.6.1). *A morphism $f: A \to B$ is* flat *if the pushout functor $A/\mathbf{A} \to B/\mathbf{A}$ along f preserves monomorphisms.*

Proposition (1.6.2).

(i) *Flat morphisms are co-universal, i.e. the pushout of a flat morphism is flat.*

(ii) *Composites of flat morphisms are flat.*

(iii) *Coproducts and filtered colimits of flat morphisms are flat.*

(iv) *For a flat epimorphism $f: A \to B$ and any morphism $g: B \to C$: g is flat $\Leftrightarrow$ gf is flat.*

(v) *A finite product of morphisms is flat if and only if each factor is flat.*

(vi) *If $(f_i : A \to B_i)_{i \in [1,n]}$ is a finite sequence of flat morphisms, the morphism $(f_i): A \to \Pi_{i=1}^n B_i$ is flat.*

Proof.

(i) and (ii) are immediate.

(iii) First, let us consider the coproduct $f_1 \amalg f_2 : A_1 \amalg A_2 \to B_1 \amalg B_2$ of two flat morphisms $f_1 : A_1 \to B_1$ and $f_2 : A_2 \to B_2$ with canonical inductions $\alpha_1 : A_1 \to A_1 \amalg A_2$, $\alpha_2 : A_2 \to A_1 \amalg A_2$, $\beta_1 : B_1 \to B_1 \amalg B_2$, and $\beta_2 : B_2 \to B_1 \amalg B_2$. Let $(\bar{f}_1, \bar{\alpha}_1)$ be the pushout of (α_1, f_1), let $(\bar{f}_2, \bar{\alpha}_2)$ be the pushout of (a_2, f_2), and let g_1 and g_2 be morphisms such that $g_1 \bar{f}_1 = f_1 \amalg f_2$, $g_1 \bar{\alpha}_1 = \beta_1$, $g_2 \bar{f}_2 = f_1 \amalg f_2$, and $g_2 \bar{\alpha}_2 = \beta_2$. Then (g_1, g_2) is the pushout of $(\bar{f}_1, \bar{f}_2)$. As flat morphisms are co-universal, $\bar{f}_1$ and $\bar{f}_2$ are flat, as are g_1 and g_2. As composites of flat morphisms are flat, $f_1 \amalg f_2 = g_1 \bar{f}_1$ is flat.

Second, let us consider a filtered diagram of flat morphisms $(f_i : A_i \to B_i)_{i \in \mathbf{I}}$, whose colimit is $f: A \to B$ with inductions $(\alpha_i : A_i \to A, \beta_i : B_i \to B)_{i \in \mathbf{I}}$. For any $i \in \mathbf{I}$, let $\bar{f}_i : A \to \bar{B}_i$ be the pushout of f_i along α_i. Then $f = \xrightarrow{lim}_{i \in \mathbf{I}} \bar{f}_i$ is a filtered colimit. Let $m : (X, x) \to (Y, y)$ be a monomorphism in $A/\mathbf{A}$. Denote the pushout of m along $\bar{f}_i$ by $\bar{m}_i : (X_i, x_i) \to (Y_i, y_i)$ and the pushout of m along f by $\bar{m} : (\bar{X}, \bar{x}) \to (\bar{Y}, \bar{y})$. As pushouts commute with colimits, we obtain a filtered colimit $\bar{m} = \xrightarrow{lim}_{i \in \mathbf{I}} \bar{m}_i$. As $\bar{f}_i$ is flat, $\bar{m}_i$ is monomorphic. As a filtered colimit of monomorphisms is a monomorphism, $\bar{m}$ is monomorphic. As a result, f is flat. As any coproduct is a filtered colimit of finite coproducts, any coproduct of flat morphisms is a flat morphism.

(iv) The implication $\Rightarrow$ follows from (ii). Conversely, if gf is flat, then g is flat as it is the pushout of gf along f.

(v) For any pair of objects A, C in $\mathbf{A}$ the co-universality of the product $A \times C$ implies that the category $A \times C/\mathbf{A}$ is equivalent to the category $A/\mathbf{A} \times C/\mathbf{A}$. Moreover, for any pair of morphisms $f: A \to B, g: C \to D$ in $\mathbf{A}$, the pushout functor $A \times C/\mathbf{A} \to B \times D/\mathbf{A}$ along $f \times g : A \times C \to B \times D$ is equivalent to the functor $A/\mathbf{A} \times C/\mathbf{A} \to B/\mathbf{A} \times D/\mathbf{A}$, which is the product of the pushout functor $A/\mathbf{A} \to B/\mathbf{A}$ along f with the pushout functor $C/\mathbf{A} \to D/\mathbf{A}$ along g. Consequently, the pushout functor along $f \times g$ preserves monomorphisms if and only if the pushout functors along f and g preserve monomorphisms, i.e. $f \times g$ is flat if and only if f and g are flat. This result extends immediately to products of finitely many morphisms.

(vi) For any object A in $\mathbf{A}$, the pushout functor along the diagonal $\Delta_A : A \to A \times A$ is equivalent to the diagonal functor $A/\mathbf{A} \to A/\mathbf{A} \times A/\mathbf{A}$. Therefore it preserves monomorphisms, and so the diagonal Δ_A is flat. It follows that, for any pair of flat morphisms $f: A \to B, g: A \to C$, the morphism $(f, g): A \to B \times C$ is flat as it is the composite of the flat morphisms Δ_A and $f \times g$. This result extends immediately to a morphism of the form $(f_i): A \to \Pi_{i=1}^{n} B_i$. ■

Proposition (1.6.3). *Any flat morphism $f: A \to B$ induces a morphism of bounded lattices $f_{*_c} : \mathrm{Cong}(A) \to \mathrm{Cong}(B)$.*

Proof. The map f_{*_c} is order-preserving and left-adjoint to f^{*_c}. Thus it is a morphism of complete join-semilattices. Let $r, s \in \mathrm{Cong}(A)$ and $t = r \wedge s$. Denote by $p: A \to P$, $q: A \to Q$, and $u: A \to U$ the quotients of A by r, s, and t respectively and let $v: U \to P$ and $w\ U \to Q$ such that $vu = p$ and $wu = q$. In QuotReg(A), we have $u = p \vee q$ (Proposition (1.5.2)). Thus, since **A** is a regular category, we have $1_U = v \vee w$ in QuotReg(U). Thus $(v, w) : U \to P \times Q$ is a monomorphism. Let $\bar{p}$, $\bar{q}$, $\bar{u}$, $\bar{v}$, and $\bar{w}$ be the morphisms obtained from p, q, u, v, and w by applying the pushout functor along f. Because products of two objects are co-universal, the image of $(v, w) : U \to P \times Q$ by the pushout functor along f is the morphism $(\bar{v}, \bar{w}) : \bar{U} \to \bar{P} \times \bar{Q}$. Because the morphism f is flat, $(\bar{v}, \bar{w})$ is monomorphic. Thus $\bar{v} \vee \bar{w} = 1_{\bar{u}}$ in QuotReg($\bar{U}$) and $\bar{p} \vee \bar{q} = \bar{u}$ in QuotReg(B). But $f_{*_c}(r)$, $f_{*_c}(s)$, and $f_{*_c}(t)$ are the kernel pairs of $\bar{p}$, $\bar{q}$, and $\bar{u}$ respectively; therefore $f_{*_c}(r) \wedge f_{*_c}(s) = f_{*_c}(t)$ (Proposition (1.5.2)). Moreover, $f_{*_c}(1_{A \times A}) = 1_{B \times B}$. As a result, f_{*_c} is a morphism of bounded lattices. ∎

Proposition (1.6.4). *The Beck–Chevalley condition for congruences along flat morphisms: if $f : A \to B$ is a flat morphism and $(m : B \to M, n : C \to M)$ is the pushout of $(f : A \to B, g : A \to C)$, then the congruence maps f_{*_c}, g^{*c}, m^{*c}, n_{*_c} satisfy the Beck–Chevalley condition $f_{*_c} \circ g^{*c} = m^{*c} \circ n_{*_c}$.*

Proof. Let $r \in \mathrm{Cong}(C)$, $s = n_{*c}(r)$, $t = g^{*c}(r)$, and $u = f_{*c}(t)$. Then

$$\begin{aligned} m_{*c}(u) &= m_{*c}(f_{*c}(t)) = (mf)_{*c}(t) = (ng)_{*c}(t) = n_{*c}(g_{*c}(t)) \\ &= n_{*c}(g_{*c}(g^{*c}(r))) \leqslant n_{*c}(r) = s. \end{aligned}$$

Let $q_r : C \to C/r$, $q_s : M \to M/s$, $q_t : A \to A/t$, and $q_u : B \to B/u$ be the quotients by r, s, t, and u respectively. There are unique morphisms $\bar{f} : A/t \to B/u$, $\bar{g} : A/t \to C/r$, $\bar{m} : B/u \to M/s$, and $\bar{n} : C/r \to M/s$ such that $\bar{f}q_t = q_u f$, $\bar{g}q_t = q_r g$, $\bar{m}q_u = q_s m$, and $\bar{n}q_r = q_s n$. Then $(\bar{f}, q_u)$ is the pushout of (q_t, f), $(\bar{n}, q_s)$ is the pushout of (q_r, n), and $\bar{g}$ is monomorphic. Consequently, $(\bar{n}, \bar{m}q_u)$ is the pushout of $(q_r g, f)$ and thus $(\bar{m}, \bar{n})$ is the pushout of $(\bar{f}, \bar{g})$. Because f is flat and $\bar{g}$ is monomorphic, $\bar{m}$ is monomorphic. Therefore $u = m^{*c}(s)$, i.e. $f_{*c}(g^{*c}(r)) = m^{*c}(n_{*c}(r))$. As a result, $f_{*c} \circ g^{*c} = m^{*c} \circ n_{*c}$. ∎

Examples

In **CRng**, **CAlg**(k), and **GradCRng** the morphisms $f : A \to B$, which make B a flat A-module are precisely flat. In **RedCRng**, **RedCAlg**(k), and **RedGradCRng** the morphisms $f : A \to B$ which make B a flat A-module are flat, but the converse is in question. In **CAlg**, those morphisms $(\varphi, f) : (A, E) \to (B, F)$ such that φ makes B a flat A-module and f makes F a flat E-module are precisely the flat morphisms. In **RlRng** and **RedRlRng**, the inclusion morphism $f : \mathbf{Q}[\sqrt{2}] \to \mathbf{R}$ is not flat because the pushout along f of the monomorphism $g : \mathbf{Q}[\sqrt{2}] \to \mathbf{R}$ defined by $g(a + b\sqrt{2}) = a - b\sqrt{2}$ is the null morphism $\mathbf{R} \to \{0\}$. Similarly, in **RlOrdRng** and **RedRlOrdRng** the morphism f is not flat if $\mathbf{Q}[\sqrt{2}]$ is equipped with the order whose positive elements are sums of squares and $\mathbf{R}$ is equipped with the usual order.

1.7 FLAT EPIMORPHISMS AND FLAT BIMORPHISMS

Proposition (1.7.1).

(i) *Flat epimorphisms are co-universal.*

(ii) *Flat bimorphisms are essential monomorphisms.*

(iii) *Colimits (filtered colimits) of flat epimorphisms (bimorphisms) are flat epimorphisms (bimorphisms).*

(iv) *Composites and co-intersections of flat epimorphisms (bimorphisms) are flat epimorphisms (bimorphisms).*

(v) *For any flat epimorphism (bimorphism) $f: A \to B$ and any morphism $g: B \to C$, we have that g is a flat epimorphism (bimorphism) $\Leftrightarrow$ gf is a flat epimorphism (bimorphism).*

(vi) *A finite product of morphisms is a flat epimorphism (bimorphism) if and only if each of its factors is a flat epimorphism (bimorphism).*

Proof.

(i) follows immediately from Proposition (1.6.2).

(ii) Let $f: A \to B$ be a flat bimorphism and let $g: B \to C$ be a morphism such that gf is monomorphic. Then $(g, 1_C)$ is the pushout of (f, gf), and hence g is monomorphic. Therefore f is an essential monomorphism ([35], § 15.2.3).

(iii) Following Proposition (1.6.2), coproducts and filtered colimits of flat epimorphisms are flat epimorphisms. As filtered colimits of monomorphisms are monomorphic, filtered colimits of flat bimorphisms are flat bimorphisms. Finally, let us consider a pair of morphisms $(m_1, m_2): A_1 \rightrightarrows A_2$ with coequalizer $q: A_2 \to A_3$, a pair of morphisms $(n_1, n_2): B_1 \rightrightarrows B_2$ with coequalizer $r: B_2 \to B_3$, and flat epimorphisms $f_1: A_1 \to B_1$, $f_2: A_2 \to B_2$ such that $f_2 m_1 = n_1 f_1$ and $f_2 m_2 = n_2 f_1$. The morphism f_3: $A_3 \to B_3$ defined by $f_3 q = r f_2$ is the pushout of f_2 along q. Thus it is a flat epimorphism. As a result, coequalizers of flat epimorphisms are flat epimorphisms.

(iv) The property of composition is immediate. The property of co-intersections of flat epimorphisms follows from (iii). Let $(f: A \to B$, $g: A \to C)$ be a pair of flat bimorphisms and let $(h: B \to D, k: C \to D)$ be the pushout of (f, g). Since f is flat and g is monomorphic, h is monomorphic. Thus $hf = kg$ is a flat bimorphism. It follows that finite co-intersections of flat bimorphisms are flat bimorphisms. Since any co-intersection is a cofiltered co-intersection of finite co-intersections and any filtered colimit of monomorphisms is monomorphic, it follows that any co-intersection of flat bimorphisms is a flat bimorphism.

(v) follows from Proposition (1.6.2) and (ii).

(vi) follows from Propositions (1.6.2) and (1.4.6). ■

Proposition (1.7.2). *For a flat epimorphism $f: A \to B$, any congruence on B is the direct image by f of its inverse image by f, i.e. $f_{*_c} \circ f^{*c} = Id$.*

Proof. The pushout of (f, f) is $(1_B, 1_B)$. According to the Beck–Chevalley condition (Proposition (1.6.4)), $f_{*c} \circ f^{*c} = (1_B)^{*c} \circ (1_B)_{*c} = Id$. ■

Flat quotient objects

Flat epimorphisms $f: A \to B$ represent *flat quotient objects* of A. Let us denote by $\mathrm{QuotFl}(A)$ the set of flat quotient objects of A. According to Proposition (1.7.1), $\mathrm{QuotFl}(A)$ is a complete lattice whose meets are the co-intersections of quotient objects of A.

Flat epimorphic extensions

A flat bimorphism $f: A \to B$ is also called a *flat epimorphic extension* of A. The full subcategory of $A/\mathbf{A}$ whose objects are the flat epimorphic extensions of A is small and is associated to a preorder. It is denoted by $\mathbf{ExtEpFl}(A)$. According to Proposition (1.7.1), $\mathbf{ExtEpFl}(A)$ is a complete prelattice whose joins are the co-intersections of epimorphisms.

Theorem (1.7.3). *Any object A has a flat epimorphic hull, i.e. a flat epimorphic extension $\varphi_A : A \to F(A)$ with the property that, for any flat epimorphic extension $f: A \to B$ of A, there exists a unique morphism $g: B \to F(A)$ such that $gf = \varphi_A$.*

Proof. Take the top object $\varphi_A : A \to F(A)$ in $\mathbf{ExtEpFl}(A)$. ■

Definition (1.7.4). *An object is* flatly complete *if any of its flat epimorphic extensions is an isomorphism.*

The category **FlComplA** *of flatly complete objects of* **A** has the flatly complete objects of **A** as objects and the flat morphisms as morphisms. We denote by $\mathbf{A}_{\mathrm{flat}}$ the category whose objects are those of **A** and whose morphisms are flat morphisms of **A**.

Theorem (1.7.5). *To any object is associated a universal flatly complete object, i.e.* **FlComplA** *is a reflective subcategory of* $\mathbf{A}_{\mathrm{flat}}$. *The reflection is precisely the flat epimorphic hull.*

Proof. Let A be an object and let $\varphi_A : A \to F(A)$ be its flat epimorphic hull. The object $F(A)$ is flatly complete because any flat epimorphic extension $m: F(A) \to M$ is such that $m\varphi_A : A \to M$ is a flat epimorphic extension of A. Thus $m\varphi_A \simeq \varphi_A$ and m is an isomorphism. Let $f: A \to B$ be a flat morphism whose codomain is a flatly complete object B. Let $(m: B \to M,\ n: F(A) \to M)$ be the pushout of (f, φ_A). Then m is a flat epimorphism. Since φ_A is monomorphic, m is monomorphic and hence is a flat bimorphism. Therefore m is an isomorphism. It follows that $f = m^{-1}n\varphi_A$, where $m^{-1}n : F(A) \to B$ is a flat morphism, factors through φ_A. This factorization is unique since φ_A is epimorphic. As a result $\varphi_A : A \to F(A)$ is the universal flatly complete object associated to A. ■

Examples

In **CRng** the flat epimorphic hull of a ring A is called the maximal flat epimorphism $f: A \to M(A)$ [30]. If A is noetherian, or coherent and reduced, it is also the total ring of fractions of A [30, 36].

1.8 SINGULAR EPIMORPHISMS

Definition (1.8.1). *A* singular epimorphism *is a morphism which is the codisjunctor of some pair of parallel morphisms.*

Proposition (1.8.2). *For any morphism f the following assertions are equivalent:*

(i) *f is a singular epimorphism;*

(ii) *f is the codisjunctor of a pair of parallel morphisms whose domain is a finitely presentable object;*

(iii) *f is the codisjunctor of a congruence;*

(iv) *f is the codisjunctor of a finitely generated congruence.*

Proof.

(iv) ⇒ (iii) ⇒ (i) is obvious.

(i) ⇒ (ii) Let $f: A \to B$ be a singular epimorphism, which is a codisjunctor of a pair of morphisms $(g, h) : C \rightrightarrows A$. The object C is a filtered colimit of finitely presentable objects. Let $(C, (\gamma_i)_{i\in \mathbf{I}}) = \xrightarrow{lim}_{i\in\mathbf{I}} C_i$. For each $i \in \mathbf{I}$, we write $g_i = g\gamma_i$, $h_i = h\gamma_i$, and $q_i : A \to Q_i$ for the coequalizer of (g_i, h_i). The coequalizer $q : A \to Q$ of (g, h) is the cofiltered co-intersection of the family of epimorphisms $(q_i)_{i\in\mathbf{I}}$. The pushout of q along f is the cofiltered co-intersection of the family of pushouts of q_i along f. But the pushout of q along f is $O_B : B \to 1$. As the terminal object 1 is finitely presentable, there exists $i \in \mathbf{1}$ such that the pushout of q_i along f is O_B. Then f codisjoints the pair (g_i, h_i). Moreover, any morphism $k : A \to K$ which codisjoints (g_i, h_i) codisjoints (g, h) and so factorizes uniquely through f. As a result, f is the codisjunctor of the pair of morphisms (g_i, h_i) whose domain is a finitely presentable object.

(ii) ⇒ (iv) Let $f: A \to B$ be the codisjunctor of a pair of morphisms $(g, h) : C \rightrightarrows A$ whose domain C is finitely presentable. Let $(p_1, p_2) : A \times A \rightrightarrows A$ be the product of A with itself. The morphism $(g, h) : C \to A \times A$ factors in a regular epimorphism $q : C \to X$ followed by a monomorphism $m : X \to A \times A$. The pair $(m_1, m_2) = (p_1 m, p_2 m) :$

$X \rightrightarrows A$ represents the relation on A generated by (g, h). The object X is finitely generated as a regular quotient of C. Let $r = (r_1, r_2) : R \rightrightarrows A$ be the congruence on A generated by (m_1, m_2). It is a finitely generated congruence (cf. Proposition (1.5.2)). According to Proposition (1.1.5), f is the codisjunctor of (m_1, m_2). It follows easily that f is also the codisjunctor of r. ■

Proposition (1.8.3). *Singular epimorphisms are flat.*

Proof. Let $f : A \to B$ be a singular epimorphism, which is the codisjunctor of a pair of morphisms $(g, h) : C \rightrightarrows A$ whose domain C is finitely presentable. By Axiom (1.2.2) of Zariski categories, there is a finite sequence $X_1, \ldots, X_n$ of flatly codisjunctable objects and an epimorphism $p : \amalg_{i \in [1,n]} X_i \to C$. For each $i \in [1, n]$, let us denote the induction indexed by i by $\alpha_i : X_i \to \amalg^{X_i}_{i \in [1,n]}$, the coequalizer of the pair $(fgp\alpha_i, fhp\alpha_i)$ by q_i, and the congruence on B kernel pair of q_i by r_i. Let $q : B \to Q$ be the co-intersection of $(q_i)_{i \in [1,n]}$. For any $i \in [1, n]$, we have $qfgp\alpha_i = qfhp\alpha_i$. Therefore we have $qfg = qfh$. Because (fg, fh) is codisjointed, we have $Q \simeq 1$. Thus $q_1 \wedge \ldots \wedge q_n = \mathrm{O}_B$ in $\mathrm{QuotReg}(B)$. By Proposition (1.5.2) it follows that $r_1 \overset{c}{\vee} \ldots \overset{c}{\vee} r_n = 1_{B \times B}$. For each $i \in [1, n]$, let $f_i : A \to B_i$ be the codisjunctor of the pair $(gp\alpha_i, hp\alpha_i)$ and let $d_i : B \to B_i$ be the morphism such that $d_i f = f_i$. Because f is epimorphic, d_i is the pushout of f_i along f. Hence d_i is the codisjunctor of the pair $(fgp\alpha_i, fhp\alpha_i)$ (Proposition (1.1.5)). Then d_i is also the codisjunctor of r_i. By Axiom (1.2.6) of Zariski categories, we have $d_i \vee \ldots \vee d_n = 1_B$ in $\mathrm{Quot}(B)$. Thus the family $(d_i : B \to B_i)_{i \in [1,n]}$ is monomorphic. Because X_i is flatly codisjunctable, the morphism f_i is flat. Let $m : A \to M$ be a monomorphism, let $(k : M \to N, n : B \to N)$ be the pushout of (m, f), and let $(k_i : N \to N_i, n_i : B_i \to N_i)$ be the pushout of (n, d_i). As f_i is flat, n_i is monomorphic. It follows that the family of morphisms $(k_i n)_{i \in I} = (n_i d_i : B \to N_i)_{i \in I}$ is monomorphic. Thus n is monomorphic. As a result, the pushout of a monomorphism along any singular epimorphism is monomorphic. Because singular epimorphisms are co-universal, it follows that they are flat. ■

Proposition (1.8.4).

(i) *Singular epimorphisms are co-universal, i.e. the pushout of a singular epimorphism is a singular epimorphism.*

(ii) *Composites of singular epimorphisms are singular epimorphisms.*

(iii) *Finite co-intersections of singular epimorphisms are singular epimorphisms.*

(iv) *Finite products of singular epimorphisms are singular epimorphisms.*

(v) *Any projection of a product is a singular epimorphism.*

Proof.

(i) follows from Proposition (1.1.5).

(ii) Let $f: A \to B$ and $g: B \to C$ be two singular epimorphisms. There is a congruence r on A whose codisjunctor is f, and a congruence s on B whose codisjunctor is g. Let us show that gf is the codisjunctor of the congruence $r \wedge f^{*c}(s)$ on A. Because f and g are flat epimorphisms (Proposition (1.8.3)) we have, by Propositions (1.6.3) and (1.7.2),

$$(gf)_{*c}(r \wedge f^{*c}(s)) = g_{*c}(f_{*c}(r \wedge f^{*c}(s))) =$$
$$g_{*c}(f_{*c}(r)) \wedge g_{*c}(f_{*c}(f^{*c}(s))) = g_{*c}(f_{*c}(r)) \wedge g_{*c}(s).$$

But f codisjoints r, and so $f_{*c}(r) = 1_{B \times B}$ and g codisjoints s so that $g_{*c}(s) = 1_{C \times C}$. Therefore $(gf)_{*c}(r \wedge f^{*c}(s)) = 1_{C \times C}$ and thus gf codisjoints $r \wedge f^{*c}(s)$. Let $h: A \to H$ be a morphism which codisjoints $r \wedge f^{*c}(s)$. Then h codisjoints r and $f^{*c}(s)$. Thus h factors through f in a morphism $k: B \to H$. We have

$$k_{*c}(s) = k_{*c}(f_{*c}(f^{*c}(s))) = (kf)_{*c}(f^{*c}(s)) =$$
$$h_{*c}(f^{*c}(s)) = 1_{H \times H}.$$

Therefore k codisjoints s, and so k factors through g in a morphism $v: C \to H$. As a result, $h = vgf$ and gf is the codisjunctor of $r \wedge f^{*c}(s)$. Thus gf is a singular epimorphism.

(iii) follows from (i) and (ii).

(iv) follows from Proposition (1.4.7).

(v) The projection $p_i: \Pi_{i \in I} A_i \to A_i$ is the product of the singular epimorphisms 1_{A_i} and $\Pi_{j \in I - \{i\}} A_j \to 1$. ■

Proposition (1.8.5). *A singular epimorphism $f: A \to B$ is a finitely presentable morphism, i.e. a finitely presentable object in the category $A/\mathbf{A}$.*

Proof. Let $f: A \to B$ be the codisjunctor of $(g, h): C \rightrightarrows A$. As the category $A/\mathbf{A}$ of objects of $\mathbf{A}$ under A is locally finitely presentable [16], there exists a filtered category $\mathbf{I}$, a diagram $(f_i: A \to B_i)_{i \in \mathbf{I}}$ of finitely presentable objects in $A/\mathbf{A}$, and an inductive cone $(\beta_i: f_i \to f)_{i \in \mathbf{I}}$ in $A/\mathbf{A}$ which is a colimit. Then $(\beta_i: B_i \to B)_{i \in \mathbf{I}}$ is a colimit in $\mathbf{A}$. Let $q: A \to Q$ be the coequalizer of (g, h) and, for any $i \in \mathbf{I}$, let $q_i: B_i \to Q_i$ be the coequalizer of $(g_i, h_i) = (f_i g, f_i h)$. Then q_i is the pushout of q along f_i. The pushout of q along f is, on the one hand, the morphism O_B and, on the other hand, the filtered colimit of the diagram $(q_i)_{i \in \mathbf{I}}$. Because 1 is a finitely presentable object, there exists $i \in \mathbf{1}$ such that $q_i = O_{B_i}$, so that f_i codisjoints (g, h). Therefore f_i factors through f in a morphism $h: B \to B_i$. The relation $\beta_i h f = \beta_i f_i = f$ implies the relation $\beta_i h = 1_B$. The object f in $A/\mathbf{A}$ is thus a split quotient object of the object f_i. For this reason, it is a finitely presentable object in $A/\mathbf{A}$ [16]. ■

Corollary (1.8.6). *If $f: A \to B$ is a singular epimorphism and A is a finitely presentable (finitely generated) object, then B is a finitely presentable (finitely generated) object.*

Singular quotient objects

Singular epimorphisms $f: A \to B$ represent singular quotient objects of A. Let us denote by $\mathrm{QuotSing}(A)$ the subordered set of $\mathrm{Quot}(A)$ whose elements are singular quotient objects of A. It is a bounded meet-semilattice. A singular epimorphism, or a singular quotient object, $f: A \to B$ is said to be *proper* if it is not isomorphic.

Example

The following assertions are equivalent for a morphism $f: A \to B$ in **CRng**.

(i) f is a singular epimorphism.

(ii) f is an epimorphism of rings which makes B a finitely presentable A-algebra and a flat A-module.

(iii) There exist $e_1, \ldots, e_k \in A$ and $b_1, \ldots, b_k \in B$ such that

(a) $b_1 f(e_1) + \ldots + b_k f(e_k) = 1$

(b) $\forall b \in B, \exists n \in \mathbb{N}, \forall i \in [1, k], bf(e_i^n) \in f(A)$

(c) $\forall a \in \mathrm{Ker}(f), \exists n \in \mathbb{N}, \forall i \in [1, k], ae_i^n = 0.$

This result can be found in [12].

1.9 THE PRESINGULAR FACTORIZATION

Notation. Given morphisms $e: E \to F$ and $f: A \to B$, we shall say that e is *left-orthogonal* to f, or alternatively that f is *right-orthogonal* to e, if for any morphisms $m: E \to A$ and $n: F \to B$ such that $fm = ne$, there exists a unique morphism $u: F \to A$ such that $ue = m$ and $fu = n$ (cf. [15], § 2).

Definition (1.9.1). *A morphism f is* prelocal *if it does not factor as $f = gd$ with d a proper singular epimorphism.*

Proposition (1.9.2). *Prelocal morphisms are precisely the morphisms which are right-orthogonal to singular epimorphisms.*

Proof. Let $f: A \to B$ be a prelocal morphism, let $d: X \to Y$ be a singular epimorphism, and let $m: X \to A$, $n: Y \to B$ be such that $nd = fm$. Let $(\delta: A \to \Delta, e: Y \to \Delta)$ be the pushout of (m, d). There exists $u: \Delta \to B$ such that $u\delta = f$ and $ue = n$. But δ is a singular epimorphism, and so it is isomorphic. It follows that the morphism $\delta^{-1}e: Y \to A$ is such that $\delta^{-1}ed = m$ and $f\delta^{-1}e = n$. As d is epimorphic, such a morphism $Y \to A$ is unique. As a result, f is right-orthogonal to d. Conversely, if $f: A \to B$ is right-orthogonal to any singular epimorphism and $f = gd$, where $d: A \to D$ is a singular epimorphism, then the relation $f1_A = gd$ implies the existence of a morphism $u: D \to A$ such that $ud = 1_A$. As d is epimorphic, it is also isomorphic. Therefore f is prelocal. ■

Corollary (1.9.3).
(i) *Regular monomorphisms are prelocal.*
(ii) *Prelocal morphisms are universal.*
(iii) *Composites of prelocal morphisms are prelocal.*
(iv) *Limits of prelocal morphisms are prelocal morphisms.*

Definition (1.9.4). *A morphism is* presingular *if it is left-orthogonal to any prelocal morphism.*

Proposition (1.9.5).
(i) *Singular epimorphisms are presingular.*
(ii) *Presingular morphisms are flat epimorphisms.*
(iii) *Presingular morphisms are co-universal.*
(iv) *The class of presingular epimorphisms is closed under compositions, co-intersections, and colimits.*

Proof. The only non-immediate point is (ii). Let $e: E \to F$ be a presingular morphism. Let $g: E \to G$ be the co-intersection of all flat epimorphisms with domain E through which e factors. Then g is a flat epimorphism (Proposition (1.7.1)) through which e factors in the form $e = hg$. Let us suppose that $h = kd$ where d is a singular epimorphism. Then dg is a flat epimorphism (Proposition (1.7.1)) through which e factors. Therefore $dg \simeq g$ and d is isomorphic. It follows that h is prelocal. Because h is right-orthogonal to e, the equality $1_F e = hg$ implies the existence of a morphism $u: F \to G$ such that $hu = 1_F$ and $ue = g$. Then u is epimorphic because g is also. Thus u and h are isomorphic. It follows that e is a flat epimorphism. ■

Theorem (1.9.6). *The presingular factorization: any morphism factors in an essentially unique way as the composite of a presingular epimorphism followed by a prelocal morphism.*

Proof. Let $f: A \to B$ be a morphism. Let $g: A \to C$ be the co-intersection of all presingular epimorphisms with domain A through which f factors. Then g is a presingular epimorphism (Proposition (1.9.5)) through which f factors in the form $f = hg$. Let $d: C \to D$ be a singular epimorphism through which h factors in the form $h = kd$. Then dg is a presingular epimorphism (Proposition (1.9.5)) with domain A through which f factors. Therefore $dg \simeq g$ and d is isomorphic. It follows that h is prelocal. Let $f = h'g'$ be another factorization of f, where g' is a presingular epimorphism and h' is a prelocal morphism. Because g is left-orthogonal to h' (Proposition (1.9.2)), there exists $v: C \to C'$ such that $vg = g'$ and $h'v = h$. Because g' is left-orthogonal to h, there exists $v': C' \to C$ such that $v'g' = g$ and $hv' = h'$. Then v and v' are inverse to each other. ■

Presingular extensions

A presingular bimorphism $f: A \to B$ is also called a *presingular extension* of A. The full subcategory of $A/\mathbf{A}$ whose objects are the presingular extensions of A is small and is associated to a preorder. It is denoted by $\mathbf{ExtPrSing}(A)$. According to Propositions (1.7.1) and (1.9.5), $\mathbf{ExtPrSing}(A)$ is a complete prelattice whose joins are the co-intersections of epimorphisms.

Theorem (1.9.7). *Any object A has a presingular hull, i.e. a presingular extension $\sigma_A: A \to S(A)$ which factors uniquely through any presingular extension $f: A \to B$ of A.*

Proof. Take the top object $\sigma_A: A \to S(A)$ in $\mathbf{ExtPrSing}(A)$. ■

2
CLASSICAL OBJECTS

In an arbitrary Zariski category **A**, we define a range of objects corresponding to the usual types of commutative ring:

Zariski category	...	**CRng**
simple	...	field
integral	...	integral domain
reduced	...	without non-null nilpotent elements
pseudo-simple	...	exactly one prime ideal
quasi-primary	...	$ab = 0 \Rightarrow$ (a or b is nilpotent)
primary	...	any zero divisor is nilpotent
singularly closed	...	total ring of quotients
irreducible	...	the ideal $\{0\}$ is irreducible with respect to $\cap$
regular	...	von Neumann regular
local	...	local

To any type of objects there corresponds a type of morphism in that enables categories to be obtained and the usual constructions to be performed as universal constructions. Different types of congruence occur naturally with the different types of object and correspond to usual ideals: proper, maximal, prime, radical, quasi-primary, primary, and irreducible.

2.1 SIMPLE OBJECTS

Definition (2.1.1). *An object is* simple *if it has exactly two congruences. The* category ***SimA*** *of* simple objects *of* **A** *is the full subcategory of* **A** *whose objects are simple objects.*

Proposition (2.1.2). *For an object K, the following statements are equivalent.*

(i) *K is simple.*

(ii) *K has exactly two regular quotient objects.*

(iii) *K is not the terminal object and any congruence on K is the equality or the unit congruence.*

(iv) *K is not the terminal object and any regular quotient object of K is an isomorphism or has codomain 1.*

(v) *K is not terminal and any pair of morphisms* $(m, n) : X \rightrightarrows K$ *is such that* m, n *are equal or codisjointed.*

Proof. The terminal object has only one congruence and thus it is not a simple object. Any non-terminal object A has at least two congruences: the equality Δ_A and the unit congruence $1_{A \times A}$. As a consequence, (i) $\Leftrightarrow$ (iii). The equivalences (i) $\Leftrightarrow$ (ii) and (iii) $\Leftrightarrow$ (iv) follow from the isomorphism $\mathrm{Cong}(K) \simeq \mathrm{QuotReg}(K)^{\mathrm{op}}$ (Proposition (1.5.2)). The equivalence (iv) $\Leftrightarrow$ (v) follows from the fact that the pair (m, n) is equal if and only if its coequalizer is an isomorphism, and that it is codisjointed if and only if its coequalizer has codomain 1. ■

Proposition (2.1.3). *Any filtered colimit or ultraproduct of simple objects is a simple object.*

Proof

(i) Let $(K_i)_{i \in \mathbf{I}}$ be a filtered diagram of simple objects, the colimit of which is $(\alpha_i : K_i \to K)_i \in \mathbf{I}$. The object K is not terminal, otherwise the unit morphism 1_K should factor through one of the morphisms α_i, which would imply that K_i is terminal. Let $(m, n) : X \rightrightarrows K$ be a pair of distinct morphisms. There exists a finitely presentable object X_0 and a morphism $f : X_0 \to X$ such that mf, nf are distinct. There exist an index $i \in \mathbf{I}$ and a pair of morphisms $(m_i, n_i) : X_0 \rightrightarrows K_i$ such that $(\alpha_i m_i, \alpha_i n_i) = (mf, nf)$. The morphisms m_i, n_i are distinct. Because the object K_i is simple, the pair (m_i, n_i) is codisjointed. It follows that the pair $(\alpha_i m_i, \alpha_i n_i) = (mf, nf)$ is codisjointed, and thus the pair (m, n) is codisjointed. Hence the object K is simple (Proposition (2.1.2)).

(ii) Let $(K_i)_{i \in I}$ be a family of simple objects, let $\mathfrak{U}$ be an ultrafilter on I, and let $(\alpha_{I_0} : \Pi_{i \in I_0} K_i \to K)_{I_0 \in \mathfrak{U}}$ be the filtered colimit in $\mathbf{A}$ which defines the ultraproduct of $(K_i)_{i \in I}$ with respect to $\mathfrak{U}$. The object K is not terminal as it is a filtered colimit of non-terminal objects. Let $q : K \to Q$ be a finitely generated regular quotient of K which is the coequalizer of a pair of morphisms $(m, n) : X \rightrightarrows K$ whose domain is finitely presentable and codisjunctable. There exist $I_0 \in \mathfrak{U}$ and $(m_0, n_0) : X \rightrightarrows \Pi_{i \in I_0} K_i$ such that $(m, n) = (\alpha_{I_0} m_0, \alpha_{I_0} n_0)$. Let q_0 be the coequalizer of (m_0, n_0) and let d_0 be its codisjunctor. Let I_1 be the set of $i \in I_0$ such that the projection $p_i : \Pi_{i \in I_0} K_i \to K_i$ factors through q_0. Let I_2 be the set of $i \in I_0$ such that this projection p_i factors through d_0. Because the objects K_i are simple, we have a disjoint union $I_0 = I_1 \cup I_2$. Because $I_0 \in \mathfrak{U}$, we have $I_1 \in \mathfrak{U}$ or $I_2 \in \mathfrak{U}$. If $I_1 \in \mathfrak{U}$, the canonical projection $\Pi_{i \in I_0} K_i \to \Pi_{i \in I_1} K_i$ factors through q_0; thus $\alpha_{I_0} m_0 = \alpha_{I0} n_0$, $m = n$, and hence q is isomorphic. If $I_2 \in \mathfrak{U}$, the projection $\Pi_{i \in I_0} K_i \to \Pi_{i \in I_1} K_i$ factors through d_0; thus α_{I_0} codisjoints (m_0, n_0), (m, n) is codisjointed, and hence q is null. As a result, K is a simple object because the finitely presentable codisjunctable objects properly generate the category. ■

Proposition (2.1.4). *Any prelocal subobject of a simple object is simple.*

Proof. Let $f: A \to K$ be a prelocal monomorphism whose codomain is simple. The object A is not terminal because K is not. Let $(m, n): X \rightrightarrows A$ be a pair of distinct morphisms. By Axiom (1.2.2) of Zariski categories there exist a codisjunctable object Y and a morphism $u: Y \to X$ such that $mu \neq nu$. Let $d: A \to D$ be the codisjunctor of (mu, nu). The pushout δ of d along f is the codisjunctor of (fmu, fnu). Because K is simple and $(fmu \neq fnu$, the pair (fmu, fnu) is codisjointed (Proposition (2.1.2)). Thus δ is isomorphic. Because f is prelocal, d is isomorphic. Thus (mu, nu) is codisjointed, as is (m, n). According to Proposition (2.1.2), K is a simple. ∎

Proposition (2.1.5). *Any morphism between simple objects is an essential monomorphism.*

Proof. Let $f: K \to L$ be such a morphism. Let $f = me$ be its regular factorization where $e: K \to E$ is a regular epimorphism and $m: E \to L$ is a monomorphism. The object E is not terminal, otherwise L should be also. Thus e is isomorphic and f is monomorphic. Let $g: L \to B$ be a morphism such that gf is monomorphic. The object B is not terminal, otherwise the object K would be also. Let $g = me$ be the regular factorization of g with $e: L \to E$. The object E is not terminal, otherwise the object B would be also. Thus e is isomorphic and g is monomorphic. ∎

Definition (2.1.6).

(i) *A* simple quotient object of *A is a regular quotient object of A whose codomain is simple.*

(ii) *A* maximal congruence *on A is a proper congruence on A which is a maximal element in the set of proper congruences on A.*

Proposition (2.1.7). *A congruence r on A is maximal if and only if the quotient of A by r is simple.*

Proof. The isomorphism $\mathrm{Cong}(A) \xrightarrow{\sim} \mathrm{QuotReg}(A)^{\mathrm{op}}$ associates to the unit congruence the zero regular quotient object $0_A: A \to 1$. Thus a congruence r on A is maximal if and only if the quotient of A by r is a minimal element among non-zero regular quotient objects of A. Let us suppose that r is maximal and let $q: A \to Q$ be the quotient of A by r. If $p: Q \to P$ is a regular epimorphism, then $pq \leqslant q$ in $\mathrm{QuotReg}(A)$ and so $pq \simeq q$ or $pq = 0_A$, i.e. p is isomorphic or the codomain of p is 1. The object Q is thus simple. Conversely, let us consider a congruence r such that the quotient $q: A \to Q$ of A by r is such that Q is a simple object. If $p: A \to P$ is a regular quotient object such that $p \leqslant q$ in $\mathrm{QuotReg}(A)$, there is a regular epimorphism $u: Q \to P$ such that $uq = p$. But u is isomorphic or is 0_Q. Thus

$p \simeq q$ or $p = O_A$. Thus q is a minimal element among non-zero elements in QuotReg(A), i.e. r is a maximal congruence. ∎

Theorem (2.1.8). *Any non-terminal object has a maximal congruence.*

Proof. Let A be non-terminal object. Let $\mathrm{Cong}^{\mathrm{p}}(A)$ be the set of proper congruences on A. It is not empty. Let C be a chain in $\mathrm{Cong}^{\mathrm{p}}(A)$. Let us suppose that $\overset{c}{\vee} C = 1_{A \times A}$. As $1_{A \times A}$ is compact in $\mathrm{Cong}(A)$ (Proposition (1.5.2)), there is a finite subset C_0 of C such that $\overset{c}{\vee} C_0 = 1_{A \times A}$. But $\overset{c}{\vee} C_0$ is an element of C. Thus we have a contradiction. As a consequence $\overset{c}{\vee} C \neq 1_{A \times A}$, i.e. $\overset{c}{\vee} C$ belongs to $\mathrm{Cong}^{\mathrm{p}}(A)$. By Zorn's lemma, $\mathrm{Cong}^{\mathrm{p}}(A)$ has a maximal element. It is a maximal congruence on A. ∎

Corollary (2.1.9). *Any non-terminal object has a simple quotient object.*

Corollary (2.1.10). *Any proper congruence is included in a maximal congruence.*

Proof. Let r be a proper congruence on A. The quotient $q_r : A \to A/r$ of A by r is such that A/r is not terminal. According to Corollary (2.1.9), A/r has a simple quotient $k : A/r \to K$. Then kq_r is a simple quotient of A, whose kernel pair is a maximal congruence containing r. ∎

Proposition (2.1.11). *A pair of morphisms $(g, h) : C \rightrightarrows A$ is codisjointed if and only if, for any morphism $f : A \to K$ whose codomain is simple, we have $fg \neq fh$.*

Proof. Let us suppose that the pair (g, h) is codisjointed. Let $f : A \to K$ be a morphism with a simple codomain K. The pair (fg, fh) is codisjointed; thus $fg \neq fh$ (Proposition (2.1.2)). Conversely, let (g, h) be such that, for any $f : A \to K$ with K simple, we have $fg \neq fh$. Let $m : A \to M$ be such that $mg = mh$. If M is not terminal, it has a simple quotient object $q : M \to K$, and we have $qmg = qmh$, which is impossible. Thus $M \simeq 1$ and (g, h) is codisjointed. ∎

Examples

In **CRng**, **RedCRng**, and **RegCRng** the simple objects are the fields. In **CAlg**(k), **RedCAlg**(k), and **RegCAlg**(k) they are the field extensions of k. In **AlgCAlg**(k) they are the algebraic field extensions of k. In **RlRng** and **RedRlRng** they are the real fields. In **RlOrdRng** and **RedRlOrdRng** they are the ordered real fields. In **RlLatRng** and **RedRlLatRng**, they are the totally ordered fields. In **Mod** and **CAlg** they are the pairs $(K, \{0\})$ where K is a field. In **GradCRng** and **RedGradCRng** they are the graded commutative rings in which any non-zero homogeneous element is invertible; they are precisely of the form K or $K\ [X, X^{-1}]$ where K is an homogeneous field of degree zero and X is an homogeneous element of degree $n \in \mathbf{N}^*$.

2.2 INTEGRAL OBJECTS

Definition (2.2.1).

(i) *An object is* integral *if it is a subobject of a simple object.*

(ii) *An* integral quotient *of an object A is a regular quotient object of A whose codomain is integral.*

The *category* **IntA** *of integral objects* of **A** is the subcategory of **A** whose objects are the integral objects and whose morphisms are the monomorphisms.

Theorem (2.2.2). *The category* **IntA** *of integral objects of* **A** *is a multireflective subcategory of* **A** [7].

Proof. Let A be an object of **A**. Let us show that the family $(f_i : A \to B_i)_{i \in I}$ of integral quotients of A is a universal family of morphisms from A to **IntA**. If $i, j \in I$ and $g_i : B_i \to B$, $g_j : B_j \to B$ are morphisms in **IntA** such that $g_i f_i = g_j f_j$, there are morphisms $h_i : B_i \to B_j$ and $h_j : B_j \to B_i$ such that $h_i f_i = f_j$ and $h_j f_j = f_i$. Thus $f_i = f_j$ and $i = j$. Moreover, any morphism $f : A \to B$ from A to an integral object B has a regular factorization $f = me$, where $m : E \to B$ is monomorphic, and so E is integral as it is a subobject of an integral object, thus $e : A \to E$ is an integral quotient of A. ■

Corollary (2.2.3). *The category* **InA** *has non-empty connected limits.*

Theorem (2.2.4). *To any integral object is associated a universal simple object. More precisely,* **SimA** *is a reflective subcategory of* **IntA**.

Proof. Let A be an integral object. Let $k : A \to K$ be a monomorphism whose codomain is simple. Let $k = gf$ be the presingular factorization of k where $f : A \to B$ is presingular and $g : B \to K$ is prelocal (Theorem (1.9.6)). Because f is a flat bimorphism (Proposition (1.9.5)), it is an essential monomorphism (Proposition (1.7.1)); thus g is monomorphic. By Proposition (2.1.4), the object B is simple. Let $h : A \to C$ be another monomorphism with a simple codomain. Let $(\bar{f} : C \to \bar{B}, \bar{h} : B \to \bar{B})$ be the pushout of (h, f). As f is flat, $\bar{h}$ is monomorphic and so $\bar{B} \not\approx 1$. According to Proposition (1.9.5), $\bar{f}$ is a presingular epimorphism. On the other hand, if $\bar{f}$ factors in the form $\bar{f} = ud$, where $d : C \to D$ is a singular epimorphism, then $D \not\approx 1$; thus, according to Proposition (2.1.2), d is an isomorphism. It follows that $\bar{f}$ is prelocal. Consequently, $\bar{f}$ is an isomorphism (Theorem

(1.9.6)). It follows that h factors through f in a unique way because f is epimorphic. As a result, $f: A \to B$ is the universal simple object associated to A. ■

Corollary (2.2.5). *The category* **SimA** *of simple objects of* **A** *is a multireflective subcategory of* **A**.

Proof. The proof follows from Theorems (2.2.2) and (2.2.4). ■

Corollary (2.2.6). *The category* **SimA** *has non-empty connected limits.*

Proposition (2.2.7). *Any flat morphism with an integral domain and a non-terminal codomain is monomorphic.*

Proof. Let $f: A \to B$ be such a morphism. There exist a simple object K and a monomorphism $k: A \to K$. Let $(h: B \to H, g: K \to H)$ be the pushout of (f, k). The morphism h is monomorphic and thus H is not terminal. Hence g is monomorphic. Therefore $gk = hf$ monomorphic and hence f is monomorphic. ■

Proposition (2.2.8). *Any essential flat extension of an integral object is integral.*

Proof. Let $f: A \to B$ be a flat essential monomorphism with an integral domain A. We use the notation of the proof of Proposition (2.2.7). According to Corollary (2.1.9), H has a simple quotient $s: H \to S$. Then $sg: K \to S$ is monomorphic (Proposition (2.1.5)). Thus $shf = sgk$ is monomorphic. Therefore $sh: B \to S$ is monomorphic and B is integral. ■

Proposition (2.2.9). *Any flat epimorphic extension of an integral object is integral.*

Proof. The proof follows from Propositions (1.7.1) and (2.2.8). ■

Proposition (2.2.10). *An object is integral if and only if its flat epimorphic hull is a simple object.*

Proof. Let A be an object and let $\psi_A: A \to F(A)$ be its flat epimorphic hull. If $F(A)$ is simple, then A is integral. Let us assume that A is integral. According to Proposition (2.2.9), the object $F(A)$ is integral. Let $k: F(A) \to K$ be the universal simple object associated to $F(A)$. According to the proof of Theorem (2.2.4), k is a presingular bimorphism. Thus $k\psi_A: A \to K$ is a flat epimorphic extension of A. According to the universal property of ψ_A, k is an isomorphism. Hence $F(A)$ is simple. ■

Definition (2.2.11). *A congruence p on A is* prime *if the quotient A/p of A by p is integral.*

Proposition (2.2.12). *Any maximal congruence is prime.*

Proof. The proof follows from proposition (2.1.7). ■

Proposition (2.2.13) *Any prime congruence p on A is a universally irreducible element in* $\mathrm{Cong}(A)$, *i.e. it satisfies the property $r \wedge s \leqslant p \Rightarrow r \leqslant p$ or $s \leqslant p$.*

Proof. Let $r, s \in \mathrm{Cong}(A)$ such that $r \nleqslant p$ and $s \nleqslant p$. Let $q_p : A \to A/p$ be the quotient of A by p and let $m : A/p \to K$ be a monomorphism with a simple codomain. Let $f = mq_p$. The morphism q_p does not coequalize r or s; thus, neither does the morphism f. Because r and s are joins of codisjunctable congruences (Proposition (1.5.3)), there exists a pair of codisjunctable congruences $r_0, s_0 \in \mathrm{Cong}(A)$ such that $r_0 \leqslant r, s_0 \leqslant s$, and f does not coequalize r_0 or s_0. According to proposition (2.1.2), f codisjoints r_0 and s_0. According to Proposition (1.5.4), f codisjoints $r_0 \wedge s_0$. Therefore f codisjoints $r \wedge s$ and hence neither f nor q_p coequalizes $r \wedge s$. It follows that $r \wedge s \nleqslant p$. ■

Examples

In **CRng** and **RedCRng**, the integral objects are the integral domains. In **CAlg**(k) and **RedCAlg**(k) they are the integral domain extensions of k. In **RlRng** and **RedRlRng** they are the real integral domains. In **RlOrdRng** and **RedRlOrdRng** they are the ordered real integral domains. In **RlLatRng** and **RedRlLatRng** they are the totally ordered integral domains. In **GradCRng** and **RedGradCRng** the integral objects are the graded integral domains. In **Mod** and **CAlg** they are the pairs $(A, \{0\})$ where A is an integral domain. In **RegCRng**, **RegCAlg**(k) and **RegGradCRng**, the integral objects are identical to the simple objects.

2.3 REDUCED OBJECTS

Definition (2.3.1). *An object is* reduced *if it is a subobject of a product of simple objects.*

The *category* **RedA** *of reduced objects of* **A** is the full subcategory of **A** whose objects are the reduced ones. For example, the terminal object and the integral objects are reduced. In categories of commutative algebras reduced objects are objects without a nilpotent element other than 0.

Theorem (2.3.2). *To any object is associated a universal reduced object. More precisely,* **RedA** *is a reflective subcategory of* $\mathbf{A}$.

Proof. Let A be an object of $\mathbf{A}$. Let us consider the family $(r_i : A \to R_i)_{i \in I}$ of regular quotients of A whose codomains are reduced. The morphism $(r_i)_{i \in I} : A \to \Pi_{i \in I} R_i$ factors in the form $(r_i)_{i \in I} = mr$ where $r : A \to R$ is a regular epimorphism and $m : R \to \Pi_{i \in I} R_i$ is a monomorphism. The object R is reduced as it is a subobject of the reduced object $\Pi_{i \in I} R_i$. Let $f : A \to B$ be a morphism with a reduced codomain B. It factors in the form $f = ng$ where $g : A \to C$ is a regular epimorphism and $n : C \to B$ is a monomorphism. The object C is reduced as it is a subobject of a reduced object. Thus g is isomorphic to the morphism $r_i : A \to R_i$ for some $i \in I$. It follows that f factors through $(r_i)_{i \in I}$ and thus through r. As a result, $r : A \to R$ is the universal reduced object associated to A. ∎

Proposition (2.3.3). *The category* **RedA** *of reduced objects of* $\mathbf{A}$ *is closed in* $\mathbf{A}$ *under filtered colimits and singular quotient objects.*

Proof.

(i) Let $(r_i : R_i \to R)_{i \in \mathbf{I}}$ be a filtered colimit in $\mathbf{A}$ such that the objects R_i are reduced. Let $(m, n) : X \rightrightarrows R$ be a pair of distinct morphisms whose domain X is a finitely presentable codisjunctable object. There exist $i_0 \in \mathbf{I}$ and a pair $(m_0, n_0) : X \rightrightarrows R_{i_0}$ such that $r_{i_0} m_0 = m$ and $r_{i_0} n_0 = n$. We shall assume that i_0 is an initial object in $\mathbf{I}$; otherwise, we replace $\mathbf{I}$ by $i_0/\mathbf{I}$. For any $i \in \mathbf{I}$, we write $\alpha_i : i_0 \to i$ in $\mathbf{I}$, $m_i = R_{\alpha_i} m_0$, *and* $n_i = R_{\alpha_i} n_0$. For $i \in \mathbf{I}$, let $d_i : R_i \to D_i$ be the codisjunctor of (m_i, n_i), and let $(d : R \to D,\ \delta_i : D_i \to D)$ be the pushout of (r_i, d_i). Then d is the codisjunctor of (m, n) (Proposition (1.1.5)). As pushouts commute with colimits, we obtain a filtered colimit $(\delta_i : D_i \to D)_{i \in \mathbf{I}}$. As the pair (m, n) is distinct, any pair (m_i, n_i) is distinct. As R_i is reduced, there exist a simple object K_i and a morphism $k_i : R_i \to K_i$ such that the pair $(k_i m_i, k_i n_i)$ is distinct. Then the pair $(k_i m_i, k_i n_i)$ is codisjointed (Proposition (2.1.2)). Hence k_i codisjoints (m_i, n_i) and factors through d_i. Consequently, any D_i is a non-terminal object. From the fact that 1 is finitely presentable, it follows that D is not terminal. By Corollary (2.1.9) there exists a simple quotient object $k : D \to K$. Thus the morphism $kd : R \to K$ is such that (kdm, kdn) is codisjointed and hence distinct. As a result, for any distinct pair $(m, n) : X \rightrightarrows R$ whose domain belongs to a generating set in $\mathbf{A}$, there is a morphism $q : R \to K$, whose codomain is simple, such that (qm, qn) is distinct. Consequently, the set of morphisms $q : R \to K$ with a simple codomain is a monomorphic family of morphisms. Hence the object R is a subobject of a product of simple objects. The object R is reduced.

(ii) Let A be a reduced object and let $f : A \to B$ be a singular epimorphism codisjunctor of a pair $(m, n) : X \rightrightarrows A$. Let $q : A \to Q$ be the coequalizer of

(m, n). Let $(f_i : A \to K_i)_{i\in I}$ be a family of morphisms, whose codomains are simple, such that $(f_i)_{i\in I} : A \to \Pi_{i\in I} K_i$ is a monomorphism. Let $I_0 = \{i\in I : f_i$ coequalizes $(m, n)\}$ and $I_1 = \{i\in I : f_i$ codisjoints $(m, n)\}$. Then $I = I_0 \cup I_1$ is a disjoint union. For $i\in I_0$, the morphism f_i factors through q, while for $i\in I_1$, it factors through f in a morphism $g_i : B \to K_i$. Therefore the morphism $(f_i)_{i\in I_0} : A \to \Pi_{i\in I_0} K_i$ factors through q, while the morphism $(f_i)_{i\in I_1} : A \to \Pi_{i\in I_1} K_i$ factors through f in the morphism $(g_i)_{i\in I_1} : B \to \Pi_{i\in I_1} K_i$. It follows that the pushout of $(f_i)_{i\in I_0}$ along f is the null morphism $0_B : B \to 1$ and that the pushout of $(f_i)_{i\in I_1}$ along f is $(g_i)_{i\in I_1} : B \to \Pi_{i\in I_1} K_i$. Because the product $\Pi_{i\in I} K_i = (\Pi_{i=I_0} K_i) \times (\Pi_{i=I_1} K_i)$ is co-universal, the pushout of $(f_i)_{i\in I}$ along f is $(g_i)_{i\in I_1}$. Because f is flat (Proposition (1.8.3)), $(g_i)_{i\in I_1}$ is monomorphic. Thus B is a reduced object. ■

Proposition (2.3.4). *The reflector $R : A \to$ **RedA** preserves finite products, monomorphisms, and codisjunctors.*

Proof. Let us denote the left-adjoint to the inclusion functor J: **RedA** → **A** by R: **A** → **RedA** and the unit of the adjunction by $r : 1_{\mathbf{A}} \to JR$.

(i) Let $(p : A \times B \to A, q : A \times B \to B)$ be the product of two objects A and B in **A**. Let $(v : R(A) \times R(B) \to R(A), w : R(A) \times R(B) \to R(B))$ be the product of $R(A)$ and $R(B)$, and let $u : R(A \times B) \to R(A) \times R(B)$ be the morphism such that $vu = R(p)$ and $wu = R(q)$. The morphisms $r_A : A \to R(A)$ and $r_B : B \to R(B)$ are regular epimorphisms, and thus $r_A \times r_B = ur_{A\times B} : A \times B \to R(A) \times R(B)$ is a regular epimorphism (Proposition (1.4.8)); hence u is a regular epimorphism. Let us show that u is monomorphic. Because $R(A \times B)$ is reduced, there exist a family of simple objects $(K_i)_{i\in I}$ and a family of morphisms $(f_i : R(A \times B) \to K_i)_{i\in I}$ such that $(f_i)_{i\in I} : R(A \times B) \to \Pi_{i\in I} K_i$ is a monomorphism. Let $i\in I$. As finite products are co-universal, the morphism $f_i r_{A\times B} : A \times B \to K_i$ factors through p or q, and thus through $r_A p$ or $r_B q$, i.e. through $R(p)\ r_{A\times B}$ or $R(q)\ r_{A\times B}$ equivalently, it factors through $vur_{A\times B}$ or $wur_{A\times B}$, and thus through $ur_{A\times B}$. Because $r_{A\times B}$ is epimorphic, f_i factors through u, and thus $(f_i)_{i\in I} : R(A \times B) \to \Pi_{i\in I} K_i$ factors through u which implies that u is monomorphic. As a result, u is isomorphic and $(R(p), R(q))$ is a product. Moreover $R(1) \simeq 1$. Therefore R preserves finite products.

(ii) Let $f : A \to B$ be a monomorphism. Let $R(f) = mq$ where $q : R(A) \to C$ is a regular epimorphism and $m : C \to R(B)$ is a monomorphism. Then $g = qr_A$ is a regular epimorphism. Let us show that $g \simeq r_A$. Let $(s, t) : X \rightrightarrows A$ be a pair of morphisms such that the domain X is codisjunctable and $gs = gt$. Let $d : A \to D$ be the codisjunctor of (s, t). The pushout of d along g is null, and so its pushout along $r_B f = mg$ is null. Let $\delta : B \to \Delta$ be the pushout of d along f. Then the pushout of δ along r_B is null. It implies that δ is null; otherwise a simple quotient $p : \Delta \to P$ of Δ would exist and $p\delta$ would factor through r_B, which is impossible. As d is flat and f is monomorphic, the pushout of f along d is monomorphic. It implies that $D \simeq 1$, i.e. d is null. Thus the codisjunctor of $(r_A s, r_A t)$ is null. Because $R(A)$ is reduced, it follows that $r_A s = r_A t$. As codisjunctable objects form a generating set, it follows that $g \simeq r_A$; thus q is isomorphic and $R(f)$ is monomorphic.

(iii) Since R preserves coequalizers and the terminal object, it preserves codisjointed pairs. Moreover, R reflects the terminal object because a non-terminal object A in **A** has a simple quotient $s: A \to S$ (Corollary (2.1.9)) which must factor through r_A so that $R(A)$ is non-terminal. It follows that R reflects codisjointed pairs. Let A be in **A**, B be in **Red A**, and $(f, g): A \rightrightarrows B$. From the preceding, we deduce the following equivalence: (f, g) is codisjointed in $\mathbf{A} \Leftrightarrow (R(f), R(g))$ is codisjointed in **RedA**. According to [11], Proposition 4.3, it follows that R preserves codisjunctors. ∎

2.4 RADICAL CONGRUENCES

Definition (2.4.1). *A congruence r on A is* radical *if the quotient of A by r is reduced.*

In categories of commutative algebras, radical congruences are congruences modulo radical ideals, i.e. nilradical ideals. The set of radical congruences on A is denoted by $\mathrm{CongRad}(A)$. It is equipped with the order induced by that of $\mathrm{Cong}(A)$.

Proposition (2.4.2). *The ordered set* $\mathrm{CongRad}(A)$ *of radical congruences on A is a reflective subset of the ordered set* $\mathrm{Cong}(A)$ *closed under filtered joins.*

Proof

(i) Let s be a congruence on A. Let $q: A \to Q$ be the quotient of A by s and let $r_Q: Q \to R(Q)$ be the universal reduced object associated to Q (Theorem (2.3.2)). The morphism r_Q is a regular epimorphism (see proof of Theorem (2.3.2)). Thus $r_Q q$ is a regular epimorphism. Let r be the congruence kernel pair of $r_Q q$. It is a radical congruence on A such that the quotient of A by it is $r_Q q$. The relation $r_Q q \leqslant q$ in QuotReg(A) implies the relation $s \leqslant r$ in Cong(A) (Proposition (1.5.2)). Let r_1 be a radical congruence on A such that $s \leqslant r_1$. The quotient $q_1: A \to Q_1$ of A by r_1 factors through q. Because Q_1 is reduced, q_1 factors through $r_Q q$. Therefore $r \leqslant r_1$. As a result, r is the universal radical congruence associated to s.

(ii) Let $(r_i)_{i \in I}$ be a filtered family of radical congruences on A and let $r = \vee_{i \in I} r_i$. For $i \in I$, let $q_i: A \to Q_i$ be the quotient of A by r_i and let $q: A \to Q$ be the quotient of A by r. Then $q = \wedge_{i \in I} q_i$ is a cofiltered cointersection in $\mathrm{QuotReg}(A)$ (Proposition (1.5.2)). Thus $Q = \xrightarrow{lim}_{i \in I} Q_i$ is a filtered colimit. Because the objects Q_i are reduced, the object Q is reduced (Proposition (2.3.3)). Thus the kernel pair r of q is radical. ∎

Corollary (2.4.3). CongRad(A) *is a complete lattice.*

Notation. Meets in CongRad(A) are denoted by $\wedge$ as they are the same as in Cong(A). Joins in CongRad(A) are denoted by $\overset{\mathrm{cr}}{\vee}$. The universal radical congruence associated to a congruence r is called the *radical of* r and is denoted by rad(r).

Proposition (2.4.4). *Radical congruences are precisely intersections of prime congruences.*

Proof. As a consequence of Proposition (2.4.2) intersections of radical congruences are radical. Because any prime congruence is radical, intersections of prime congruences are radical. Conversely, let r be a radical congruence on A and let $q: A \to Q$ be the quotient of A by r. Because Q is reduced, there exists a family of morphisms $(f_i : Q \to K_i)_{i\in I}$ with simple codomains such that $f = (f_i)_{i\in I} : Q \to \Pi_{i\in I} K_i$ is monomorphic. For any $i \in I$, let $f_i q = m_i q_i$, where $q_i : A \to Q_i$ is a regular epimorphism and $m_i : Q_i \to K_i$ is monomorphic, be the regular factorization of $f_i q$. There exists a morphism h_i: $Q \to Q_i$ such that $h_i q = q_i$ and $m_i h_i = f_i$. Let $h = (h_i)_{i\in I} : Q \to \Pi_{i\in I} Q_i$ and $m = \Pi_{i\in I} m_i : \Pi_{i\in I} Q_i \to \Pi_{i\in I} K_i$. Then $f = mh$ and thus h is monomorphic. It follows that $q = \vee_{i\in I} q_i$ in QuotReg(A). Let p_i be the kernel pair of q_i. Then $(p_i)_{i\in I}$ is a family of prime congruences on A such that $\wedge_{i\in I} p_i = r$. ∎

Proposition (2.4.5).

(i) *For any family* $(r_i)_{i\in I}$ *of congruences on* A,

$$\mathrm{rad}\Big(\overset{\mathrm{c}}{\bigvee_{i\in I}} r_i\Big) = \overset{\mathrm{cr}}{\bigvee_{i\in I}} \mathrm{rad}(r_i).$$

(ii) *For any pair* r, s *of congruences on* A,

$$\mathrm{rad}(r \wedge s) = \mathrm{rad}(r) \wedge \mathrm{rad}(s).$$

Proof.

(i) follows from the fact that the functor rad : Cong(A) → CongRad(A) is a left-adjoint.

(ii) The relation $r \wedge s \leqslant \mathrm{rad}(r) \wedge \mathrm{rad}(s)$ and the fact that $\mathrm{rad}(r) \wedge \mathrm{rad}(s)$ is radical imply that $\mathrm{rad}(r \wedge s) \leqslant \mathrm{rad}(r) \wedge \mathrm{rad}(s)$. Let p be a prime congruence on A such that $\mathrm{rad}(r \wedge s) \leqslant p$. Then $r \wedge s \leqslant p$. Because p is universally irreducible (Proposition (2.2.13)), we have that $r \leqslant p$ or $s \leqslant p$. Because p is radical, we have that $\mathrm{rad}(r) \leqslant p$ or $\mathrm{rad}(s) \leqslant p$. Thence

$\mathrm{rad}(r) \wedge \mathrm{rad}(s) \leqslant p$. According to Proposition (2.4.4), it follows that $\mathrm{rad}(r \wedge s) = \mathrm{rad}(r) \wedge \mathrm{rad}(s)$. ■

Proposition (2.4.6). *Inverse image congruence maps preserve radicals, i.e. for any $f: A \to B$ and $s \in \mathrm{Cong}(B)$, $f^{*c}(\mathrm{rad}(s)) = \mathrm{rad}(f^{*c}(s))$.*

Proof. Let $f: A \to B$ be a morphism in **A**. Let s be a congruence on B, let $q_s : B \to B/s$ be the quotient of B by s, and let $q_{\mathrm{rad}(s)} : B \to B/\mathrm{rad}(s)$ be the quotient of B by $\mathrm{rad}(s)$. Let $t = f^{*c}(s)$, $q_t : A \to A/t$, and $q_{\mathrm{rad}(t)} : A \to A/\mathrm{rad}(t)$. The morphism $q_s f$ factors in the form $q_s f = g q_t$ where $g : A/t \to B/s$ is monomorphic. The morphism $q_{\mathrm{rad}(s)} f$ factors in the form $q_{\mathrm{rad}(s)} f = h q_{\mathrm{rad}(t)}$ with $h : A/\mathrm{rad}(t) \to B/\mathrm{rad}(s)$. By the construction of $\mathrm{rad}(t)$ and $\mathrm{rad}(s)$ (Proposition (2.4.2)), the morphism h is the image by the reflector $R : \mathbf{A} \to \mathbf{RedA}$ of the morphism f. According to Proposition (2.3.4), $h = R(f)$ is monomorphic. Therefore the kernel pair $\mathrm{rad}(t)$ of $q_{\mathrm{rad}(t)}$ is the inverse image by f of the kernel pair $\mathrm{rad}(s)$ of $q_{\mathrm{rad}(s)}$, i.e. $f^{*c}(\mathrm{rad}(s)) = \mathrm{rad}(f^{*c}(s))$. ■

Proposition (2.4.7).

(i) *A congruence is codisjunctable if and only if its radical is codisjunctable.*

(ii) *For a pair r, s of codisjunctable congruences on A, the following assertions are equivalent.*

(a) *r and s have the same codisjunctor;*

(b) *r and s have the same radical.*

(iii) *Any radical congruence is the join of codisjunctable radical congruences.*

Proof.

(i) Let r be a congruence on A and let $f: A \to B$ be any morphism. If f codisjoints r, it codisjoints $\mathrm{rad}(r)$ for $r \leqslant \mathrm{rad}(r)$. Let us assume that f does not codisjoint r. Then $f_{*c}(r)$ is a proper congruence on B. According to Corollary (2.1.10) there exists a maximal congruence m on B such that $f_{*c}(r) \leqslant m$. Thence $r \leqslant f^{*c}(m)$. Because $f^{*c}(m)$ is radical, $\mathrm{rad}(r) \leqslant f^{*c}(m)$ and thus $f_{*c}(\mathrm{rad}(r)) \leqslant m$. Therefore $f_{*c}(\mathrm{rad}(r))$ is proper, and thus f does not codisjoint $\mathrm{rad}(r)$. As a result, f codisjoints r if and only if it codisjoints $\mathrm{rad}(r)$. Consequently, f is a codisjunctor for r if and only if f is a codisjunctor for $\mathrm{rad}(r)$.

(ii) (b) ⇒ (a) By the proof of (i), $\mathrm{codis}(r) = \mathrm{codis}(\mathrm{rad}(r))$ and $\mathrm{codis}(s) = \mathrm{codis}(\mathrm{rad}(s))$. Therefore $\mathrm{codis}(r) = \mathrm{codis}(s)$.

(a) ⇒ (b) Let d be the common codisjunctor or r and s. Let p be a prime congruence on A such that $r \leqslant p$. Let $q : A \to Q$ be the quotient of A by p and $k : Q \to K$ be a monomorphism with a simple codomain. Then kq coequalizes r and thus it does not factor through d. Hence it does not codisjoint s and thus it coequalizes s. Consequently, q coequalizes s and $s \leqslant p$. As a result, for any prime congruence p on A, $r \leqslant p \Leftrightarrow s \leqslant p$. According to Proposition (2.4.4), it follows that $r = s$. ■

(iii) Let r be a radical congruence on A. According to Proposition (1.5.3), there exists a family of codisjunctable congruences $(r_i)_{i \in I}$ on A such that $r = \overset{c}{\vee}_{i \in I} r_i$. According to (i), the radical of r_i is codisjunctable for any $i \in I$. Then the family $(\mathrm{rad}(r_i))_{i \in I}$ is such that $r = \mathrm{rad}(r) = \mathrm{rad}(\overset{c}{\vee}_{i \in I} r_i) = \overset{cr}{\vee}_{i \in I} \mathrm{rad}(r_i)$. ■

Theorem (2.4.8). *The set* CongRad(A) *of radical congruences on* A *is a distributive algebraic lattice.*

Proof. Let s be a compact element in Cong(A) and let $(r_i)_{i \in I}$ be a filtered family of elements in CongRad(A) such that $\mathrm{rad}(s) \leqslant \overset{cr}{\vee}_{i \in I} r_i$. By Proposition (2.4.2), $\overset{cr}{\vee}_{i \in I} r_i = \overset{c}{\vee}_{i \in I} r_i$. Then $s \leqslant \mathrm{rad}(s) \leqslant \vee_{i \in I} r_i$. Thus there exists $i_0 \in I$ such that $s \leqslant r_{i_0}$ and so $\mathrm{rad}(s) \leqslant r_{i_0}$. As a result, $\mathrm{rad}(s)$ is compact in CongRad(A). Let r be an element in CongRad(A). Then $r = \overset{c}{\vee}_{i \in I} s_i$, where the s_i are compact elements in Cong(A) (Proposition (1.5.2). It follows that $r = \mathrm{rad}(r) = \mathrm{rad}(\overset{c}{\vee}_{i \in I} s_i) = \overset{cr}{\vee}_{i \in I} \mathrm{rad}(s_i)$. Thus r is the join of compact elements in CongRad(A). As a result, CongRad(A) is an algebraic lattice [17]. Let us show that it is a distributive lattice. Let $r, s, t \in$ CongRad(A). The relations $r \leqslant r \overset{cr}{\vee} s$ and $s \leqslant r \overset{cr}{\vee} s$ imply $r \wedge t \leqslant (r \overset{cr}{\vee} s) \wedge t$ and $s \wedge t \leqslant (r \overset{cr}{\vee} s) \wedge t$. Thus $(r \wedge t) \overset{cr}{\vee} (s \wedge t) \leqslant (r \overset{cr}{\vee} s) \wedge t$. Let p be a prime congruence on A. Let us assume that $(r \wedge t) \overset{cr}{\vee} (s \wedge t) \leqslant p$. Then $r \wedge t \leqslant p$ and $s \wedge t \leqslant p$. Because p is universally irreducible (Proposition (2.2.13)), it follows that ($r \leqslant p$ or $t \leqslant p$) and ($s \leqslant p$ or $t \leqslant p$). If $t \leqslant p$, then $(r \overset{cr}{\vee} s) \wedge t \leqslant p$. If $t \nleqslant p$, then $r \leqslant p$ and $s \leqslant p$; thus $r \overset{cr}{\vee} s \leqslant p$ and hence $(r \overset{cr}{\vee} s) \wedge t \leqslant p$. If follows that, in any case, $(r \overset{cr}{\vee} s) \wedge \leqslant p$. Because any element of CongRad(A) is an intersection of prime congruences (Proposition (2.4.4.)), we have the equality $(r \wedge t) \overset{cr}{\vee} (s \wedge t) = (r \overset{cr}{\vee} s) \wedge t$. As a result, the lattice CongRad(A) is distributive. ■

The radical congruence functor

Let $f : A \to B$ be a morphism in **A**. Let r be a congruence on B, let $q_r : B \to B/r$ be the quotient of B by r, let $s = f^{*c}(r)$, and let $q_s : A \to A/s$ be the quotient of A by s. Then A/s is a subobject of B/r. If r is prime (radical), then B/r is integral (reduced); thus A/s is integral (reduced) and hence s is prime (radical). It follows that the map $f^{*c} : \mathrm{Cong}(B) \to \mathrm{Cong}(A)$ induces an order-preserving map $f^{*cr} : \mathrm{CongRad}(B) \to \mathrm{CongRad}(A)$. On the other hand, let us define the map $f_{*cr} : \mathrm{CongRad}(A) \to \mathrm{CongRad}(B)$ by $f_{*cr}(t) = \mathrm{rad}(f_{*c}(t))$. It is order-preserving and, for any $t \in \mathrm{CongRad}(A)$ and $r \in \mathrm{CongRad}(B)$, we have $f_{*cr}(t) \leqslant r \Leftrightarrow \mathrm{rad}(f_{*cr}(t)) \leqslant r \Leftrightarrow f_{*c}(t) \leqslant r \Leftrightarrow t \leqslant f^{*c}(r) \Leftrightarrow t \leqslant f^{*cr}(r)$. It follows that f_{*cr} is left-adjoint to f^{*cr}.

Consequently, $f_{*\mathrm{cr}}$ preserves joins and $f^{*\mathrm{cr}}$ preserves meets. Let us prove that $f_{*\mathrm{cr}}$ preserves finite meets. Let $s, t \in \mathrm{CongRad}(A)$. Because $f_{*\mathrm{cr}}$ is order-preserving, $f_{*\mathrm{cr}}(s \wedge t) \leqslant f_{*\mathrm{cr}}(s) \wedge f_{*\mathrm{cr}}(t)$. Let p be a prime congruence on B such that $f_{*\mathrm{cr}}(s \wedge t) \leqslant p$. Then $s \wedge t \leqslant f^{*\mathrm{cr}}(p)$. Because $f^{*\mathrm{cr}}(p)$ is prime, it is universally irreducible (Proposition (2.2.13)). Thus $s \leqslant f^{*\mathrm{cr}}(p)$ or $t \leqslant f^{*\mathrm{cr}}(p)$. Thence $f_{*\mathrm{cr}}(s) \leqslant p$ or $f_{*\mathrm{cr}}(t) \leqslant p$. Hence $f_{*\mathrm{cr}}(s) \wedge f_{*\mathrm{cr}}(t) \leqslant p$. Because any radical congruence on B is the meet of prime congruences (Proposition (2.4.4), it follows that $f_{*\mathrm{cr}}(s \wedge t) = f_{*\mathrm{cr}}(s) \wedge f_{*\mathrm{cr}}(t)$. As a result, $f_{*\mathrm{cr}} : \mathrm{CongRad}(A) \to \mathrm{CongRad}(B)$ is a morphism of bounded join-complete lattices. Let **LatAlgDist** denote the *category of distributive algebraic lattices* whose morphisms are morphisms of bounded join-complete lattices. Then the *radical congruence functor* RadCong : **A** $\to$ **LatAlgDist** assigns RadCong(A) to an object A and $f_{*\mathrm{cr}}$ to a morphism f. ■

Notation. A *finitely generated radical congruence* is a congruence of the form $\mathrm{rad}(r)$ where r is a finitely generated congruence. They are precisely the compact elements of CongRad(A) (proof of Theorem (2.4.8).

Proposition (2.4.9). *Any codisjunctable radical congruence is finitely generated.*

Proof. Let r be a codisjunctable radical congruence on A whose codisjunctor is d. According to Proposition (1.8.2), there exists a finitely generated congruence s on A whose codisjunctor is d. According to Proposition (2.4.7), $r = \mathrm{rad}(r) = \mathrm{rad}(s)$. Thus r is finitely generated. ■

Theorem (2.4.10). *The set* $\mathrm{CongRad}_0(A)$ *of finitely generated radical congruences on A is a bounded distributive lattice.*

Proof. As $\mathrm{CongRad}_0(A)$ is the set of compact elements in CongRad(A), it is closed in CongRad(A) under finite joins ([17], proof of Theorem 1.3). The top element $1_{A \times A}$ of CongRad(A) is compact, because it is compact in Cong(A). Let r, s be a pair of compact radical congruences on A. According to Proposition (2.4.7), there are two finite families $(r_i)_{i \in I}$, $(s_i)_{j \in J}$ of codisjunctable radical congruences such that $r = \overset{\mathrm{cr}}{\bigvee}_{i \in I} r_i$ and $s = \overset{\mathrm{cr}}{\bigvee}_{j \in J} s_j$. Then $r \wedge s = (\overset{\mathrm{cr}}{\bigvee}_{i \in I} r_i) \wedge \overset{\mathrm{cr}}{\bigvee}_{j \in J} s_j) = (\overset{\mathrm{cr}}{\bigvee}_{(i,j) \in I \times J} r_i \wedge s_j$. According to Proposition (1.5.4), $r_i \wedge s_j$ is codisjunctable, and according to Proposition (2.4.9), $r_i \wedge s_j$ is compact. As finite joins of compact elements are compact, $r \wedge s$ is compact. As a result, $\mathrm{CongRad}_0(A)$ is a bounded sub lattice of CongRad(A), and thus is a bounded distributive lattice. ■

The finitely generated radical congruence functor

Let $f: A \to B$ be a morphism of **A**. Let r be a congruence on **A**, let $q_r: A \to A/r$ be the quotient of A by r, let $s = f_{*c}(r)$, and let $q_s: B \to B/s$ be the quotient of B by s. Then q_s is the pushout of q_r along f. If r is finitely generated, q_r is a finitely generated regular quotient and thus q_s is a finitely generated regular quotient. Hence s is finitely generated. Let t be a finitely generated radical congruence on A. There exists a finitely generated congruence r on A such that $t = \mathrm{rad}(r)$. Then $r \leqslant t$ and thus $\mathrm{rad}(f_{*c}(r)) \leqslant \mathrm{rad}(f_{*c}(t))$. On the other hand, $r \leqslant f^{*c}(f_{*c}(r))$ and thus $t \leqslant \mathrm{rad}(f^{*c}(f_{*c}(r)))$. According to Proposition (2.4.6), $t \leqslant f^{*c}(\mathrm{rad}(f_{*c}(r)))$. Thus $f_{*c}(t) \leqslant \mathrm{rad}(f_{*c}(r))$ and hence $\mathrm{rad}(f_{*c}(t)) \leqslant \mathrm{rad}(f_{*c}(r))$. It follows that $\mathrm{rad}(f_{*c}(r)) = \mathrm{rad}(f_{*c}(t))$. Therefore $f_{*cr}(t) = \mathrm{rad}(f_{*c}(r))$ is a finitely generated radical congruence. As a result, and according to Theorem (2.4.10), the radical congruence functor CongRad : **A** $\to$ **LatAlgDist** induces a functor $\mathrm{CongRad}_0$: **A** $\to$ **LatDist** called the *finitely generated radical congruence functor.*

Proposition (2.4.11) *The finitely generated radical congruence functor* $\mathrm{CongRad}_o$: ***A*** $\to$ ***LatDist*** *preserves filtered colimits.*

Proof. The category **RedA** of reduced objects of **A** is a Zariski category (cf. Theorem (12.3.2) below). According to Proposition (1.5.5) applied to the category **Red A**, the finitely generated congruence functor Cong_o : **RedA** $\to$ $\vee$-**SemLat** preserves filtered colimits. As the reflector R : **A** $\to$ **RedA** is a left-adjoint, it preserves filtered colimits. Thence the composite functor $\mathrm{Cong}_0 R$: **A** $\to$ $\vee$-**SemLat** preserves filtered colimits. Let us prove that this functor is isomorphic to the finitely generated radical congruence functor $\mathrm{CongRad}_0$: **A** $\to$ $\vee$-**SemLat**. Let us define the natural transformation $\alpha : \mathrm{CongRad}_0 \to \mathrm{Cong}_0 R$ at the object A by the map $\alpha_A : \mathrm{CongRad}_0(A) \to \mathrm{CongRad}_0(R(A))$ which assigns to s the direct image of s by the unit morphism $r_A : A \to R(A)$ of the reflection of A in **RedA**. This map is injective because, for any $s \in \mathrm{CongRad}_o(A)$, $s = (r_A)^{*c}((r_A)_{*c}(s)) = r_A^{*c}(\alpha_A(s))$. Let $t \in \mathrm{CongRad}_0(R(A))$. Let $s = r_A^{*c}(t)$. Then $s \in \mathrm{CongRad}(A)$. Let us prove that $s \in \mathrm{CongRad}_o(A)$. Let $(s_i)_{i \in I}$ be a family of elements in $\mathrm{CongRad}(A)$ such that $s \leqslant \overset{\mathrm{cr}}{\vee}_{i \in I} s_i$. Then $t = (r_A)_{*c}(s) \leqslant \overset{\mathrm{cr}}{\vee}_{i \in I}(r_A)_{*c}(s_i)$. Because $t \in \mathrm{CongRad}_0(R(A))$, there exists a finite subset I_0 of I such that $t \leqslant \overset{\mathrm{cr}}{\vee}_{i \in I_0}(r_A)_{*c}(s_i) = (r_A)_{*c}(\overset{\mathrm{cr}}{\vee}_{i \in I_0} s_i)$. Thence $s = (r_A)^{*c}(t) \leqslant (r_A^{*c})((r_A)_{*c}(\overset{\mathrm{cr}}{\vee}_{i \in I_0} s_i)) = \overset{\mathrm{cr}}{\vee}_{i \in I_0} s_i$. It follows that $s \in \mathrm{CongRad}_0(A)$. The relation $t = (r_A)_{*c}((r_A)^{*c}(t)) = (r_A)_{*c}(s) = \alpha_A(s)$ implies that α_A is bijective, so that α is an isomorphism. As a result, the functor $\mathrm{CongRad}_0$: **A** $\to$ **LatDist** preserves filtered colimits. ■

Theorem (2.4.12). *The set* CongRadCodis(A) *of codisjunctable radical congruences on A is a bounded meet-semilattice.*

Proof. According to Proposition (2.4.9), CongRadCodis(A) is a subset of $\text{CongRad}_0(A)$. The bounds Δ_A and $1_{A\times A}$ are codisjunctable and, according to Proposition (1.5.4), CongRadCodis(A) is closed under finite meets in CongRad(A) and thus is also closed in $\text{CongRad}_0(\text{A})$. It follows that CongRadCodis(A) is a sub-bounded meet-semilattice of $\text{CongRad}_0(A)$. ■

The codisjunctable radical congruence functor

Let $f: A \to B$ be a morphism in **A**. Let r be a codisjunctable radical congruence on A. Then $f_{*c}(r)$ is a codisjunctable congruence (cf. § 1.5, p. 32). Thus $f_{*cr}(r) = \text{rad}(f_{*c}(r))$ is codisjunctable (Proposition (2.4.7). We denote by $\wedge$-**SemLat** the category of bounded meet-semilattices. The *codisjunctable radical congruence functor* CongRadCodis : **A** $\leftarrow$ $\wedge$-**SemLat** assigns the set CongRadCodis(A) of codisjunctable radical congruences on A to an object A, and assigns the restriction of f_{*cr} to a morphism f.

Proposition (2.4.13). *A congruence is unit if and only if its radical is unit.*

Proof. Let r be a congruence on A. If $r = 1_{A\times A}$ then $\text{rad}(r) = 1_{A\times A}$. If $r \neq 1_{A\times A}$, r is proper. Thus there exist a maximal congruence m on A such that $r \leqslant m$ (Corollary (2.1.10). Then $\text{rad}(r) \leqslant m$ and hence $\text{rad}(r) \neq 1_{A\times A}$. ■

Definition (2.4.14). *The* radical rad(A) of an object A *is the radical of the equality of A, i.e.* $\text{rad}(A) = \text{rad}(\Delta_A)$.

Proposition (2.4.15). *The quotient* $q_{\text{rad}(A)} : A \to A/\text{rad}(A)$ *of an object A by its radical is precisely the universal reduced object associated to A.*

Proof. The proof follows from the fact that the universal reduced object $r_A : A \to R(A)$ is a regular epimorphism. ■

Corollary (2.4.16). *An object is reduced if and only if its radical is the equality.*

2.5 CONJOINT MORPHISMS

Definition (2.5.1). *A pair of parallel morphisms* $(f, g) : A \rightrightarrows B$ *is* conjoint *if it has, as a codisjunctor, the morphism* $O_B : B \to 1$.

Alternatively, we say that f and g are conjoint and this is denoted by $f \approx g$.

Proposition (2.5.2). *For a pair of parallel morphisms $(f, g) : A \rightrightarrows B$, the following assertions are equivalent:*

(i) *f and g are conjoint.*

(ii) *For any morphism $h : B \to K$ with a simple codomain, $hf = hg$.*

(iii) *$r_B f = r_B g$ where $r_B : B \to R(B)$ is the universal reduced object associated to B.*

(iv) *The pair (f, g) factors through the radical of B.*

(v) *$R(f) = R(g)$ where $R : \mathbf{A} \to \mathbf{RedA}$ is the reflector.*

Proof.

(i) ⇒ (ii) The pair (hf, hg) is conjoint. Thus it is not codisjointed and hence $hf = hg$ (Proposition (2.1.2)).

(ii) ⇒ (i) Let $d : B \to D$ be a morphism which codisjoints (f, g). If D were not terminal there would exist a simple quotient $k : D \to K$ and (kdf, kdg) would be codisjointed, which contradicts the equality $kdf = kdg$. Thence D is terminal. Consequently $O_B : B \to 1$ is the codisjunctor of (f, g).

(ii) ⇒ (iii) As $R(B)$ is reduced, it is a subobject of a product of simple objects, say $m = (m_i)_{i \in I} : R(B) \to \Pi_{i \in I} K_i$. For each $i \in I$, we have $m_i r_B f = m_i r_B g$. Thus $r_B f = r_B g$.

(iii) ⇒ (ii) Any morphism $h : B \to K$ with a simple codomain factors through r_B and thus satisfies $hf = hg$.

(iii) ⇔ (iv) follows from Proposition (2.4.15).

(iii) ⇔ (v) The relations $R(f) r_A = r_B f$ and $R(g) r_A = r_B g$, and the fact that r_A is epimorphic, imply the equivalences $R(f) = R(g) \Leftrightarrow R(f) r_A = R(g) r_A \Leftrightarrow r_B f = r_B g$. ■

Corollary (2.5.3). *The binary relation $\approx$ on morphisms of* **A** *is a congruence on* **A** *([31], p. 52).*

The equivalence class of a morphism $f : A \to B$ is denoted by $\overset{\approx}{f} : A \to B$, and the quotient category of **A** by $\approx$ ([31], p. 51) is denoted by $\mathbf{A}/\approx$.

Corollary (2.5.4). *A congruence on A is conjoint if and only if it is contained in the radical of A.*

Corollary (2.5.5). *An object A is reduced if and only if conjoint morphisms with codomain A are equal.*

Proof. The proof follows from Corollary (2.4.16). ■

Definition (2.5.6). *A* premonomorphism *is a morphism* $f: A \to B$ *such that, for any pair of morphisms* $(g, h) : C \rightrightarrows A$, *we have* $fg \approx fh \Rightarrow g \approx h$.

Proposition (2.5.7). *For a morphism* $f: A \to B$, *the following statements are equivalent.*

(i) f *is a premonomorphism.*

(ii) $\overset{\approx}{f}$ *is a monomorphism in* $\mathbf{A}/\approx$.

(iii) *The kernel pair of* f *is conjoint.*

(iv) $f^{*c}(\mathrm{rad}(B))$ *is conjoint.*

(v) $f^{*c}(\mathrm{rad}(B)) = \mathrm{rad}(A)$.

(vi) *The morphism* $R(f) : R(A) \to R(B)$ *is monomorphic where* R: $\mathbf{A} \to \mathbf{RedA}$ *is the reflector.*

(vii) *Pushouts along* f *of non-null singular epimorphisms are not null.*

Proof. Let $k = (k_1, k_2) : K \rightrightarrows A$ be the kernel pair of f. Then $k = f^{*c}(\Delta_B)$ and, according to Proposition (2.4.6), $\mathrm{rad}(k) = \mathrm{rad}(f^{*c}(\Delta_B)) = f^{*c}(\mathrm{rad}(\Delta_B)) = f^{*c}(\mathrm{rad}(B))$. Thus $\mathrm{rad}(A) \leqslant f^{*c}(\mathrm{rad}(B))$.

(i) ⇔ (ii) is obvious.

(i) ⇒ (iii) The equality $fk_1 = fk_2$ implies that (k_1, k_2) is conjoint.

(iii) ⇒ (iv) If k is conjoint, then $f^{*c}(\mathrm{rad}(B)) = \mathrm{rad}(k)$ is conjoint (Proposition (2.4.7).

(iv) ⇒ (v) follows from the relation $f^{*c}(\mathrm{rad}(B)) \leqslant \mathrm{rad}(A)$ (Proposition (2.5.4)).

(v) ⇒ (vi) follows from the relation $R(f)r_A = r_B f$, where $r_A : A \to A/\mathrm{rad}(A)$ and $r_B : B \to B/\mathrm{rad}(B)$ are the quotient morphisms.

(vi) ⇒ (vii) Let $d: A \to D$ be a non-null singular epimorphism codisjunctor of a pair $(g, h) : C \rightrightarrows A$. Then (g, h) is not conjoint. Thus $r_A g \neq r_A h$, and hence $r_B fg = R(f)r_A g \neq R(f)r_A h = r_B fh$. Therefore (fg, fh) is not conjoint, i.e. the codisjunctor of (fg, fh) is not null. But it is precisely the pushout of d along f.

(vii) ⇒ (i) Let $(g, h) : C \rightrightarrows A$ be a pair of morphisms which is not conjoint. Then $r_A g \neq r_A h$. According to Axiom (1.2.2) of Zariski categories, there exist a codisjunctable object X and a morphism $m: X \to C$ such that $r_A gm \neq r_A hm$. Then (gm, hm) is not conjoint. Thus the codisjunctor d: $A \to D$ of (gm, hm) is not null. Therefore its pushout $\delta : B \to \Delta$ along f is not null. But δ is the codisjunctor of (fgm, fhm). Thus (fgm, fhm) is not conjoint, therefore (fg, fh) is not conjoint. As a result, f is a premonomorphism. ∎

Corollary 2.5.8 *Monomorphisms are premonomorphisms.*

Proof. The proof follows from Proposition (2.3.4). ∎

Examples

In categories of commutative algebras, conjoint pairs of morphisms are pairs of morphisms which are equal up to nilpotent elements, and premonomorphisms are the morphisms which are injective with respect to nilpotent elements.

2.6 PSEUDO-SIMPLE OBJECTS

Definition (2.6.1). *An object is* pseudo-simple *if it has exactly two singular quotient objects.*

For example the object $k[\varepsilon] = k[X]/(X^2)$ is pseudo-simple in $\mathbf{CAlg}(k)$. The *category* **PsdSimA** *of pseudo-simple objects* of **A** is the full subcategory of **A** whose objects are the pseudo-simple ones.

Proposition (2.6.2). *For any object A, the following statements are equivalent.*

(i) *A is pseudo-simple.*

(ii) *A is not terminal and any singular epimorphism $f: A \to B$ is isomorphic or null.*

(iii) *A is not terminal and any presingular epimorphism $f: A \to B$ is isomorphic or null.*

(iv) *The universal reduced object associated to A is simple.*

(v) *The radical of A is a maximal congruence.*

(vi) *A has exactly one integral quotient object.*

(vii) *A has exactly one prime congruence.*

(viii) *A is not terminal and any pair of parallel morphisms with codomain A is conjoint or codisjointed.*

Proof

(i) $\Leftrightarrow$ (ii) The terminal object has exactly one singular quotient object. Thus it is not pseudo-simple. Thus any pseudo-simple object is not terminal. For any non-terminal object, the morphisms $1_A : A \to A$, $0_A : A \to 1$ are two distinct singular quotients. Therefore a non-terminal object A is pseudo-simple if and only if any singular quotient of A is isomorphic or null.

(ii) $\Rightarrow$ (iii) Let A be pseudo-simple and let $f: A \to B$ be a non-null presingular epimorphism. If $f = gd$, with d singular, then d is not null. Thus it is isomorphic. It follows that f is prelocal (Definition (1.9.1)) and thence isomorphic (Theorem (1.9.6)).

(iii) $\Rightarrow$ (ii) follows from the fact that any singular epimorphism is presingular.

(ii) $\Rightarrow$ (iv) In what follows, let $r: A \to R$ be the universal reduced object associated to A. As A is not terminal, there is a morphism $k: A \to K$ with

K simple (Corollary (2.1.9)). This morphism k factors through r and hence R is not terminal. Let $q: R \to Q$ be a proper regular quotient of R. Then $qr: A \to Q$ is a regular quotient object of A different from r. Hence there exist a pair of morphisms $(m, n): X \rightrightarrows A$ whose domain X is condisjunctable such that $rm \neq rn$ and $qrm = qrn$. As the object R is reduced, there exists a simple object S and a morphism $s: R \to S$ such that $srm \neq srn$, and, consequently, such that sr codisjoints the pair (m, n) (Proposition (2.1.2)). Let $d: A \to D$ be the codisjunctor of (m, n). The morphism sr factors through d and so D is not terminal, i.e. d is not null. Thus d is isomorphic and (m, n) is codisjointed, as is (qrm, qrn). The equality $qrm = qrn$ implies that Q is terminal, i.e. q is null. As a result, R is simple.

(iv) ⇒ (ii) As R is not terminal A is not terminal. Let $d: A \to D$ be a non-null singular quotient object of A. Let $(m, n): X \rightrightarrows A$ be a pair of morphisms whose codisjunctor is d. Let $\delta: R \to \Delta$ be the pushout of d along r. As D is not terminal, there is a morphism $k: D \to K$ with K simple. The morphism kd factors through r in a morphism which factors through δ. Thus δ is not null. As R is simple and δ is a singular epimorphism, δ is isomorphic. Thus r factors through d and hence r codisjoints (m, n). Any morphism $s: A \to S$ with S simple factors through r and so it codisjoints (m, n). It follows that the pair (m, n) is codisjointed (Proposition (2.1.11)) and that d is isomorphic. As a result, the object A satisfies (ii).

(iv) ⇔ (v) follows from Propositions (2.1.7) and (2.4.15).

(iv) ⇒ (vi) If R is simple, then $r: A \to R$ is an integral quotient of A. Moreover, any integral quotient $p: A \to P$ of A factors through r in a regular epimorphism which is necessarily isomorphic. Therefore A has exactly one integral quotient object.

(iv) ⇒ (vii) The proof is immediate.

(vii) ⇒ (v) According to Proposition (2.4.4), A has exactly one radical congruence. According to Proposition (2.2.12) A has exactly one maximal congruence. Thus the radical of A is maximal.

(iv) ⇒ (viii) Let $(m, n): X \rightrightarrows A$. Let $q: A \to Q$ be the coequalizer of (m, n). If Q is terminal, (m, n) is codisjointed. Otherwise there is a simple quotient $k: Q \to K$. Then kq is a simple quotient of A which factors through r. Thus kq is isomorphic to r. Thus $rm = rn$. By Proposition (2.5.2), (m, n) is conjoint.

(viii) ⇒ (ii) Let $f: A \to B$ be a singular epimorphic codisjunctor of a pair $(m, n): X \rightrightarrows A$. Because (m, n) is conjoint or codisjointed, f is null or isomorphic. ■

Corollary (2.6.3). *An object is simple if and only if it is pseudo-simple and reduced.*

Theorem (2.6.4). *To any pseudo-simple object is associated a universal simple object. More precisely,* **SimA** *is a reflective subcategory of* **PsdSimA.**

Proof. It is the universal reduced object associated to it (Theorem (2.3.2). ■

Proposition (2.6.5). *Any filtered colimit of pseudo-simple objects is a pseudo-simple object.*

Proof. Let $(\alpha_i : A_i \to A)_{i \in \mathbf{I}}$ be a filtered colimit in **A**. If we denote the reflector by $R : \mathbf{A} \to \mathbf{RedA}$, we obtain a filtered colimit $(R(\alpha_i) : R(A_i) \to R(A)_{i \in \mathbf{I}})$ in **RedA**, as in **A** (Proposition (2.3.3)). If the objects A_i are pseudo-simple, the objects $R(A_i)$ are simple (Proposition (2.6.2)). Thus $R(A)$ is simple (Proposition (2.1.3)) and hence A is pseudo-simple (Proposition (2.6.2)). ■

2.7 QUASI-PRIMARY OBJECTS

Definition (2.7.1).

(i) *An* object A *is* quasi-primary *if it is not terminal and, for any pair of singular epimorphisms,* $d : A \to D$, $\delta : A \to \Delta$,

$$d \wedge \delta = O_A \Rightarrow (d = O_A \text{ or } \delta = O_A).$$

(ii) *A* morphism $f : A \to B$ *is* quasi-primary *if A and B are quasi-primary objects and f is a premonomorphism (Definition (2.5.6)).*

The *category* **QuaPrimA** of *quasi-primary objects of* **A** has quasi-primary objects as its objects and quasi-primary morphisms as its morphisms. It contains the category **PsdSimA** as a full subcategory.

Theorem (2.7.2). *To any quasi-primary object is associated a universal pseudo-simple object, i.e.* **PsdSimA** *is a reflective subcategory of* **QuaPrimA**.

Proof. Let A be a quasi-primary object. As the set of singular quotient objects of A is stable under finite co-intersections, the set of non-null singular quotient objects of A is stable under finite co-intersections. Let $f : A \to B$ be the co-intersection of all non-null singular quotients of A. As 1 is finitely presentable, the object B is not terminal. The morphism f is flat (Proposition (1.9.5)). Let us show that B is pseudo-simple. Let $\delta : B \to \Delta$ be a non-null singular quotient of B which is the codisjunctor of a congruence r on B (Proposition (1.8.2)). According to Proposition (1.5.3), the inverse image $f^{*c}(r)$ of r by f is the join of a family of codisjunctable congruences on A, i.e. $f^{*c}(r) = \overset{c}{\vee}_{i \in I} r_i$. The family $((\delta f)_{*c}\, r_i))_{i \in I}$ of direct images of r_i by δf is such that

$$\overset{c}{\vee}_{i \in I} (\delta f)_{*c}(r_i) = \delta_{*c}(f_{*c}(\overset{c}{\vee}_{i \in I} r_i)) = \delta_{*c}(f_{*c}(f^{*c}(r))) = \delta_{*c}(r) = 1_{\Delta \times \Delta}.$$

Because $1_{\Delta X\Delta}$ is compact in Cong (Δ), there is a finite subset $I_0 \subset I$ such that $\overset{c}{\vee}_{i\in I_0}(\delta f)_{*c}(r_i) = 1_{\Delta\times\Delta}$. For each $i \in I_0$, let $d_i : A \to D_i$ be the codisjunctor of r_i and let $\delta_i : \Delta \to \Delta_i$ be the pushout of d_i along δf. Then δ_i is the codisjunctor of $(\delta f)_{*c}(r_i)$. By Axiom (1.2.6) of Zariski categories, $v_{i\in I_0}\delta_i = 1_\Delta$. It follows that there exists some $i_0 \in I_0$ for which $\Delta_{i_0} \neq 1$ and thence $D_{i_0} \neq 1$. Then d_{i_0} is a non-null singular quotient of A. Thus f factors through d_{i_0} and codisjoints r_{i_0}. Consequently, f codisjoints $f^{*c}(r)$. Thus $r = f_{*c}(f^{*c}(r)) = 1_{B\times B}$, and δ is isomorphic. As a result, B is pseudo-simple. By construction the morphism f is quasi-primary (proposition (2.5.7)). Let $g : A \to C$ be a quasi-primary morphism with a pseudo-simple codomain C. For any non-null singular epimorphism $d : A \to D$, the pushout of d along g is not null (Proposition (2.5.7) and thus it is isomorphic. Hence g factors through d. Therefore g factors through f. As a result $f: A \to B$ is the universal pseudo-simple object associated to A. ■

Proposition (2.7.3). *IntA is a full subcategory of QuaPrimA.*

Proof. Let A be an integral object. Then A is not terminal. Let $k : A \to K$ be a monomorphism whose codomain is simple. Let $d : A \to D$, $\delta : A \to \Delta$ be a pair of non-null singular epimorphisms. Let (g, v) be the pushout of (d, k), and let (h, w) be the pushout of (δ, k). As k is monomorphic, and d and δ are flat, g and h are monomorphic. Thus v and w are not null. As v and w are singular epimorphisms, they are isomorphic. It follows that k factors through d and δ. Consequently, $d \wedge \delta \neq O_A$. As a result, A is quasi-primary. Let $f: A \to B$ be a morphism of **IntA**, i.e. f is a monomorphism and A and B are integral. According to Corollary (2.5.8), the morphism f is quasi-primary. Let A and B be two integral objects and let $f: A \to B$ be a quasi-primary morphism. According to Proposition (2.5.7), the kernel pair r of f is included in the radical of A. But the object A is reduced and thus its radical is Δ_A (Proposition (2.4.16). It follows that $r = \Delta_A$ and that f is monomorphic. As a result, **IntA** is a full subcategory of **QuaPrimA.** ■

Proposition (2.7.4). *For any object A, the following statements are equivalent.*

(i) *A is quasi-primary.*

(ii) *The universal reduced object associated to A is integral.*

(iii) *The radical of A is prime.*

Proof

(i) ⇒ (ii) Let $p : A \to P$ be the universal pseudo-simple object associated to A (Theorem (2.7.2)). If $R \cdot \mathbf{A} \to \mathbf{RedA}$ denotes the reflector, then $R(p)$ is monomorphic (Proposition (2.5.7). According to Proposition (2.6.2), the object $R(P)$ is simple. Thus $R(A)$ is integral.

(ii) $\Rightarrow$ (iii) follows from Proposition (2.4.15).

(iii) $\Rightarrow$ (i) The fact that rad (A) is prime implies that $A \neq 1$. Let $d : A \to D$, $\delta : A \to \Delta$ be a pair of singular epimorphisms such that $d \wedge \delta = O_A$. Let r and s be congruences on A with codisjunctors d and δ respectively. According to Proposition (1.5.4), the codisjunctor of $r \wedge s$ is O_A, i.e. $r \wedge s$ is conjoint. According to Corollary (2.5.4), $r \wedge s \leqslant \text{rad}(A)$. According to Proposition (2.2.13), $r \leqslant \text{rad}(A)$ or $s \leqslant \text{rad}(A)$. According to Corollary (2.5.4), r is conjoint or s is conjoint. Thus $d = O_A$ or $\delta = O_A$. As a result, A is quasi-primary. ∎

Corollary (2.7.5). *An object is integral if and only if it is quasi-primary and reduced.*

Theorem (2.7.6). *To any quasi-primary object is associated a universal integral object, i.e.* **IntA** *is a reflective subcategory of* **QuaPrimA**.

Proof. Proposition (2.7.4) implies that the reflector $R : \mathbf{A} \to \mathbf{Red\ A}$ induces a left-adjoint to the inclusion functor **IntA** $\to$ **QuaPrimA**. ∎

2.8 PRIMARY OBJECTS

Definition (2.8.1). *An object is* primary *if it is not terminal and any singular quotient of it is monomorphic or null.*

Proposition (2.8.2). *Any primary object is quasi-primary.*

Proof. Let A be a primary object and $d : A \to D$, $\delta : A \to \Delta$ be a pair of non-null singular epimorphisms. Then d and δ are monomorphic, and thus they are flat bimorphisms (Proposition (1.8.3)). Therefore $d \wedge \delta$ is a flat bimorphism (Proposition (1.7.1)). Thus $d \wedge \delta$ is not null. ∎

The *category* **PrimA** *of primary objects of* **A** is the full subcategory of **QuaPrimA** whose objects are primary objects.

Proposition (2.8.3). *An object is primary if and only if it is a subobject of a pseudo-simple object.*

Proof. Let A be a primary object. As A is quasi-primary, by Theorem (2.7.2) there exists a universal pseudo-simple object associated to it, say $p_A : A \to P(A)$. By the proof of Theorem (2.7.2), p_A is the co-intersection of all non-null singular quotient objects of A. Because these singular quotients are monomorphic, they are flat bimorphisms (Proposition (1.8.3)). Thus p_A is a flat bimorphism (Proposition (1.7.1)) and hence p_A is monomorphic. It follows that A is a subobject of the pseudo-simple object $P(A)$.

Conversely, let A be an object such that there exists a monomorphism $p: A \to P$ whose codomain is pseudo-simple. Let $d: A \to D$ be a non-null singular epimorphism. Let $(\delta : P \to \Delta, m : D \to \Delta)$ be the pushout of (p, d). Because d is flat, m is monomorphic and thus $\Delta \neq 1$. Then $\delta \neq O_P$, and thus δ is isomorphic. Then p factors through d and hence d is monomorphic. It follows that A is pseudo-simple. ∎

Theorem (2.8.4). *To any quasi-primary object is associated a universal primary object, i.e.* **PrimA** *is a reflective subcategory of* **QuaPrimA.**

Proof. Let A be a quasi-primary object, let $p_A : A \to P(A)$ be the universal pseudo-simple object associated to A, and let $p_A = m_A q_A$ with $q_A : A \to Q(A)$ a regular epimorphism and m_A a monomorphism. Then $Q(A)$ is a primary object (Proposition (2.8.3). Let $g: A \to C$ be a quasi-primary morphism whose codomain is primary. There exists a monomorphism $p: C \to P$ where P is pseudo-simple. The morphism p is quasi-primary, as is the composite pg. Thus there exists a morphism $h: P(A) \to P$ such that $hp_A = pg$. The relation $pg = hm_A q_A$ implies the existence of a morphism $f: Q(A) \to C$ such that $fq_A = g$ and $hm_A = pf$. This morphism f is quasi-primary because the morphism $pf = hm_A$ is. It follows that $q_A : A \to Q(A)$ is the universal primary object associated to A. ∎

Theorem (2.8.5). *To any primary object is associated a universal pseudo-simple object, i.e.* **PsdSimA** *is a reflective subcategory of* **PrimA.**

Proof. The proof is a consequence of Theorem (2.7.2). ∎

Definition (2.8.6). *A congruence r on A is* primary (quasi-primary) *if the quotient of A by r is a primary (quasi-primary) object.*

2.9 SINGULARLY CLOSED OBJECTS

Definition (2.9.1).

(i) *A* singular extension *of an object A is a singular bimorphism $m: A \to M$.*

(ii) *An object is* singularly closed *if any singular extension of it is isomorphic.*

The *category* **SingClA** of *singularly closed object of* **A** is the full subcategory of **A** whose objects are the singularly closed ones.

Proposition (2.9.2). *For an object A, the following assertions are equivalent.*

(i) *A is singularly closed.*

(ii) *Any monomorphism $m: A \to M$ is prelocal.*

(iii) *Any presingular bimorphism $m: A \to M$ is isomorphic.*

Proof.
(i) ⇒ (ii) Let $m: A \to M$ be a monomorphism. If $d: A \to D$ is a singular morphism through which m factors, then d is a singular extension of A. Thus it is isomorphic. It follows that m is prelocal.
(ii) ⇒ (iii) Any presingular monomorphism $m: A \to M$ is prelocal and thus is isomorphic.
(iii) ⇒ (i) Any singular extension $m: A \to M$ is a presingular monomorphism. Hence it is isomorphic. ■

Definition (2.9.3). *A morphism $f: A \to B$ is* preflat *if the pushout along f of any presingular bimorphism is monomorphic.*

The category $\mathbf{A}_{\text{Prflat}}$ has the objects of $\mathbf{A}$ as its objects, and preflat morphisms as its morphisms. It contains the category **SingClA** as a full subcategory.

Theorem (2.9.4). *To any object is associated a universal singularly closed object, i.e. SingClA is a reflective subcategory of $\mathbf{A}_{\text{Prflat}}$. The reflector is precisely the presingular hull.*

Proof. Let A be an object. Let $m: A \to \bar{A}$ be the presingular hull of A (Theorem (1.9.7). If $n: \bar{A} \to N$ is a singular extension of $\bar{A}$, then $nm: A \to N$ is a presingular extension of A. Thus $nm = m$ and hence n is an isomorphism. It follows that $\bar{A}$ is singularly closed. Moreover, the morphism $m: A \to \bar{A}$ is preflat as it is flat (Proposition (1.9.5)). Let $f: A \to B$ be a preflat morphism whose codomain is singularly closed. Let $(\bar{m}: B \to \bar{B}, \bar{f}: \bar{A} \to \bar{B})$ be the pushout of (f, m). The morphism $\bar{m}$ is presingular and monomorphic. By Proposition (2.9.2), $\bar{m}$ is isomorphic. Thus f factors through m in the form $f = hm$. The morphism h is preflat because the pushout of a presingular monomorphism $g: \bar{A} \to C$ along h is equal to the pushout of the presingular monomorphism gm along f. Thus it is monomorphic. As a result, $m: A \to \bar{A}$ is the universal singularly closed object associated to A. ■

Notation. The universal singularly closed object $m: A \to \bar{A}$ associated to A is also called *the singular closure of A*.

Proposition (2.9.5). *An object is pseudo-simple if and only if it is primary and singularly closed.*

Proof. The necessity is immediate. Let A be primary and singularly closed. By the proof of Theorem (2.7.2), the universal pseudo-simple object associated to A is a presingular epimorphism $p_A: A \to P(A)$ and thus it is an isomorphism. Thus A is pseudo-simple. ■

2.10 IRREDUCIBLE OBJECTS

Definition (2.10.1). *An object A is* irreducible *if it is not terminal and, for any pair of congruences r, s on A,*

$$r \wedge s = \Delta_A \Rightarrow (r = \Delta_A \quad \text{or} \quad s = \Delta_A).$$

The *category* **IrrA** of *irreducible objects* of **A** has the irreducible objects of **A**, as its objects and the monomorphisms of **A** as its morphisms.

Proposition (2.10.2). *For any non-terminal object A, the following assertions are equivalent.*

(i) *A is irreducible.*

(ii) *For any pair of regular epimorphisms $p: A \to P$, $q: A \to Q$, we have in* QuotReg(A)

$$p \vee q = 1_A \Rightarrow (p = 1_A \quad \text{or} \quad q = 1_A).$$

(iii) *For any pair $p: A \to P$, $q: A \to Q$ of regular epimorphisms such that the morphism $(p, q): A \to P \times Q$ is monomorphic, one of the morphisms p, q must be an isomorphism.*

Proof.

(i) $\Leftrightarrow$ (ii) follows from the isomorphism Cong $(A) \simeq$ QuotReg $(A)^{0p}$ (Proposition (1.5.2)).

(ii) $\Rightarrow$ (iii) Let $u: A \to U$ be a regular epimorphism such that $p \leqslant u$ and $q \leqslant u$. Let $r_1: U \to P$, $r_2: U \to Q$ such that $r_1 u = p$ and $r_2 u = q$. Then $(r_1, r_2): U \to P \times Q$ is such that $(r_1, r_2)u = (p, q)$. Since (p, q) is monomorphic, u is isomorphic. It follows that $p \vee q = 1_A$ in QuotReg(A). Thus $p \simeq 1_A$ or $q \simeq 1_A$. Equivalently, p or q is isomorphic.

(iii) $\Rightarrow$ (ii) Let us denote by f the morphism $(p, q): A \to P \times Q$. This morphism factorizes in the form $f = me$ where e is a regular epimorphism and m is a monomorphism. We have $p \leqslant e$ and $q \leqslant e$. Thus $e = 1_A$ and hence f is monomorphic. It follows that p or q is isomorphic, i.e. $p = 1_A$ or $q = 1_A$. ■

Proposition (2.10.3). *An object is integral if and only if it is reduced and irreducible.*

Proof. Let A be an integral object. The congruence Δ_A is prime. According to Proposition (2.2.13), Δ_A is a universally irreducible element in Cong (A), i.e. for any pair of congruences r, s on $A: r \wedge s \leqslant \Delta_A \Rightarrow r \leqslant \Delta_A$ or $s \leqslant \Delta_A$. Thus A is irreducible. Conversely, let A be a reduced and irreducible object. Let us show that A is quasi-primary. Let $d: A \to D$, $\delta: A \to \Delta$ be a pair of singular epimorphisms such that $d \wedge \delta = O_A$. There are congruences r, s on A such that codis $(r) = d$ and codis $(s) = \delta$. Then codis

$(r \wedge s) =$ codis$(r) \wedge$ codis$(s) = d \wedge \delta = O_A$. Because A is reduced, it implies $r \wedge s = \Delta_A$ (Proposition (2.5.5)). Because A is irreducible, it implies $r = \Delta_A$ or $s = \Delta_A$. Thus $d = O_A$ or $\delta = O_A$. It follows that A is quasi-primary and, by corollary (2.7.5), A is integral. ■

Theorem (2.10.4). *Any object is a subobject of a product of irreducible objects.*

Proof. The proof is standard and uses only a few of the structures of Zariski categories. Let A be an object. Let $(m, n) : X \rightrightarrows A$ be a pair of distinct morphisms whose domain is finitely presentable. Let S be the set of regular quotients q of A such that $qm \neq qn$. It is a co-inductive set in QuotReg (A). By the Zorn lemma, S has a minimal element $q : A \to Q$. The object Q is irreducible because any pair of regular epimorphisms $(q_1 : Q \to Q_1,\ q_2 : Q \to Q_2)$ such that $q_1 \vee q_2 = \Delta_Q$ is monomorphic. Thus $q_1 qm \neq q_1 qn$ or $q_2 qm \neq q_2 qn$ and so $q_1 q \simeq q$ or $q_2 q \simeq \omega$, i.e. q_1 or q_2 is isomorphic. Consequently, the family $(q_i : A \to Q_i)_{i \in I}$ of regular quotients of A whose codomains are irreducible is such that the morphiss $(q_i)_{i \in I}$: $A \to \Pi_{i \in I} Q_i$ is monomorphic. Therefore A is a subobject of a product of irreducible objects. ■

Definition (2.10.5). *A congruence r on A is* irreducible *if it is not unit and, for any pair s, t of congruences on A,*

$$s \wedge t = r \Rightarrow (s = r \quad \text{or} \quad t = r).$$

Proposition (2.10.6). *A congruence r on A is irreducible if and only if the quotient of A by r is an irreducible object.*

Proof. Let r be a congruence on A and let $f : A \to B$ be the quotient of A by r. Let us suppose that r is irreducible. If s, t is a pair of congruences on B such that $s \wedge t = \Delta_B$, then $f^{*c}(s) \wedge f^{*c}(t) = f^{*c}(s \wedge t) = f^{*c}(\Delta_B) = r$. Thus $f^{*c}(s) = r$ or $f^{*c}(t) = r$, which implies $s = \Delta_B$ or $t = \Delta_B$. Therefore B is irreducible. Conversely, let us assume that B is irreducible. Let s, t be a pair of congruences on A such that $s \wedge t = r$. Let $q_s : A \to A/s$ and $q_t : A \to A/t$ be the quotients of A by s and t respectively, and let u: $B \to A/s$ and $v : B \to A/t$ be regular epimorphisms such that $uf = q_s$ and $vf = q_t$. According to Proposition (1.5.2), $f = q_s \vee q_t$. Then $u \vee v = 1_B$ and therefore $u = 1_B$ or $v = 1_B$, which implies $q_s \simeq f$ or $q_t \simeq f$, or equivalently $s = r$ or $t = r$. As a result, r is irreducible. ■

Proposition (2.10.7). *A congruence is prime if and only if it is radical and irreducible.*

Proof. The proposition is a consequence of Propositions (2.10.3) and (2.10.6). ■

2.11 REGULAR OBJECTS

Definition (2.11.1) ([9], § 2.0). *A* direct factor morphism *is a morphism* $\delta : A \to B$ *such that there exists a morphism* $\delta' : A \to B'$ *composing with* δ *a product* $(\delta : A \to B, \delta' : A \to B')$.

According to Propositions (1.4.2) and (1.8.4), direct factor morphisms are both regular and singular epimorphisms.

Definition (2.11.2). *An* object *is* regular *if any finitely generated regular quotient of it is a direct factor.*

The *category* **Reg A** *of regular objects* of **A** is the full subcategory of **A** whose objects are the regular ones.

Proposition (2.11.3). *A regular object is reduced.*

Proof. Let A be a regular object. Let $(g, h) : X \rightrightarrows A$ be a pair of distinct morphisms whose domain is finitely presentable. The coequalizer $q : A \to Q$ of (g, h) is a direct factor, and so there is an epimorphism $d : A \to D$ such that $(q : A \to Q, d : A \to D)$ is a product. Because q is not isomorphic, d is not null. Thus there is a simple quotient $s : D \to S$ of D. Because d codisjoints (g, h), sd codisjoints (g, h). Thence $sdg \neq sdh$. It follows that the family of quotient objects $(k_i : A \to K_i)_{i \in I}$ of A with a simple codomain is such that the morphism $(k_i)_{i \in I} : A \to \Pi_{i \in I} K_i$ is monomorphic. Therefore A is reduced. ■

Proposition (2.11.4). *For a regular object A, the following assertions are equivalent.*

(i) *A is irreducible.*
(ii) *A is quasi-primary.*
(iii) *A is primary.*
(iv) *A is integral.*
(v) *A is pseudo-simple.*
(vi) *A is simple.*

Proof.
(iv) ⇒ (iii) ⇒ (ii) and (vi) ⇒ (v) ⇒ (iii) are immediate.
(i) ⇔ (iv) By Proposition (2.11.3) A is reduced, and by Proposition (2.10.3) A is integral ⇔ A is irreducible.
(ii) ⇒ (iv) follows from Proposition (2.11.3) and Corollary (2.7.5).

(iv) $\Rightarrow$ (vi) Let $k : A \to K$ be a monomorphism with a simple codomain. Let $q : A \to Q$ be a non-null finitely generated regular quotient of A. It is a direct factor and thus it is flat. Thus the pushout $(\bar{k}, \bar{q})$ of (q, k) is such that $\bar{k}$ is monomorphic; thus $\bar{q}$ is not null. Because K is simple, $\bar{q}$ is isomorphic and thus q is isomorphic. It follows that any finitely generated regular quotient of A is isomorphic and therefore that A is simple. ■

Proposition (2.11.5). *The category* **RegA** *of regular objects of* **A** *is closed in* **A** *under products, filtered colimits, regular quotient objects, flat quotient objects, and extremal subobjects.*

Proof.

(i) Let $(p_i : A \to A_i)_{i \in I}$ be the product of a family $(A_i)_{i \in I}$ of regular objects. Let $q : A \to Q$ be a finitely generated regular epimorphism. First, let us assume that q is the coequalizer of a pair $(m, n) : X \rightrightarrows A$ with a finitely presentable codisjunctable domain. Let $d : A \to D$ be the codisjunctor of (m, n). For each $i \in I$, let $(r_i : Q \to Q_i,\ q_i : A_i \to Q_i)$ be the pushout of (q, p_i) and let $(t_i : D \to D_i,\ d_i : A_i \to D_i)$ be the pushout of (d, p_i). Then q_i is the coequalizer of $(p_i m, p_i n)$ while d_i is the codisjunctor of $(p_i m, p_i n)$. Because A_i is regular, $(q_i,\ d_i)$ is a product. Thus $(\Pi_{i \in I} q_i : A \to \Pi_{i \in I} O_i$, $\Pi_{i \in I} d_i : A \to \Pi_{i \in I} D_i)$ is a product. Let r be the morphism $(r_i)_{i \in I} : Q \to \Pi_{i \in I} Q_i$ and let t be the morphism $(t_i)_{i \in I} : D \to \Pi_{i \in I} D_i$. Then $rq = \Pi_{i \in I} q_i$ and $td = \Pi_{i \in I} d_i$. Then the pushout of $\Pi_{i \in I} q_i$ along q is r, while the pushout of $\Pi_{i \in I} d_i$ along q is null. Because finite products are co-universal, the pair (r, O_Q) is a product. Thus r is isomorphic and q is a direct factor. Second, let us consider any finitely generated regular epimorphism $q : A \to Q$. Following Axiom (1.2.2) of Zariski categories, q is a finite intersection of a sequence $(q_k)_{k \in [1,\, n]}$ of finitely generated regular epimorphisms which are coequalizers of pairs of parallel morphisms with a finitely presentable codisjunctable domain. Because each q_k is a direct factor and finite products are co-universal, $q = \wedge_{k \in [l, n]} q_k$ is a direct factor. As a result, A is regular.

(ii) Let $(\alpha_i : A_i \to A)_{i \in \mathbf{I}}$ be the colimit of a filtered diagram $(A_i)_{i \in \mathbf{I}}$ of regular objects. Let $q : A \to Q$ be a finitely generated regular quotient of A. It is the coequalizer of a pair $(m, n) : X \rightrightarrows A$ where X is finitely presentable. There exist $i \in \mathbf{I}$ and $(m_i,\ n_i) : X \rightrightarrows A_i$ such that $(\alpha_i m_i,\ \alpha_i n_i) = (m, n)$. Let $q_i : A_i \to Q_i$ be the coequalizer of $(m_i,\ n_i)$. Then q is the pushout of q_i along α_i. Because A_i is regular, q_i is a direct factor. Because direct factors are co-universal, q is a direct factor. As a result, A is regular.

(iii) Let A be a regular object and $f : A \to B$ be a finitely generated regular epimorphism. Let $q : B \to Q$ be a finitely generated regular epimorphism. Then $qf : A \to Q$ is a finitely generated regular epimorphism and thus it is a direct factor. Thus q is a direct factor as it is the pushout of qf along f. Thence B is regular. Let $g : A \to C$ be any regular epimorphism. It is a cofiltered co-intersection of finitely generated regular epimorphisms. Thus C is a regular object, as it is a filtered colimit of regular objects.

(iv) Let A be a regular object and let $f: A \to B$ be a flat epimorphism. Let $q: B \to Q$ be a finitely generated regular epimorphism. It is the coequalizer of a finitely generated congruence s on B. The congruence $f^{*c}(s)$ on A, which is the inverse image of s by f, is the join of a family of finitely presentable congruences on A, say $f^{*c}(s) = \overset{c}{\vee}_{i\in I} r_i$. Then $\overset{c}{\vee}_{i\in I} f_{*c}(r_i) = f_{*c}(\overset{c}{\vee}_{i\in I} r_i) = f_{*c}(f_{*c}(s)) = s$ (Proposition 1.7.2). Because s is compact, there is a finite subset I_0 of I such that $\overset{c}{\vee}_{i\in I_0} f_{*c}(r_i) = s$. Thus $f_{*c}(\overset{c}{\vee}_{i\in I_0} r_i) = s$. Because A is regular, the quotient p of A by the compact congruence $\overset{c}{\vee}_{i\in I_0} r_i$ is a direct factor. But q is the pushout of p along f. Thus q is a direct factor. As a result, B is regular.

(v) Let B be a regular object and let $f: A \to B$ be an extremal monomorphism ([35], p. 351). Let $q: A \to Q$ be a finitely generated regular epimorphism. As seen previously in (i), we can assume that q is the coequalizer of a pair of morphisms $(g, h): X \rightrightarrows A$ whose domain is finitely presentable and codisjunctable. Let $d: A \to D$ be the codisjunctor of (g, h). Let $(r, \bar{q})$ be the pushout of (q, f) and let $(s, \bar{d})$ be the pushout of (d, f). Then $\bar{q}$ is the coequalizer of (fg, fh) while $\bar{d}$ is its codisjunctor. Because B is regular, $(\bar{q}, \bar{d})$ is a product. Thus the morphism f factors in the form $f = (r \times s)(q, d)$ where $(q, d): A \to Q \times D$ and $r \times s: Q \times D \to B$. By Proposition (1.4.9), (q, d) is the co-union of q and d. Because f is an extremal monomorphism, (q, d) is isomorphic. Thus the pair $(q: A \to Q, d: A \to D)$ is a product and q is a direct factor. As a result, A is regular. ■

Theorem (2.11.6). *To any object is associated a universal regular object, i.e.* ***RegA*** *is a reflective subcategory of* ***A****.*

Proof. We shall apply the Freyd adjoint functor theorem ([31], p. 117) to the inclusion functor **RegA** $\to$ **A**. By Proposition (2.11.5), the category **RegA** is complete and the inclusion functor is continuous. Let A be an object in **A**. Let $(f_i: A \to B_i)_{i\in I}$ be the family of all quotient objects of A whose codomains are regular objects. Let B be a regular object and let $f: A \to B$ be any morphism. As the category **A** is cocomplete and co-well-powered, the morphism f factors in the form $f = me$, where $e: A \to E$ is an epimorphism and $m: E \to B$ is an extremal monomorphism ([35], Proposition 21.6.2). By Proposition (2.11.5), the object E is regular. Thus e is isomorphic to the morphism f_i for some $i \in I$. It follows that the solution set condition is satisfied ([31], p. 117). As a result, the inclusion functor has a left-adjoint. ■

2.12 LOCAL OBJECTS

Definition (2.12.1). *An object is* local *if it has exactly one maximal congruence.*

The maximal congruence on a local object A is denoted by m_A, the quotient object of A by m_A is called the *residue simple object* of A and is denoted by K_A, and the quotient morphism is denoted by $k_A : A \to K_A$.

Proposition (2.12.2). *For an object A, the following assertions are equivalent.*

(i) *A is local.*

(ii) *A is not terminal and, for any pair r,s of congruences on A,*

$$r \overset{c}{\vee} s = 1_{A\times A} \Rightarrow (r = 1_{A\times A} \quad \text{or} \quad s = 1_{A\times A}).$$

(iii) *A is not terminal and, for any pair r, s of finitely generated congruences on A,*

$$r \overset{c}{\vee} s = 1_{A\times A} \Rightarrow (r = 1_{A\times A} \quad \text{or} \quad s = 1_{A\times A}).$$

(iv) *A has exactly one simple quotient.*

(v) *A is not terminal and, for any pair p, q of regular quotients of A,*

$$p \wedge q = O_A \Rightarrow (p = O_A \quad \text{or} \quad q = O_A).$$

(vi) *A is not terminal and, for any pair p, q of finitely generated regular quotients of A,*

$$p \wedge q = O_A \Rightarrow (p = O_A \quad \text{or} \quad q = O_A).$$

Proof.

(i) $\Rightarrow$ (ii) The terminal object has no proper congruence and thus no maximal congruence. Thence A is not terminal. Let (r, s) be a pair of proper congruences on A. Then $r \leqslant m_A$ and $s \leqslant m_A$, and thus $r \overset{c}{\vee} s \leqslant m_A$. Consequently, $r \overset{c}{\vee} s$ is proper.

(ii) $\Rightarrow$ (i) As A is not terminal, A has at least one maximal congruence m (Theorem (2.1.8)). If n is another maximal congruence on A, then $m \overset{c}{\vee} n = 1_{A\times A}$. Thus $m = 1_{A\times A}$ or $n = 1_{A\times A}$, which is impossible. Therefore A has exactly one maximal congruence.

(ii) $\Rightarrow$ (iii) is obvious.

(iii) $\Rightarrow$ (ii) Let (r, s) be a pair of congruences on A such that $r \overset{c}{\vee} s = 1_{A\times A}$. As any congruence on A is the join of finitely generated congruences (Proposition (1.5.2)), $r = \overset{c}{\bigvee}_{i\in I} r_i$ and $s = \overset{c}{\bigvee}_{j\in J} s_j$ where r_i and s_j are finitely generated. Then $(\overset{c}{\bigvee}_{i\in I} r_i) \overset{c}{\vee} (\overset{c}{\bigvee}_{j\in k} s_j) = 1_{A\times A}$. Because $1_{A\times A}$ is compact in Cong (A) (Proposition (1.5.2)), there are finite subsets $I_0 \subset I$ and $J_0 \subset J$ such that $(\overset{c}{\bigvee}_{i\in I_0} r_i) \overset{c}{\vee} (\overset{c}{\bigvee}_{j\in J_0} s_j) = 1_{A\times A}$. By (iii), $r_i = 1_{A\times A}$ for some $i \in I_0$ or $s_j = 1_{A\times A}$ for some $j \in J_0$. Thus $r = 1_{A\times A}$ or $s = 1_{A\times A}$.

(i) ⇔ (iv), (ii) ⇔ (v), and (iii) ⇔ (vi) follow from Proposition (1.5.2). ■

Definition (2.12.3). *A morphism* $f: A \to B$ *is* local *if* A *and* B *are local objects and the inverse image by* f *of the maximal congruence on* B *is the maximal congruence on* A. *Then the unique morphism* $\bar{f}: K_A \to K_B$ *such that* $\bar{f}k_A = k_B f$ *is called the* residue morphism of f.

Proposition (2.12.4). *For a morphism* $f: A \to B$, *the following assertions are equivalent.*

(i) f *is local.*

(ii) B *is local and* f_{*c} *preserves proper congruences.*

(iii) B *is local and* f *is prelocal.*

Proof.

(i) ⇒ (ii) Let m_A and m_B be the maximal congruences on A and B respectively. The relation $m_A = f^{*c}(m_B)$ implies $f_{*c}(m_A) \leqslant m_B$. Thus $f_{*c}(m_A)$ is proper. If r is a proper congruence on A, then $r \leqslant m_A$. Thus $f_{*c}(r) \leqslant f_{*c}(m_A)$ and hence $f_{*c}(r)$ is proper.

(ii) ⇒ (iii) Let $f = gd$ where d is a singular epimorphic codisjunctor of a congruence r. Because f codisjoints r, we have $f_{*c}(r) = 1_{B\times B}$. Therefore $r = 1_{A\times A}$ and thus d is isomorphic. As result, f is prelocal.

(iii) ⇒ (i) A is not terminal because B is not. Let r be a codisjunctable congruence on A such that $f_{*c}(r) = 1_{B\times B}$. Then f codisjoints r and thus f factors through the codisjunctor d of r. Consequently, d is isomorphic and thus $r = 1_{A\times A}$. Let r be a proper congruence on A. Then $r = \overset{c}{\vee}_{i\in[l,n]} r_i$ where the r_i are proper codisjunctable congruences on A (Proposition (1.5.3)). Then $f_{*c}(r) = \overset{c}{\vee}_{i\in[l,n]} f_{*c}(r_i)$. As shown above, $f_{*c}(r_i)$ are proper congruences on B. Because B is local, $\overset{c}{\vee}_{i\in[l,n]} f_{*c}(r_i)$ is also proper and thus $f_{*c}(r)$ is proper. Let r, s be two proper congruences on A. Then $f_{*c}(r \overset{c}{\vee} s) = f_{*c}(r) \overset{c}{\vee} f_{*}(s)$ is a proper congruence on B. Hence $r \overset{c}{\vee} s$ is a proper congruence on A. It follows that A is local. Let m_A and m_B be the maximal congruences on A and B respectively. Then $f_{*c}(m_A)$ is proper and so $f_{*c}(m_A) \leqslant m_B$. Thus $m_A \leqslant f^{*c}(m_B)$. But $f^{*c}(m_B)$ is proper, and so $m_A = f^{*c}(m_B)$ and f is a local morphism. ■

The *category* **LocA** *of local objects of* **A** is the subcategory of **A** with the local objects as its objects and the local morphisms as its morphisms.

Proposition (2.12.5). *SimA and PsdSimA are full subcategories of* **LocA**.

Proof. As **SimA** is a full subcategory of **PsdSimA**, it is enough to prove that **PsdSimA** is a full subcategory of **LocA**. Let A be a pseudo-simple object. By Proposition (2.6.2), the universal reduced object $r_A: A \to R(A)$

associated to A is simple. It is a simple quotient of A. If $q: A \to L$ is another simple quotient of A, then q factors through r_A in a quotient $p: R(A) \to L$. As $R(A)$ is simple, p is isomorphic. Therefore r_A is the unique simple quotient of A and thus A is local. Let $f: A \to B$ be a morphism between two pseudo-simple objects. The image $R(f)$ of the morphism f by the reflector $R: \mathbf{A} \to \mathbf{RedA}$ is monomorphic. Because the maximal congruence m_A (m_B) on A (B) is the kernel pair of r_A (r_B), the relation $R(f)r_A = r_B f$ implies $f^{*c}(m_B) = m_A$. Thus f is a local morphism. ■

Theorem (2.12.6). *To any local object is associated a universal simple object, i.e.* **SimA** *is a reflective subcategory of* **LocA**.

Proof. Let A be a local object, let m_A be its maximal congruence, and let $k: A \to K$ be its residual simple object. Any local morphism $f: A \to L$ with a simple codomain is such that $f^{*c}(\Delta_L) = m_A$. Thus $f_{*c}(m_A) = \Delta_L$ and so f factors through k. Therefore $k: A \to K$ is the universal simple object associated to A. ■

Proposition (2.12.7). *A local object is simple if and only if it is regular.*

Proof. Let A be a regular local object. Let $q: A \to Q$ be a finitely generated regular epimorphism. Because A is regular, q is a direct factor. Thus there exists a regular epimorphism $q': A \to Q'$ composing with q a product $(q: A \to Q, q': A \to Q')$. Because A is local, the relation $q \wedge q' = O_A$ implies $q = O_A$ or $q' = O_A$. Alternatively, q is null or isomorphic. It follows that A is simple. ■

Proposition (2.12.8). *The category* **LocA** *has filtered colimits preserved by the inclusion functor* **LocA** $\to$ **A**.

Proof. Let $(A_i)_{i \in \mathbf{I}}$ be a filtered diagram of **LocA** and let $(\alpha_i: A_i \to A)_{i \in \mathbf{I}}$ be its colimit in **A**. The object A is not terminal for it is a filtered colimit of non-terminal objects.

(i) Let $s, r \in \mathrm{Cong}_0\ (A)$ such that $r \overset{c}{\vee} s = 1_{A \times A}$. According to Proposition (1.5.5), there exist $i \in \mathbf{1}$ and $r_i, s_i \in \mathrm{Cong}_0\ (A_i)$ such that $(\alpha_i)_{*c}(r_i) = r$, $(\alpha_i)_{*c}(s_i) = s$ and $r_i \overset{c}{\vee} s_i = 1_{A_i \times A_i}$. Since A_i is local, $r_i = 1_{A_i \times A_i}$ or $s_i = 1_{A_i \times A_i}$. Thus $r = 1_{A \times A}$ or $s = 1_{A \times A}$. Therefore A is local.

(ii) Let $i \in \mathbf{I}$. Let us assume that α_i factors in the form $\alpha_i = \beta d$ where d is a singular epimorphism codisjunctor of a congruence $r \in \mathrm{Cong}_0\ (A_i)$. Then $(\alpha_i)_{*c}(r) = 1_{A \times A}$. According to Proposition (1.5.5), there exists a morphism $u: i \to j$ in **I** such that $(A_u)_{*c}(r) = 1_{A_j \times A_j}$. Then A_u codisjoints r and thus factors through d. Since A_u is a local morphism, d is an isomorphism (Proposition (2.12.4)) It follows that α_i is prelocal and therefore local (Proposition (2.12.4)).

(iii) Let $(\beta_i : A_i \to B)_{i\in\mathbf{I}}$ be an inductive cone in **LocA**. There exists a morphism $f: A \to B$ in **A** such that $f\alpha_i = \beta_i$ for any $i \in \mathbf{I}$. Let us assume that f factors in the form $f = gd$ where d is a singular epimorphic codisjunctor of a congruence $r \in \mathrm{Cong}_0\,(A)$. According to Proposition (1.5.5), there exists $i \in \mathbf{I}$ and $r_i \in \mathrm{Cong}_0\,(A_i)$ such that $(\alpha_i)_{*c}(r_i) = r$. Then $(\beta_i)_{*c}(r_i) = (f_{*c})((\alpha_i)_{*c}(r_i)) = f_{*c}(r) = 1_{B\times B}$. Since β_i is a local morphism, it follows that $r_i = 1_{A_i\times A_i}$ (Proposition (2.12.4)). Therefore $r = 1_{A\times A}$ and d is isomorphic. Consequently, f is prelocal and thus local (Proposition (2.12.4)). As a result, $(\alpha_i : A_i \to A)_{i\in\mathbf{I}}$ is a colimit in **LocA**. ■

Examples

In **CRng**, **RedCRng**, **CAlg**(k), **RedCAlg**(k), **RlRng**, **RedRlRng**, **RlOrdRng**, and **RedRlOrdRng** local objects are the objects which are local rings. In **RegCRng** and **RegCAlg**(k) local objects are identical to simple objects, i.e. they are fields. In **GradCRng** and **RedGradCRng** local objects are the objects whose set of non-invertible homogeneous elements is stable under addition. In **Mod** and **CAlg** local objects are the pairs (L, E) where L is a local ring.

2.13 LOCALIZATIONS

The quotient of an object A by a prime congruence p is denoted by $q_p : A \to A/p$ and the universal simple object associated to A/p is denoted by $j_p : A/p \to K(p)$.

Definition (2.13.1). *The* localization of an object A at a prime congruence p on A *is the quotient* $l_p : A \to A_p$ *of* A *arising in the presingular factorization* $k_p l_p$ *of the morphism* $j_p q_p : A \to K(p)$ (cf. Theorem (1.9.6)).

Proposition (2.13.2). *The object* A_p *is local, the morphism* $k_p : A_p \to K(p)$ *is its residue simple object, and* (k_p, j_p) *is the pushout of* (l_p, q_p).

Proof. According to Proposition (2.12.4), the morphism k_p is local and thus A_p is local. Let $k : A_p \to K$ be the residue simple object of A_p. It is the universal simple object associated to A_p (Theorem (2.12.6)). Thus the local morphism $k_p : A_p \to K(p)$ factors through k in the form $k_p = uk$ where u is monomorphic. The relations $ukl_p = k_p l_p = j_p q_p$ imply the existence of a monomorphism $n : A/p \to K$ such that $un = j_p$. Because $j_p : A/p \to K(p)$ is the universal simple object associated to A/p, there exists $r : K(p) \to K$ such that $rj_p = n$. Then r is inverse to u, u is isomorphic, and k_p is the residue simple object of A_p. Let m be the maximal congruence on A_p. Because j_p is monomorphic, the equality $j_p q_p = k_p l_p$ implies that $l_p^{*c}(m) = p$. Because

l_p is a flat epimorphism (Proposition (1.9.5)), $(l_p)_{*c}(l_p^{*c}(m)) = m$ (Proposition (1.7.2)). Thus $(l_p)_{*c}(p) = m$. Therefore k_p is the pushout of q_p along l_p, and hence (k_p, j_p) is the pushout of (l_p, q_p). ■

Notation. The object A_p is called the *local object of A at p* and its maximal congruence is denoted by m_p. Then $p = l_p^{*c}(m_p)$.

Theorem (2.13.3). *The category* **LocA** *of local objects of A is a multireflective subcategory of A, where the family of localizations of an object A at prime congruences is a universal family of morphisms from A to* **LocA** [7].

Proof. Let p, p' be a pair of prime congruences on A, and let $g : A_p \to B$, $g' : A_{p'} \to B$ be a pair of local morphisms such that $gl_p = g'l_{p'}$. According to the essential uniqueness of the presingular factorization (Theorem (1.9.6)), there is an isomorphism $u : A_p \to A_{p'}$ such that $ul_p = l_{p'}$. Then $p' = l_{p'}^{*c}(m_{p'}) = (ul_p)^{*c}(\mathrm{m}_{p'}) = l_p^{*c}(u^{*c}(m_{p'})) = l_p^{*c}(m_p) = p$. Let f: $A \to B$ be a morphism in **A** with a local codomain. It factorizes in the form $f = ge$ where $e : A \to E$ is a presingular epimorphism and $g : E \to B$ is a prelocal morphism (Theorem (1.9.6)). By Proposition (2.12.4), the morphism g is local and thus E is local. Let $k : E \to K$ be the residue simple object of E, and let nq be the regular factorization of ke with $q : A \to Q$ be a regular epimorphism and n be a monomorphism. Let p be the kernel pair of q. Then p is prime and $p = e^{*c}(m_E)$. Because e is a flat epimorphism (Proposition (1.9.5)), $e_{*c}(p) = e_{*c}(e^{*c}(m_E)) = m_E$ (Proposition (1.7.2)). Therefore k is the pushout of q along e. Thus n is the pushout of e along q, and hence n is a presingular epimorphism (Proposition (1.9.5)). It follows that n is the universal simple object associated to Q. Because $q \simeq q_p$, e is the localization of A at p. Then f factors through the localization of A at p by a local morphism. As a result, the family of localizations of A at prime congruences is a universal family of morphisms from A to **LocA**, and **LocA** is a multireflective subcategory of **A** [7]. ■

Let $f : A \to B$ be a morphism and let p be a prime congruence on B. According to Theorem (2.13.3), the morphism $l_p f : A \to B_p$ factors in a unique way in the form $l_p f = gl_q$ where q is a prime congruence on A and $g : A_q \to B_p$ is a local morphism. Then $q = l_q^{*c}(m_q) = l_q^{*c}(g^{*c}(m_p)) = (gl_q)^{*c}(m_p) = (l_p f)^{*c}(m_p) = f^{*c}(l_p^{*c}(m_p)) = f^{*c}(p)$. The morphism g will be denoted by f_p.

Definition (2.13.4). *The* local morphism of a morphism $f : A \to B$ at a prime congruence *p on B is the unique local morphism $f_p : A_{f^{*c}(p)} \to B_p$ such that $f_p l_{f^{*c}(p)} = l_p f$.*

Proposition (2.13.5). *The localization of a quasi-primary object at its radical is the universal pseudo-simple object associated to it.*

Proof. Let A be a quasi-primary object, let $p = \operatorname{rad}(A)$ be its radical (Proposition (2.7.4)), and let $f: A \to B$ be the universal pseudo-simple object associated to it (Theorem (2.7.2)). Let q be a prime congruence on A_p. Then $q \leqslant m_p$ and thus $l_p^{*c}(q) \leqslant l_p^{*c}(m_p) = p$. Since $p = \operatorname{rad}(A) \leqslant l_p^{*c}(q)$, it follows that $l_p^{*c}(q) = l_p^{*c}(m_p)$ and hence $q = m_p$ (Proposition (1.7.2)). Consequently the object A_p is pseudo-simple (Proposition (2.6.2)). The relation $\ker(l_p) = l_p^{*c}(\Delta_{A_p}) \leqslant l_p^{*c}(m_p) = p = \operatorname{rad}(A)$ imply that l_p is a premonomorphism (Proposition (2.5.7)). Therefore the morphism l_p factors through f in the form $l_p = gf$ where $g: B \to A_p$ is a local morphism (Proposition (2.12.5)). Since f is a presingular epimorphism (Proof of Theorem (2.7.2)), g must be an isomorphism. As a result $l_p: A \to A_p$ is the universal pseudo-simple object associated to A. ■

Proposition (2.13.6). *Any localization of an integral object is a subobject of the universal simple object associated to it.*

Proof. Let A be an integral object, let $k: A \to K$ be the universal simple object associated to A, and let $l_p: A \to A_p$ be a localization of A. Let $(f: A_p \to B,\ g: K \to B)$ be the pushout of (l_p, k). Since l_p is a presingular epimorphism, f is a monomorphism. Therefore $B \neq 1$ and g is a non-null presingular epimorphism. Since K is pseudo-simple (Corollary (2.6.3)), g is an isomorphism. Consequently, l_p is a subobject of k in $A/\mathbf{A}$. ■

3

SPECTRA

The prime spectrum of an object A is the set of prime congruences on A with a topology called the Zariski topology. The open sets are defined by congruences on A, the closed sets by regular quotients of A, the affine open sets by codisjunctable congruences on A or singular quotients of A, and the affine subsets by simultaneously codisjunctable families of congruences on A or semisingular quotients of A. The maximal spectrum of A is the subspace of the prime spectrum whose elements are the maximal congruences on A. The patch spectrum of A is the patch space of the prime spectrum of A or the co-universal boolean space associated to it. Many properties of objects and morphisms in Zariski categories have nice topological interpretations via their spectra. In particular, the study of direct factors of objects, and of indecomposable objects, is considered here. The prime spectrum of an object A is canonically equipped with a sheaf of local objects whose object of global sections is precisely A. This is the structure sheaf on the prime spectrum. Many properties of objects and morphisms in Zariski categories can then be localized via the stalks of the structure sheaf. Indeed, this is because in a Zariski category the class of local objects is a strong cogenerating class.

3.1 PRIME SPECTRA

Notations. We denote the set of prime congruences on an object A by Spec (A) and, for any congruence r on A, we denote the set $\{p \in \operatorname{Spec}(A) : r \not\leqslant p\}$ by $D(r)$ and the set $\{p \in \operatorname{Spec}(A) : r \leqslant p\}$ by $V(r)$.

Proposition (3.1.1). *For any congruence r on A and any* $p \in \operatorname{Spec}(A)$,

(i) $D(r) = \{p \in \operatorname{Spec}(A) : l_p$ codisjoints $r\} = \{p \in \operatorname{Spec}(A) : k_p l_p$ codisjoints $r\} = D(\operatorname{rad}(r))$,

(ii) $V(r) = \{p \in \operatorname{Spec}(A) : k_p l_p$ coequalizes $r\} = V(\operatorname{rad}(r))$,

(iii) $l_p = \wedge\{\operatorname{codis}(r) : r$ is codisjunctable and $p \in D(r)\}$.

Proof.

(i) and (ii) In the notation used in § 2.13, $l_p : A \to A_p$ is the localization of A at p, $k_p : A_p \to K(p)$ is the residual simple object of A_p, $q_p : A \to A/p$ is the quotient of A by p, and $j_p : A/p \to K(p)$ is the universal simple object associated to A/p. These morphisms satisfy $j_p q_p = k_p l_p$. For any $p \in \operatorname{Spec}(A)$, we have that $r \not\leqslant p \Leftrightarrow q_p$ does not coequalize $r \Leftrightarrow j_p q_p$ does not coequalize $r \Leftrightarrow k_p l_p$ does not coequalize $r \Leftrightarrow k_p l_p$ codisjoints r (Proposition (2.1.2)). Because k_p is a prelocal morphism, it is local (Proposition (2.12.4)) and therefore we have that $k_p l_p$ codisjoints $r \Leftrightarrow l_p$ codisjoints r. As a result, we have that $r \nleqslant p \Leftrightarrow k_p l_p$ does not coequalize $r \Leftrightarrow k_p l_p$ codisjoints $r \Leftrightarrow l_p$ codisjoints r. On the other hand, the equalities $D(r) = D(\text{rad}(r))$ and $V(r) = V(\text{rad}(r))$ follow from the fact that prime congruences are radical.

(iii) Let $l: A \to L$ be $\wedge\{\text{codis}(r) : r \in \text{CongCodis}(A)$ and $p \in D(r)\}$. According to (i), the relation $p \in D(r)$ implies that l_p codisjoints r. Thus l_p factors through l in the form $l_p = ul$. Let us prove that u is prelocal. Let $u = vd$ where d is a singular epimorphic codisjunctor of a congruence s on L. Let $r = l^{*c}(s)$. According to Proposition (1.5.3), $r = \overset{c}{\vee}_{i \in I} r_i$ where the r_i are codisjunctable. Thus $(l_p)_{*c}(r)) = u_{*c}(l_{*c}(r)) = u_{*c}(s) = 1_{A_p \times A_p}$. Because $p \notin D(p)$, l_p does not codisjoint p and so $(l_p)_{*c}(p) \neq 1_{A_p \times A_p}$. Consequently $r \nleq p$. Thus there exists $i \in I$ such that $r_i \nleqslant p$; hence $p \in D(r_i)$ and l codisjoints r_i. It follows that l codisjoints r; thus $s = 1_{L \times L}$ and d is isomorphic. As a result, u is prelocal and, according to Theorem (1.9.6), u is isomorphic. Thus $l_p \simeq l$. ■

Let us denote by $D: \text{Cong}(A) \to \mathcal{P}(\text{Spec}(A))$ the map which assigns to a congruence r on A the subset $D(r)$ of $\text{Spec}(A)$.

Proposition (3.1.2). *The map $D: \text{Cong}(A) \to \mathcal{P}(\text{Spec}(A))$ is a homomorphism of bounded sup complete lattices.*

Proof.

(i) $D(\Delta_A) = \{p \in \text{Spec}(A) : \Delta_A \nleqslant p\} = \varnothing$.

(ii) $D(1_{A \times A}) = \{p \in \text{Spec}(A) : 1_{A \times A} \nleqslant p\} = \text{Spec}(A)$.

(iii)
$$\begin{aligned} D(\overset{c}{\bigvee_{i \in I}} r_i) &= \{p \in \text{Spec}(A) : \overset{c}{\bigvee_{i \in I}} r_i \nleqslant p\} \\ &= \{p \in \text{Spec}(A) : \exists i \in I,\ r_i \nleqslant p\} \\ &= \bigcup_{i \in I} \{p \in \text{Spec}(A) : r_i \nleqslant p\} \\ &= \bigcup_{i \in I} D(r_i). \end{aligned}$$

(iv) Let $r, s \in \text{Cong}(A)$ and $p \in \text{Spec}(A)$. By Proposition (2.2.13), the congruence p is universally irreducible, i.e. $r \wedge s \leqslant p \Leftrightarrow (r \leqslant p$ or $s \leqslant p)$, which implies that $r \wedge s \nleqslant p \Leftrightarrow (r \nleqslant p$ and $s \nleqslant p)$. It follows that

$$D(r \wedge s) = \{p \in \text{Spec}(A) : r \wedge s \nleqslant p\} = \{p \in \text{Spec}(A) : r \nleqslant p \text{ and } s \nleqslant p\} = \{p \in \text{Spec}(A) : r \nleqslant p\} \cap \{p \in \text{Spec}(A) : s \nleqslant p\} = D(r) \cap D(s).$$ ■

Corollary (3.1.3). *The set $\{D(r) : r \in \text{Cong}(A)\}$ is a topology on* Spec(A).

Definition (3.1.4). *The* prime spectrum *of an object A is the topological space* Spec(A) *whose elements are the prime congruences on A and whose* *topology, called the* Zariski topology, *has as its open sets the subsets of the form* $D(r) = \{p \in \mathrm{Spec}(A) : r \nleqslant p\}$, *where* r *runs through the set of congruences on* A.

We denote by $\Omega(\mathrm{Spec}(A))$ the *lattice of open subsets of* $\mathrm{Spec}(A)$ and by $\Omega_0(\mathrm{Spec}(A))$ *the subset of compact open subsets.*

Theorem (3.1.5). *The map* $D : \mathrm{CongRad}(A) \to \Omega(\mathrm{Spec}(A))$ *which assigns to a radical congruence* r *on* A *the set* $D(r) = \{p \in \mathrm{Spec}(A) : r \nleqslant p\}$ *is an isomorphism of lattices.*

Proof. The map D is order-preserving. Let r, s be a pair of radical congruences on A such that $D(r) \subset D(s)$. Then, for any $p \in \mathrm{Spec}(A)$, $s \leqslant p \Rightarrow r \leqslant p$. Because r and s are intersections of prime congruences (Proposition (2.4.4)), it follows that $r \leqslant s$. Therefore D is injective and induces an isomorphism of ordered sets between $\mathrm{CongRad}(A)$ and its image. Let $D(t) \in \Omega(\mathrm{Spec}(A))$. Then $\mathrm{rad}(t)$ is a radical congruence on A such that $D(\mathrm{rad}(t)) = D(t)$. Thus D is surjective. As a result, D is an isomorphism of ordered sets and thus is an isomorphism of lattices. ■

Corollary (3.1.6). *The map* $D_0 : \mathrm{CongRad}_0(A) \to \Omega_0(\mathrm{Spec}(A))$ *induced by* D *is an isomorphism of bounded distributive lattices.*

Proof. According to Theorem (2.4.10), $\mathrm{CongRad}_0(A)$ is a bounded distributive lattice. Moreover, compact open sets of $\mathrm{Spec}(A)$ are precisely compact elements in $\Omega(\mathrm{Spec}(A))$. ■

Proposition (3.1.7). *The set* $\{V(r) : r \in \mathrm{Cong}(A)\}$ *is the set of closed sets of* $\mathrm{Spec}(A)$ *and, for any* $p \in \mathrm{Spec}(A)$, $V(p)$ *is the closure of* $\{p\}$ *in* $\mathrm{Spec}(A)$.

Proof. The first assertion follows from the fact that the subset $V(r)$ is precisely the complement of $D(r)$ in $\mathrm{Spec}(A)$. Let $p \in \mathrm{Spec}(A)$. Then $V(p)$ is a closed subset containing p. Any closed subset containing p is of the form $V(r)$ with $r \leqslant p$, so that $V(p) \subset V(r)$. Therefore $V(p)$ is the closure of $\{p\}$. ■

Proposition (3.1.8). *For a pair* p, q *of prime congruences on* A, *the following assertions are equivalent.*

(i) $p \leqslant q$ *in* $\mathrm{Cong}(A)$.
(ii) p *is a generization of* q *in* $\mathrm{Spec}(A)$.
(iii) $l_p \leqslant l_q$ *in* $\mathrm{Quot}(A)$.

Proof.

(i) $\Leftrightarrow$ (ii) According to proposition (3.1.7), the closure of $\{p\}$ is $^-\{p\} = V(p)$. Thus $p \leqslant q \Leftrightarrow q \in V(p) \Leftrightarrow q \in \overline{\{p\}} \Leftrightarrow q$ is a specialization of $p \Leftrightarrow p$ is a generization of q.

(i) $\Rightarrow$ (iii) According to Proposition (3.1.1), $l_p = \wedge\{\text{codis}(r) : r$ is codisjunctable and $p \in D(r)\}$, and $l_q = \wedge\{\text{codis}(r) : r$ is codisjunctable and $q \in D(r)\}$. Since $q \in D(r) \Rightarrow r \nleqslant q \Rightarrow r \nleqslant p \Rightarrow p \in D(r)$, we have $l_p \leqslant l_q$.

(iii) $\Rightarrow$ (ii) For an arbitrary open set $D(r)$ of $\text{Spec}(A)$, we have $q \in D(r)$ $\Rightarrow l_q$ codisjoints $r \Rightarrow l_p$ codisjoints $r \Rightarrow p \in D(r)$ (Proposition (3.1.1)). Thus p is a generization of q. ■

Following Hochster [25], a topological space is said to be spectral if it is a T_0 - space, if its set of compact open subsets form an open basis closed under finite intersections (including the empty one), and if any irreducible closed subset has a generic point.

Theorem (3.1.9). *Prime spectra are spectral spaces.*

Proof. Let p_1, p_2 be a pair of distinct elements in $\text{Spec}(A)$. Then $p_1 \nleqslant p_2$ or $p_2 \nleqslant p_1$. If $p_1 \nleqslant p_2$, then $p_2 \in D(p_1)$ and $p_1 \notin D(p_1)$. If $p_2 \nleqslant p_1$, then $p_1 \in D(p_2)$ and $p_2 \notin D(p_2)$. It follows that $\text{Spec}(A)$ is a T_0 - space. According to Corollary (3.1.6), the set of compact open sets of $\text{Spec}(A)$ is closed under finite intersections. According to Theorems (2.4.8) and (3.1.5), any open set is the union of a family of compact open sets. Thus compact open sets form an open basis for $\text{Spec}(A)$. Let us consider an irreducible closed set of $\text{Spec}(A)$. It is of the form $V(r) = \{p \in \text{Spec}(A) : r \leqslant p\}$ with $r \in \text{CongRad}(A)$. Following Theorem (3.1.5), r is irreducible. According to Proposition (2.10.7), r is prime. Therefore r is an element of $\text{Spec}(A)$ with closure $V(r)$, i.e. r is a generic point of $V(r)$. As a result, the space $\text{Spec}(A)$ is spectral in the sense of Hochster. ■

Definition (3.1.10). *An* affine open set *of* $\text{Spec}(A)$ *is a set of the form* $D(r) = \{p \in \text{Spec}(A) : r \nleqslant p\}$ *where* r *is a codisjunctable congruence on* A.

Proposition (3.1.11). *A congruence* r *on* A *is codisjunctable if and only if* $D(r)$ *is an affine open set of* $\text{Spec}(A)$.

Proof. The necessity follows from Definition (3.1.10). Conversely, if $D(r)$ is affine, then there exists a codisjunctable congruence s on A such that $D(r) = D(s)$. According to Theorem (3.1.5), the relations $D(\text{rad}(r)) = D(r) = D(s) = D(\text{rad}(s))$ imply $\text{rad}(r) = \text{rad}(s)$. According to Proposition (2.4.7), $\text{rad}(s) = \text{rad}(r)$ is codisjunctable and r is codisjunctable. ■

Proposition (3.1.12). *An affine open set is compact.*

Proof. Let r be a codisjunctable congruence on A. Then $\mathrm{rad}(r)$ is codisjunctable (Proposition (2.4.7)). According to Proposition (2.4.9), $\mathrm{rad}(r)$ is a finitely generated radical congruence. Following Corollary (3.1.6), $D(r) = D(\mathrm{rad}(r))$ is a compact open subset of $\mathrm{Spec}(A)$. ■

Proposition (3.1.13). *Finite intersections of affine open sets are affine.*

Proof. The empty set is affine because it is $D(\Delta_A)$. If r and s are codisjunctable congruences on A, then $r \wedge s$ is codisjunctable (Proposition (1.5.4)) and $D(r \wedge s) = D(r) \cap D(s)$ (Proposition (3.1.2)). Therefore $D(r) \cap D(s)$ is affine. ■

Proposition (3.1.14). *Affine open sets form an open basis for the Zariski topology.*

Proof. Any open set of $\mathrm{Spec}(A)$ is of the form $D(r)$ with $r \in \mathrm{Cong}(A)$. According to Proposition (1.5.3), there exists a family $(r_i)_{i\in I}$ of codisjunctable congruences on A such that $r = \overset{c}{\vee}_{i\in I} r_i$. According to Proposition (3.1.2), $D(r) = \cup_{i\in I} D(r_i)$. Therefore any open set of $\mathrm{Spec}(A)$ is the union of affine open sets. ■

We denote by $\Omega_a(\mathrm{Spec}(A))$ the *set of affine* open sets of Spec(A).

Theorem (3.1.15). *The map* $D_a : \mathrm{CongRadCodis}(A) \to \Omega_a(\mathrm{Spec}(A))$ *induced by D is an isomorphism of bounded meet semilattices.*

Proof. The theorem is a consequence of Theorem (3.1.5) and Proposition (3.1.13). ■

Proposition (3.1.16). *For any congruence r on A,*

(i) $D(r) = \varnothing \Leftrightarrow$ *r is conjoint,*

(ii) $D(r) = \mathrm{Spec}(A) \Leftrightarrow r = 1_{A\times A}$.

Proof. According to Proposition (3.1.2), $D(\Delta_A) = \varnothing$ and $D(1_{A\times A}) = \mathrm{Spec}(A)$. According to Theorem (3.1.5), for any pair of congruences r, s on A, we have $D(r) = D(s) \Leftrightarrow D(\mathrm{rad}(r)) = D(\mathrm{rad}(s)) \Leftrightarrow \mathrm{rad}(r) = \mathrm{rad}(s)$. It follows that $D(r) = \varnothing \Leftrightarrow D(r) = D(\Delta_A) \Leftrightarrow \mathrm{rad}(r) = \mathrm{rad}(A) \Leftrightarrow r \leqslant \mathrm{rad}(A) \Leftrightarrow r$ is conjoint (Corollary (2.5.4)). Similarly, $D(r) = \mathrm{Spec}(A) \Leftrightarrow D(r) = D(1_{A\times A}) \Leftrightarrow \mathrm{rad}(r) = 1_{A\times A} \Leftrightarrow r = 1_{A\times A}$ (Proposition (2.4.13)). ■

The prime spectrum functor Spec : $\mathbf{A}^{\mathrm{op}} \to$ TopSpec

Following Hochster [25], a *map* $f: E \to F$ is said to be *spectral* is E and F are spectral spaces and the inverse image by f of any compact open subset of F is compact open. Then it is continuous.

Let **TopSpec** be the *category of spectral spaces* and spectral maps. Let $f: A \to B$ be a morphism in **A**. For any prime congruence p on B, $f^{*c}(p)$ is a prime congruence on A (cf. §2.4, p. 57). Let us define the map $\mathrm{Spec}(f)$: $\mathrm{Spec}(B) \to \mathrm{Spec}(A)$ by $\mathrm{Spec}(f)(p) = f^{*c}(p)$. Let us consider a compact open subset of $\mathrm{Spec}(A)$. It is of the form $D_0(r)$ where $r \in \mathrm{CongRad}_0(A)$ (Corollary (3.1.6)). According to § 2.4, p. 59, $f_{*cr}(r) \in \mathrm{CongRad}_0(B)$. Thence $D_0(f_{*cr}(r))$ is a compact open set in $\mathrm{Spec}(B)$ (Corollary (3.1.6)). But

$$D_0(f_{*cr}(r)) = \{p \in \mathrm{Spec}(B) : f_{*cr}(r) \nleqslant p\} =$$
$$\{p \in \mathrm{Spec}(B) : r \nleqslant f^{*c}(p)\} = \{p \in \mathrm{Spec}(B) : f^{*c}(p) \in D_0(r)\} =$$
$$(\mathrm{Spec}(f))^{-1}(D_0(r)).$$

Therefore $\mathrm{Spec}(f)$ is a spectral map. It follows that, by assigning $\mathrm{Spec}(A)$ to an object A and $\mathrm{Spec}(f)$ to a morphism $f: A \to B$, we define a functor Spec : $\mathbf{A}^{\mathrm{op}} \to$ **TopSpec** called the *prime spectrum functor.*

Examples

Congruences are identified with ideals in any category of commutative algebras. In **CRng**, **RedCRng**, **CAlg**(k), and **RedCAlg**(k) the prime spectrum of an object A is the prime spectrum of the ring A. In **RlRng** and **RedRlRng** the prime spectrum of an object A is the subspace of the prime spectrum of the ring A whose points are the real prime ideals I of A, i.e. they are such that $x_1^2 + \ldots + x_n^2 \in I \Rightarrow x_1 \in I$. In **RlOrdRng** and **RedRlOrdRng** the prime spectrum of an object A is the subspace of the prime spectrum of the ring A whose points are the convex prime ideals I of A, i.e. they are such that $0 \leqslant x \leqslant y \in I \Rightarrow x \in I$. In **GradCRng** and **RedGradCRng** the prime spectrum of an object A is the subspace of the prime spectrum of the ring A whose points are the graded prime ideals of A, i.e. they are generated by homogeneous elements. In **Mod** and **CAlg** the prime spectrum of an object (A, E) is the prime spectrum of the ring A.

3.2 MAXIMAL AND PATCH SPECTRA

Proposition (3.2.1). *Maximal congruences on A are precisely the prime congruences on A which are closed points in* $\mathrm{Spec}(A)$.

Proof. Let m be a maximal congruence on A. Then m is prime (Proposition (2.2.12)) and so $m \in \mathrm{Spec}(A)$. According to Proposition (3.1.7),

$\overline{\{m\}} = V(m) = \{m\}$. Thus m is a closed point. Conversely, let p be a closed point in Spec(A). Because p is a proper congruence, there is a maximal congruence m on A such that $p \leqslant m$ (Corollary (2.1.10)). According to Proposition (3.1.8), m is a specialization of p and thus $m = p$. As a result, p is a maximal congruence. ■

Definition (3.2.2). *The* maximal spectrum *of an object A is the topological space* Spec$_{\max}(A)$ *whose elements are the maximal congruences on A and whose topology is induced by the Zariski topology on* Spec(A).

Proposition (3.2.3). *For any regular object, the prime spectrum is identical to the maximal spectrum.*

Proof. Let A be a regular object and let p be a prime congruence on A. Let $q_p : A \to A/p$ be the quotient of A by p. According to Proposition (2.11.5), the object A/p is regular. According to Proposition (2.11.4), A/p is simple. According to Proposition (2.1.7), p is maximal. As a result, Spec$(A) =$ Spec$_{\max}(A)$. ■

Following Hochster [25], the *patch space* of a spectral space X is the topological space $T(X)$ whose elements are those of X and whose topology has the compact open sets of X and their complements as an open sub-basis. It is a boolean space, i.e. a compact Hausdorff space in which the closed open (clopen) sets form an open basis. In fact, it is the co-universal boolean space associated to X [25].

Definition (3.2.4). *The* patch spectrum Spec$_{\text{Pa}}(A)$ *of an object A is the patch space of the prime spectrum of A.*

Theorem (3.2.5). *The patch spectrum of an object is homeomorphic to the maximal spectrum of the universal regular object associated to it.*

Proof. Let $f : A \to B$ be the universal regular object associated to A. Let $p \in$ Spec(A). Because the object $K(p)$ is regular, the morphism $j_p q_p : A \to K(p)$ factors through f in the form $j_p q_p = r_p f$. Let $r_p = uq$ be the regular factorization of r_p where $q : B \to Q$ is a regular epimorphism and $u : Q \to K(p)$ is a monomorphism. The object Q is regular (Proposition (2.11.5)). Because it is integral, it is simple (Proposition (2.11.4)). The relation $uqf = r_p f = j_p q_p$ implies the existence of a monomorphism $v : A/p \to Q$ fulfilling $uv = j_p$ and $vq_p = qf$. Because j_p is the universal simple object associated to A/p, there exists a morphism $w : K(p) \to Q$ fulfilling $wj_p = v$. Because f is epimorphic, v is epimorphic and thus w is epimorphic. Moreover, $uwj_p = uv = j_p$. Because j_p is epimorphic, $uw = 1_{K(p)}$. Thus u is isomorphic and r_p is a regular epimorphism. It follows that if p_1 is the kernel pair of r_p then $f^{*c}(p_1) = p$, i.e. $\mathrm{Spec}(f)(p_1) = p$. Let p_2 be any element of $\mathrm{Spec}(B)$ such that $\mathrm{Spec}(f)(p_2) = p$. Let $q_2 : B \to Q_2$ be the quotient of B by p_2 and let $g : A/p \to Q_2$ be the monomorphism such that $gq_p = q_2 f$. The integral object Q_2 is regular (Proposition (2.11.5)) and thus simple (Proposition (2.11.4)). Because j_p is the universal simple object associated to A/p, there exists a monomorphism $k : K(p) \to Q_2$ such that $kj_p = g$. Then $kr_p f = kj_p q_p = gq_p = q_2 f$. Thus $kr_p = q_2$ and hence k is a regular epimorphism. It follows that k is isomorphic. Thence $p_1 = p_2$. If follows that $\mathrm{Spec}(f)$ is one to one. As a result, $\mathrm{Spec}(f)$ is a continuous one-to-one map between compact Hausdorff spaces and thus it is a homeomorphism. ■

3.3 TOPOLOGICAL INTERPRETATIONS

Proposition (3.3.1). *An object is pseudo-simple if and only if its prime spectrum is a singleton.*

Proof. The proof follows from Proposition (2.6.2). ■

Proposition (3.3.2).
(i) *An object is local if and only if its prime spectrum has a unique closed point.*
(ii) *A morphism between local objects is local if and only if its prime spectrum preserves the closed point.*

Proof. The proof follows from Proposition (3.2.1). ■

Proposition (3.3.3). *If* $r_A : A \to R(A)$ *is the universal reduced object associated to* A, *then* $\mathrm{Spec}(r_A) : \mathrm{Spec}(R(A)) \to \mathrm{Spec}(A)$ *is a homeomorphism.*

Proof. The map $\mathrm{Spec}(r_A)$ is injective because if $p_1, p_2 \in \mathrm{Spec}(R(A))$ are such that $\mathrm{Spec}(r_A)(p_1) = \mathrm{Spec}(r_A)(p_2)$ then the quotients q_1 *and* q_2 of $R(A)$ by p_1 and p_2 respectively are such that $q_1 r_A \simeq q_2 r_A$. Thus $q_1 \simeq q_2$ and $p_1 = p_2$. The map $\mathrm{Spec}(r_A)$ is surjective because, for any $p \in \mathrm{Spec}(A)$, the quotient q of A by p factors through r_A in the form $q = gr_A$ and the kernel pair of g is a prime congruence s on $R(A)$ such that $\mathrm{Spec}(r_A)(s) = p$. The map $\mathrm{Spec}(r_A)$ is a homeomorphism because it is continuous and, for any congruence s on $R(A)$, $\mathrm{Spec}(r_A)(D(s)) = D(r_A^{*c}(s))$. ■

Proposition (3.3.4). *An object is quasi-primary if and only if its prime spectrum is irreducible.*

Proof. According to Theorem (2.7.4), A is quasi-primary if and only if the universal reduced object $R(A)$ associated to it is integral. According to Proposition (2.10.3), $R(A)$ is integral if and only if $R(A)$ is irreducible. According to Definition (2.10.1), if $R(A)$ is irreducible, then $\Delta_{R(A)}$ is an irreducible element in $\mathrm{CongRad}(R(A))$. But the converse is also true because, if $\Delta_{R(A)}$ is an irreducible element in $\mathrm{CongRad}(R(A))$, then, for any $r,s \in \mathrm{Cong}(R(A))$ such that $r \wedge s = \Delta_{R(A)}$, we have $\mathrm{rad}(r) \wedge \mathrm{rad}(s) = \Delta_{R(A)}$ (Proposition (2.4.5)). Thus $\mathrm{rad}(r) = \Delta_{R(A)}$ or $\mathrm{rad}(s) = \Delta_{R(A)}$. Hence $r = \Delta_{R(A)}$ or $s = \Delta_{R(A)}$, and so $R(A)$ is irreducible. According to Theorem (3.1.5), $\Delta_{R(A)}$ is an irreducible element in $\mathrm{CongRad}(R(A))$ if and only if $\mathrm{Spec}(R(A))$ is an irreducible topological space. But $\mathrm{Spec}(R(A)) \simeq \mathrm{Spec}(A)$ (Proposition (3.3.3)). Thus A is quasi-primary if and only if $\mathrm{Spec}(A)$ is irreducible. ■

Proposition (3.3.5). *For any morphism $f: A \to B$, the image of* $\mathrm{Spec}(f)$ *is* $\{p \in \mathrm{Spec}(A) : f^{*c}(f_{*c}(p)) = p\}$.

Proof.

(i) Let $r \in \mathrm{Spec}(B)$, $p = \mathrm{Spec}(f)(r) = f^{*c}(r)$, and $s = f_{*c}(p)$. Let $q_p: A \to A/p$ be the quotient of A by p and let $q_r: B \to B/r$ and $q_s: B \to B/s$ be the quotients of B by r and s respectively. There is a monomorphism $m: A/p \to B/r$ such that $mq_p = q_r f$, there is a morphism $n: A/p \to B/s$ such that (n, q_s) is the pushout of (q_p, f), and there is a morphism $t: B/s \to B/r$ such that $tq_s = q_r$ and $tn = m$. Thus n is monomorphic. Therefore $p = f^{*c}(s) = f^{*c}(f_{*c}(p))$.

(ii) Let $p \in \mathrm{Spec}(A)$ such that $f^{*c}(f_{*c}(p)) = p$. Let $r = f_{*c}(p)$, and let $q_p: A \to A/p$ and $q_r: B \to B/r$ be the quotients of A and B by p and r respectively. Because $p = f^{*c}(r)$, there exists a monomorphism $m: A/p \to B/r$ such that $mq_p = q_r f$. Because $r = f_{*c}(p)$, (m, q_r) is the pushout of (q_p, f). Let $j_p: A/p \to K(p)$ be the universal simple object associated to A/p. Let $(h: K(p) \to H, u: B/r \to H)$ be the pushout of (j_p, m). Because j_p is presingular (proof of Theorem (2.2.4)), it is flat (Proposition (1.9.5)) and thus h is monomorphic. Hence H is not terminal. According to Corollary (2.1.9), there exists a simple quotient $q: H \to L$ of H. Let s be the kernel pair of quq_r. As L is simple, s is prime. The morphism quq_r factors in the form $quq_r = nq_s$ where $q_s: B \to B/s$ is the quotient of B by s and $n: B/S \to L$ is monomorphic. The relations $nq_s f = quq_r f = qumq_p = qhj_p q_p$ imply the existence of a morphism $v: A/p \to B/s$ such that $vq_p = q_s f$ and $nv = qhj_p$. Because $K(p)$ and L are simple, qh is monomorphic. Thus qhj_p is monomorphic and hence v is monomorphic. It follows that $p = f^{*c}(s)$ and that p belongs to the image of $\mathrm{Spec}(f)$. ■

Proposition (3.3.6). *A morphism $f: A \to B$ codisjoints a congruence r on A if and only if the image of the map* $\mathrm{Spec}(f)$ *is included in $D(r)$.*

Proof. Let us denote by I the image of Spec(f). Let us assume that f codisjoints r and let $p \in I$. Then $f_{*c}(r) = 1_{B\times B}$ and $f_{*c}(p)$ is proper according to the relation $p = f^{*c}(f_{*c}(p))$ (Proposition (3.3.5)). Thus $f_{*c}(r) \not\leqslant f_{*c}(p)$; hence $r \not\leqslant f^{*c}(f_{*c}(p)) = p$ and so $p \in D(r)$. As a result, $I \subset D(r)$. Let us assume that f does not codisjoint r. Then $f_{*c}(r)$ is proper. According to Corollary (2.1.10), there exists a maximal congruence m on B such that $f_{*c}(r) \leqslant m$. Then $r \leqslant f^{*c}(m)$ and hence $f^{*c}(m) \notin D(r)$. As a result $I \not\subset D(r)$. ■

Proposition (3.3.7). *For any flat morphism $f: A \to B$ the image of* Spec(f) *is* $\{p \in \mathrm{Spec}(A) : f_{*c}(p)$ *is proper*$\}$ *and is closed under generizations.*

Proof. Let $p \in \mathrm{Spec}(A)$. According to Proposition (3.3.5), p belongs to the image of Spec(f) if and only if $f^{*c}(f_{*c}(p)) = p$. Thus if p is in this image, $f_{*c}(p)$ is proper. Conversely, let us assume that $f_{*c}(p)$ is proper. Let $r = f_{*c}(p)$ and let $q_p : A \to A/p$ and $q_r : B \to B/r$ be the quotients by p and r respectively. There is a morphism $g : A/p \to B/r$ such that (g, q_r) is the pushout of (q_p, f). Because flat morphisms are co-universal, g is flat. Because r is proper, B/r is not terminal. Because A/p is integral, g is monomorphic (Proposition (2.2.7)). It follows that $p = f^{*c}(r)$ and thus $p = f^{*c}(f_{*c}(p))$. Hence p belongs to the image of Spec(f). Furthermore, if q is a generization of p then $q \leqslant p$ (Proposition (3.1.8)). Thus $f_{*c}(q)$ is proper and q belongs to the image of Spec(f). ■

Proposition (3.3.8). *If $f: A \to B$ is a flat epimorphism, the map* Spec(f) : Spec$(B) \to$ Spec(A) *is a topological embedding, i.e. it induces an homeomorphism between* Spec(B) *and its image in* Spec(A).

Proof. According to Proposition (1.7.2), the inverse congruence map f^{*c} : Cong$(B) \to$ Cong(A) satisfies $f_{*c} \circ f^{*c} = Id$ and thus is injective. Because Spec(f) is induced by f^{*c}, it is injective. Let X be the image of Spec(f) in Spec(A). The map Spec(f) : Spec$(B) \to X$ is continuous. Let $D(r)$, with $r \in \mathrm{Cong}(B)$, be an open set of Spec(B). Then $D(f^{*c}(r))$ is an open set of Spec(A). For any $p \in \mathrm{Spec}(B)$,

$$\mathrm{Spec}(f)(p) \in D(f^{*c}(\mathrm{r})) \Leftrightarrow f^{*c}(p) \not\leqslant f^{*c}(r) \Leftrightarrow p \not\leqslant r \Leftrightarrow p \in D(r).$$

Consequently, Spec$(f)(D(r)) = X \cap D(f^{*c}(r))$ is an open set of X. If follows that the map Spec(f) : Spec$(B) \to X$ is open and therefore is a homeomorphism. ■

Proposition (3.3.9). *If $l_p : A \to A_p$ is the localization of A at p, the map* Spec(l_p) : Spec$(A_p) \to$ Spec(A) *induces a homeomorphism between* Spec(A_p) *and the subspace of* Spec(A) *whose points are the generizations of p.*

Proof. The morphism l_p is presingular (Definition (2.13.1)) and thus is a flat epimorphism. According to Proposition (3.3.8), the map $\mathrm{Spec}(l_p)$ is a topological embedding. The prime congruence p belongs to the image of $\mathrm{Spec}(l_p)$ for $p = (l_p)^{*c}(m_p) = \mathrm{Spec}(l_p)(m_p)$. According to Proposition (3.3.7), any generization of p belongs to this image. Moreover, for any $q \in \mathrm{Spec}(A_p)$, we have $q \leqslant m_p$ and thus $\mathrm{Spec}(f)(q) = f^{*c}(q) \leqslant f^{*c}(m_p) = p$. Hence $\mathrm{Spec}(f)(q)$ is a generization of p. ■

Proposition (3.3.10). *If $f: A \to B$ is a singular epimorphism, which is codisjunctor of a congruence r, the map* $\mathrm{Spec}(f): \mathrm{Spec}(B) \to \mathrm{Spec}(A)$ *induces a homeomorphism between* $\mathrm{Spec}(B)$ *and its open image $D(r)$ in* $\mathrm{Spec}(A)$.

Proof. According to Proposition (1.8.3), f is a flat epimorphism, and according to Proposition (3.3.8), the map $\mathrm{Spec}(f): \mathrm{Spec}(B) \to \mathrm{Spec}(A)$ induces a homeomorphism between $\mathrm{Spec}(B)$ and its image $\mathrm{Im}(\mathrm{Spec}(f))$ in $\mathrm{Spec}(A)$. Let $q \in \mathrm{Spec}(B)$, $p = f^{*c}(q)$, $l_q: B \to B_q$, and $l_p: A \to A_p$ be the localization morphisms, and let $f_q: A_p \to B_q$ be the local morphism of f at q. Because f and l_q are presingular epimorphisms, the morphism $l_q f = f_q l_p$ is a presingular epimorphism. Thus f_q is isomorphic (Theorem (1.9.6)). It follows that l_p factors through f, and thus l_p codisjoints r. According to Proposition (3.1.1), $p \in D(r)$. Conversely, let us consider any $p \in D(r)$. According to Proposition (3.1.1), l_p codisjoints r and thus l_p factors through f in the form $l_p = hf$ with $h: B \to A_p$. If m_p is the maximal congruence on A_p, then the prime congruence $q = h^{*c}(m_p)$ on B is such that $f^{*c}(q) = f^{*c}(h^{*c}(m_p)) = l_p^{*c}(m_p) = p$. Thus $p \in \mathrm{Im}(\mathrm{Spec}(f))$. As a result, $\mathrm{Im}(\mathrm{Spec}(f)) = D(r)$. ■

Proposition (3.3.11). *If $f: A \to B$ is a regular epimorphism, which is the coequalizer of a congruence r, the map* $\mathrm{Spec}(f): \mathrm{Spec}(B) \to \mathrm{Spec}(A)$ *induces a homeomorphism between* $\mathrm{Spec}(B)$ *and its closed image $V(r)$ in* $\mathrm{Spec}(A)$.

Proof. The map $\mathrm{Spec}(f)$ is injective because it is the restriction of the map f^{*c} which satisfies the relation $f_{*c} \circ f^{*c} = Id$. Let $m \in \mathrm{Spec}(B)$ and $q_m: B \to B/m$ be the quotient of B by m, and let $p = f^{*c}(m) = \mathrm{Spec}(f)(m)$. Then the coequalizer of A by p is the morphism $q_m f$ which coequalizes r. Therefore $r \leqslant p$ and thus $p \in V(r)$. Conversely, if $p \in V(r)$, the quotient $q_p: A \to A/p$ factors through f in a morphism $q: B \to A/p$ which provides a prime congruence $m = q^{*c}(\Delta)$ on B such that $f^{*c}(m) = f^{*c}(q^{*c}(\Delta)) = (qf)^{*c}(\Delta) = q_p^{*c}(\Delta) = p$. As a result, the image of $\mathrm{Spec}(f)$ is the closed set $V(r)$. Let us prove that the map $\mathrm{Spec}(f)$ preserves closed sets. Let $V(s)$, where $s \in \mathrm{Cong}(B)$, be a closed set in $\mathrm{Spec}(B)$. Let $q_s: b \to B/s$ be the quotient of B by s and let $t = f^{*c}(s)$. According to the preceding,

$$\text{Spec}(f)(V(s)) = \text{Spec}(f)((\text{Spec}(q_s)(\text{Spec}(B/s))) =$$
$$\text{Spec}(q_s f)(\text{Spec}(B/s)) = V(f^{*c}(s)) = V(t).$$

It follows that Spec$(f)(V(s))$ is a closed set of Spec(A). Therefore the map Spec(f) : Spec$(B) \to V(r)$ is one to one, continuous, and preserves closed subsets. It is a homeomorphism. ■

Proposition (3.3.12). *If $q_p : A \to A/p$ is the quotient of A by a prime ideal p, the map* Spec(q_p) : Spec$(A/p) \to$ Spec(A) *induces a homeomorphism between* Spec(A/p) *and the subspace of* Spec(A) *whose points are the specializations of p.*

Proof. According to Proposition (3.3.11), Spec(f) induces a homeomorphism between Spec(A/p) and the subspace $V(p)$ of Spec(A). According to Proposition (3.1.7), $V(p)$ is the closure of $\{p\}$, i.e. the set of specializations of p. ■

Let us recall that a continuous map $u : E \to F$ is said to be *dominant* if its image $u(E)$ is a dense subset of F.

Proposition (3.3.13). *A morphism $f: A \to B$ is a premonomorphism if and only if the map* Spec(f) : Spec$(B) \to$ Spec(A) *is dominant.*

Proof. Let us consider an arbitrary affine open set in Spec(A). It is of the form $D(s)$ with $s \in$ CongRadCodis(A). Its inverse image by Spec(f) is Spec$(f)^{-1}(D(s)) = D(f_{*\text{cr}}(s))$ (cf. § 3.1, p. 86). Let us assume that f is a premonomorphism and that Spec$(f)^{-1}(D(s)) = \varnothing$. Then $D(f_{*\text{c}}(s)) = D(f_{*\text{cr}}(s)) = \varnothing$. According to Proposition (3.1.16), $f_{*\text{c}}(s)$ is conjoint. Then $f_{*\text{c}}(s) \leqslant \text{rad}(B)$ and hence $s \leqslant f^{*\text{c}}(\text{rad}(B))$. According to Proposition (2.5.7), $f^{*\text{c}}(\text{rad}(B)) = \text{rad}(A)$. Thus $s \leqslant \text{rad}(A)$ and s is conjoint. Therefore $D(s) = \varnothing$. It follows that the map Spec(f) is dominant. Conversely, let us assume that the map Spec(f) is dominant. Let r be the kernel pair of f. Then Spec$(f)^{-1}(D(r)) = D(f_{*\text{c}}(r)) = D(\Delta_B) = \varnothing$. Thus $D(r) = \varnothing$. By Proposition (3.1.16), r is conjoint. According to Proposition (2.5.7), f is a premonomorphism. ■

3.4 STRUCTURE SHEAF ON THE PRIME SPECTRUM

Let A be an object in **A**. The ordered set CongRadCodis(A) of codisjunctable radical congruences on A has finite meets (Theorem (2.4.12)). It can be considered as a finitely complete category denoted by $\mathbf{C}(A)$. For any object r of $\mathbf{C}(A)$, let us denote by Cov(r) the set of finite families of morphisms $(r_i \to r)_{i \in I}$ in $\mathbf{C}(A)$ such that r is the join of $(r_i)_{i \in I}$ in CongRad(A), i.e. $r = \overset{\text{cr}}{\bigvee}_{i \in I} r_i$. The set $\{\text{Cov}(r)\}_{r \in \mathbf{C}(A)}$ contains the singletons

$\{1_r\}_{r\in \mathbf{C}(A)}$ and is stable under compositions. Moreover, it is stable under change of bases, for, if $(r_i \to r)_{i\in I} \in \mathrm{Cov}(r)$ and $s \to r$ is a morphism, then the family $(r_i \wedge s \to s)_{i\in I}$ of pullbacks of $(r_i \to r)_{i\in I}$ along $s \to r$ belongs to $\mathrm{Cov}(s)$ since $s = s\wedge r = s\wedge (\overset{\mathrm{cr}}{\vee}_{i\in I} r_i) = \overset{\mathrm{cr}}{\vee}_{i\in I}(s\wedge r_i)$ (Theorem (2.4.8)). Consequently, the set $\{Cov(r)\}_{r\in \mathbf{C}(A)}$ is a Grothendieck pretopology on $\mathbf{C}(A)$ ([21], II Definition 1.3). This pretopology generates a Grothendieck topology on $\mathbf{C}(A)$ ([21], II, Definition 1.3.1) which gives the category $\mathbf{C}(A)$ the structure of a site ([21], II, Definition 1.2). According to [21], II, Definition 6.1, the notion of sheaves defined on $\mathbf{C}(A)$ with values in $\mathbf{A}$ is well defined. Let us define the functor $\Delta : \mathbf{C}(A)^{\mathrm{op}} \to \mathbf{A}$ by assigning to r the codomain of a codisjunctor $\delta(r) : A \to \Delta(r)$ of r, and to a morphism $r \to s$ the unique morphism $\Delta(r \to s) : \Delta(s) \to \Delta(r)$ such that $\Delta(s \to s)\delta(s) = \delta(r)$.

Theorem (3.4.1). *The functor $\Delta : C(A)^{\mathrm{op}} \to A$ is a sheaf with values in A.*

Proof.

(i) Let us prove that any flat extremal co-universal monomorphism $f: A \to B$ in $\mathbf{A}$ is a regular monomorphism, i.e. is the equalizer of its cokernel pair $(m, n) : B \rightrightarrows M$. Let $k : K \to B$ be the equalizer of (m, n) and let $g : A \to K$ be such that $kg = f$. Let $\bar{f}, \bar{m}, \bar{n}, \bar{k}, \bar{g}$ be the images of f, m, n, k, and g respectively by the pushout functor along f. The morphism $\bar{f}$ is a split monomorphism, and thus it is a regular monomorphism. Because $(\bar{m}, \bar{n})$ is its cokernel pair, $\bar{f}$ is the equalizer of $(\bar{m}, \bar{n})$. Because f is flat, $\bar{k}$ is monomorphic. The relation $\bar{k}\bar{g} = \bar{f}$ implies $\bar{f} \leqslant \bar{k}$ and the relation $\bar{m}\bar{k} = \bar{n}\bar{g}$ implies that $\bar{k}$ factors through $\bar{f}$. Thus $\bar{k} \leqslant \bar{f}$. It follows that $\bar{k} \simeq \bar{f}$ and that $\bar{g}$ is isomorphic. Let $(r, s) : K \rightrightarrows N$ be the cokernel pair of g. Let $\bar{r}$ and $\bar{s}$ be the images of r and s respectively by the pushout functor along f, and let v, w be morphisms such that $vg = \bar{g}f$, $wr = \bar{r}v$, and $ws = \bar{s}v$. Because $\bar{g}$ is isomorphic, $\bar{r} = \bar{s}$ and thence $wr = \bar{r}v = \bar{s}v = ws$. Because f is a co-universal monomorphism, w is monomorphic. Thus $r = s$ and thence g is epimorphic. Because f is an extremal monomorphism, g is isomorphic. It follows that f is the equalizer of (m, n).

(ii) Let us consider a finite family $(r_i)_{i\in I}$ of elements in CongRad-Codis(A) and a morphism $f: A \to B$ which codisjoints $r = \overset{\mathrm{cr}}{\vee}_{i\in I} r_i$. We shall prove that the family $(\delta_i : B \to \Delta_i)_{i\in I}$ of codisjunctors of the congruences $f_{*\mathrm{c}}(r_i)$ is co-universal effective monomorphic ([21], I, Definition 10.4) i.e. an effective co-union. According to Proposition (2.4.5) and § 2.4, p. 57),

$$\mathrm{rad}(\overset{\mathrm{c}}{\underset{i\in I}{\vee}} f_{*\mathrm{c}}(r_i)) = \overset{\mathrm{cr}}{\underset{i\in I}{\vee}} \mathrm{rad}(f_{*\mathrm{c}}(r_i)) = \overset{\mathrm{cr}}{\underset{i\in I}{\vee}} f_{*\mathrm{cr}}(r_i) = f_{*\mathrm{cr}}(\overset{\mathrm{cr}}{\underset{i\in I}{\vee}} r_i)$$

$$= f_{*\mathrm{cr}}(r) = 1_{B\times B}.$$

According to Proposition (2.4.13), $\overset{\mathrm{cr}}{\vee}_{i\in I} f_{*\mathrm{c}}(r_i) = 1_{B\times B}$. According to Axiom (1.2.6) of Zariski categories, $\vee_{i\in I}\delta_i = 1_B$ is a co-union in Quot(B). This co-union is co-universal because, for any morphism $g : B \to C$, the morphism $gf : A \to C$ codisjoints r and thus the pushouts $\overline{\delta_i}$ of δ_i along g, which are the codisjunctors of $(gf)_{*c}(r_i)$, satisfy the relation $\vee_{i \in I}\overline{\delta_i} = 1_C$. Let $(p_i : \Delta \to \Delta_i)_{i \in I}$ be the product of $(\Delta_i)_{i \in I}$ and let $\delta : B \to \Delta$ be the morphism defined by $p_i \delta = \delta_i$ for any $i \in I$. Then δ is a co-universal extremal monomorphism. Moreover, according to Proposition (1.6.2), δ is flat. It follows from (i) that δ is a co-universal regular monomorphism, i.e. a co-universal effective monomorphism. On the other hand, the family $(p_i : \Delta \to \Delta_i)_{i \in I}$ is a co-universal effective monomorphic family because finite products are co-universal. Consequently, the composite family $(\delta_i : B \to \Delta_i)_{i \in I}$ is co-universal effective monomorphic ([21], II, Definition 2.5).

(iii) Let $(r_i \to r)_{i \in I} \in \mathrm{Cov}(r)$. Then the morphism $\delta(r) : A \to \Delta(r)$ codisjoints $r = \overset{\mathrm{cr}}{\vee}_{i \in I} r_i$. For any $i \in I$, the morphism $\Delta(r_i \to r) : \Delta(r) \to \Delta(r_i)$ is the pushout of $\delta(r_i)$ along $\delta(r)$ and thus is the codisjunctor of $\delta(r)_{*c}(r_i)$. It follows from (ii) that $(\Delta(r_i \to r))_{i \in I}$ is a co-universal effective monomorphic family. On the other hand, according to Proposition (1.5.4), for $i, j \in I$, $(\Delta(r_i \wedge r_j \to r_i), \Delta(r_i \wedge r_j \to r_j))$ is the pushout of $(\Delta(r_i \to r), \Delta(r_j \to r))$. It follows that the sheaf property is satisfied for the covering $(r_i \to r)_{i \in I}$. As a result, the presheaf $\Delta : \mathbf{C}(A)^{\mathrm{op}} \to \mathbf{A}$ is a sheaf with values in $\mathbf{A}$. ■

The ordered set $\Omega(\mathrm{Spec}(A))$ of open subsets of $\mathrm{Spec}(A)$ can be considered as a category. Equipped with the canonical Grothendieck topology, it becomes a site such that the sheaves on it are precisely the sheaves on $\mathrm{Spec}(A)$. The assignment $r \to D(r)$ defines a functor $D : \mathbf{C}(A) \to \Omega(\mathrm{Spec}(A))$.

Lemma (3.4.2). *The functor $F \to F \circ D^{\mathrm{op}}$ induces an equivalence of categories between the category $\mathit{Sh}[\mathrm{Spec}(A), A]$ of sheaves on* Spec(A) *with values in* $\mathbf{A}$ *and the category $\mathit{Sh}[\mathbf{C}(A), A]$ of sheaves on $\mathbf{C}(A)$ with values in A.*

Proof.

(i) Let us prove that the covering families in **C**(A) are precisely the families $(r_i \to r)_{i \in I}$ such that $r = \overset{\mathrm{cr}}{\vee}_{i \in I} r_i$. Let $(r_i \to r)_{i \in I}$ be a covering family. Let $S = \{s \in \mathbf{C}(A) : \exists\, i \in I, s \leqslant r_i\}$. Then the family $(s \to r)_{s \in S}$ can be identified with the seave ([21], § I.4) generated by $(r_i \to r)_{i \in I}$ and is covering. According to [21], II, Proposition 1.4, there exists a family $(s_i \to r)_{i \in I_0} \in \mathrm{Cov}(r)$ which generates a seave included in $(s \to r)_{s \in S}$. Then the relation $r = \overset{\mathrm{cr}}{\vee}_{i \in I_0} s_i$ implies $r = \overset{\mathrm{cr}}{\vee}_{s \in S} s$ and $r = \overset{\mathrm{cr}}{\vee}_{i \in I} r_i$. Conversely, let us consider a family $(r_i \to r)_{i \in I}$ such that $r = \overset{\mathrm{cr}}{\vee}_{i \in I} r_i$. According to Proposition (2.4.9), r is a finitely generated radical congruence and thus is a compact element in $\mathrm{CongRad}(A)$. Therefore there exists a finite subset $I_0 \subset I$ such that $r = \overset{\mathrm{cr}}{\vee}_{i \in I_0} r_i$. Then $(r_i \to r)_{i \in I_0} \in \mathit{Cov}(r)$ and $(r_i \to r)_{i \in I}$ is a covering family.

(ii) The functor D is obviously full and faithful. According to (i) and Theorem (3.1.5), D preserves and reflects covering families. Therefore the topology on $\mathbf{C}(A)$ is the topology induced by the topology on $\Omega(\mathrm{Spec}(A))$. Moreover, according to Proposition (3.1.14), any object of $\Omega(\mathrm{Spec}(A))$ can be covered by objects which are images of objects of $\mathbf{C}(A)$ by D. Consequently, according to the comparison lemma ([21], II, Theorem 4.1), the functor $F \to F \circ D^{\mathrm{op}}$ induces an equivalence of categories $\mathbf{Sh}[\mathrm{Spec}(A), \mathbf{A}] \sim \mathbf{Sh}[\mathbf{C}(A), \mathbf{A}]$. ■

It follows from Lemma (3.4.2) that up to isomorphism there is a unique sheaf F on $\mathrm{Spec}(A)$ such that $F \circ D^{\mathrm{op}}$ is the sheaf Δ defined in Theorem (3.4.1), i.e. the structure sheaf on $\mathrm{Spec}(A)$.

Definition (3.4.3). *The* structure sheaf on $\mathrm{Spec}(A)$ *is the sheaf* $\tilde{A}$: $\Omega(\mathrm{Spec}(A))^{\mathrm{op}} \to \mathbf{A}$ *based on* $\mathrm{Spec}(A)$ *with values in* $\mathbf{A}$ *whose value, for an affine open set* $D(r)$ *of* $\mathrm{Spec}(A)$, *is* $\tilde{A}(D(r)) = \Delta(r)$, *where* $\delta(r)$: $A \to \Delta(r)$ *is the codisjunctor of* r, *and whose restriction morphism along* $D(r) \to D(s)$ *is the morphism* $\tilde{A}(D(r) \to D(s)) : \tilde{A}(s) \to \tilde{A}(r)$ *such that* $\tilde{A}(D(r) \to D(s))\delta(s) = \delta(r)$.

Theorem (3.4.4). *The structure sheaf* $\tilde{A}$ *on* $\mathrm{Spec}(A)$ *has the localized object of* A *at* p *as its stalk at* p *and the object* A *as its object of global sections.*

Proof. According to Proposition (3.1.1), the localization morphism l_p: $A \to A_p$ of A at p is the co-intersection of the family of codisjunctors $(\mathrm{codis}(r))$ where r runs through $\mathrm{CongCodis}(A)$ with $p \in D(r)$. According to Proposition (2.4.7), it is sufficient to make r run through CongRad-Codis(A) with $p \in D(r)$. As this co-intersection is cofiltered, l_p is the filtered colimit of the diagram $(\mathrm{codis}(r))$ where r runs through the ordered set $\mathrm{CongRadCodis}(A)$ with $p \in D(r)$. Therefore A_p is precisely the stalk of $\tilde{A}$ at p. On the other hand, the unit congruence $1_{A \times A}$ on A is codisjointed; thus its codisjunctor is 1_A and $\tilde{A}(\mathrm{Spec}(A)) = \tilde{A}(D(1_{A \times A})) = A$. ■

3.5 THE COGENERATING CLASS OF LOCAL OBJECTS

Proposition (3.5.1). *For any object* A, *the morphism* $(l_p)_{p \in \mathrm{Spec}(A)} : A \to \Pi_{p \in \mathrm{Spec}(A)} A_p$ *is a co-universal regular monomorphism.*

Proof.

(i) We use the notation of Van Osdol ([38], III). Let $X = \mathrm{Spec}(A)$, let $\mathbf{Sh}[X, \mathbf{A}]$ be the category of sheaves on X with values in $\mathbf{A}$, and let $\mathbf{A}^X$ be the category of families of objects of $\mathbf{A}$ indexed by the set X. The stalk functor $S : \mathbf{Sh}[X, \mathbf{A}] \to \mathbf{A}^X$ is defined by $S(F) = (F_x)_{x \in X}$ and $S(\alpha) = (\alpha_x)_{x \in X}$, while the functor $\Pi : \mathbf{A}^X \to \mathbf{Sh}[X, \mathbf{A}]$ is defined by

$(\Pi(A_x))(U) = \Pi_{x \in U} A_x$ and $(\Pi(f_x))_U = \Pi_{x \in U} f_x$. The functor S is left-adjoint to Π and the unit η of the adjunction has, as it value at F, the morphism $\eta_F : F \to \Pi S F$ defined by $(\eta_F)_U = (\rho_x^U)_{x \in U} : F(U) \to \Pi_{x \in U} F_x$ where the symbol ρ denotes any restriction morphism. A result of Van Osdol ([38], Theorem III.3) asserts that S is cotriplable. However, we cannot use this result here because the assumptions made in obtaining it are not necessarily satisfied in a Zariski category. Thus we adopt a different method of proving that S is cotriplable, i.e. comonadic.

(ii) We prove that the morphism $(\eta_F)_U : F(U) \to \Pi_{x \in U} F_x$ is monomorphic. Let $(m, n) : C \rightrightarrows F(U)$ be a pair of morphisms with a finitely presentable domain such that $(\eta_F)_U m = (\eta_F)_U n$, i.e. $\rho_x^U m = \rho_x^U n$ for any $x \in U$. Because $F_x = \xrightarrow{lim}_{x \in U} F(U)$ is a filtered colimit, there exists an open neighbourhood U_x of x in U such that $\rho_{U_x}^X m = \rho_{U_x}^X n$. We obtain an open covering $(U_x)_{x \in U}$ of U and the separation property of the sheaf F implies the equality $m = n$. If follows that $(\eta_F)_U$ is monic.

(iii) We prove that the functor S reflects isomorphisms. Let $\alpha : G \to F$ be a morphism in $\mathbf{Sh}[X, A]$ such that $S(\alpha)$ is an isomorphism, i.e. α_x is an isomorphism for any $x \in X$. Let us assume that $S(\alpha)$ is an identity, i.e. α_x is an identity for any $x \in X$. Then $\eta_F \alpha = \eta_G$. According to (ii), η_G is a pointwise monomorphism. Thus α is a pointwise monomorphism. Let $U \in \Omega(X)$. Let $f : C \to F(U)$ be a morphism with a finitely presentable domain. We prove that f factors through α_U. Let $x \in U$. Because $F_x = \xrightarrow{lim}_{x \in V} G(V)$ is a filtered colimit, there exist an open neighbourhood V_x of x in U and a morphism $h_x : C \to G(V_x)$ such that $\rho_x^{V_x} h_x = \rho_x^U f$. Then $\rho_x^{V_x} \alpha_{V_x} h_x = \rho_x^{V_x} \rho_{V_x}^U f$. Furthermore, there exists an open neighbourhood U_x of x in V_x such that $\rho_{U_x}^{V_x} \alpha_{V_x} h_x = \rho_{U_x}^{V_x} \rho_{V_x}^U f$. Let us denote by g_x the morphism $\rho_{U_x}^{V_x} h_x$: $C \to G(U_x)$. Then

$$\alpha_{U_x} g_x = \alpha_{U_x} \rho_{U_x}^{V_x} h_x = \rho_{U_x}^{V_x} \alpha_{V_x} h_x = \rho_{U_x}^U f.$$

In this way we obtain an open covering $(U_x)_{x \in U}$ of U and a family of morphisms $(g_x : C \to G(U_x))_{x \in U}$. Let $y \in U$. Then we have the relations

$$\alpha_{U_x \cap U_y} \rho_{U_x \cap U_y}^{U_y} g_y = \rho_{U_x \cap U_y}^{U_y} \alpha_{U_y} g_y = \rho_{U_x \cap U_y}^{U_y} \rho_{U_y}^U f = \rho_{U_x \cap U_y}^U f =$$
$$\rho_{U_x \cap U_y}^{U_x} \rho_{U_x}^U f = \rho_{U_x \cap U_y}^{U_x} \alpha_{U_x} g_x = \alpha_{U_x \cap U_y} \rho_{U_x \cap U_y}^{U_x} g_x.$$

Since $\alpha_{U_x \cap U_y}$ is monomorphic, it follows that

$$\rho_{U_x \cap U_y}^{U_y} g_y = \rho_{U_x \cap U_y}^{U_x} g_x.$$

From the gluing property of the sheaf G, it follows that there exists a morphism $g : C \to GU$ such that $\pi_{U_x}^U g = g_x$ for any $x \in U$. Then

$$\rho_{U_x}^U \alpha_U g = \alpha_{U_x} \pi_{U_x}^U g = \alpha_{U_x} g_x = \pi_{U_x}^U f.$$

The equality $\alpha_U g = f$ follows from the separation property of the sheaf F. The result that the monomorphism α_U is an isomorphism follows from the fact that finitely presentable objects form a proper generating class of objects in **A**. Therefore α is an isomorphism. Thus S reflects isomorphisms.

(iv) The functor S preserves equalizers since equalizers commute with filtered colimits in **A**. According to Beck's theorem ([31], § VI.7, Theorem 1, p. 147), the functor S is cotriplable, i.e. comonadic.

(v) The fact that $\eta_F: F \to \Pi SF$ is a regular monomorphism in $\mathbf{Sh}[X, \mathbf{A}]$, and thus is a pointwise regular monomorphism, follows from the canonical presentation of the co-algebra F ([31], § VI.7). In particular, $(\eta_F)_X: FX \to \Pi_{x \in X} F_x$ is a regular monomorphism. For $F = \tilde{A}$, we obtain the result that $(l_p)_{p \in X}: A \to \Pi_{p \in X} A_p$ is a regular monomorphism.

(vi) Let $f: A \to B$ be any morphism and let $(g: B \to G, h: \Pi_{p \in X} A_p \to G)$ be the pushout of $(f: A \to B, (l_p)_{p \in X}: A \to \Pi_{p \in X} A_p)$. Let $q \in \mathrm{Spec}(B)$ and $p = \mathrm{Spec}(f)(q)$, and let $\alpha_p: \Pi_{p \in X} A_p \to A_p$ be the canonical projection. The relation $l_q f = f_q l_p = f_q \alpha_p (l_p)_{p \in X}$ implies the existence of a morphism $m_q: G \to B_q$ such that $m_q g = l_q$ and $m_q h = f_q \alpha_p$. It follows that the morphism $(l_q)_{q \in \mathrm{Spec}(B)}: B \to \Pi_{q \in \mathrm{Spec}(B)} B_q$ factors through the morphism g. According to (ii), (l_q) is monomorphic; thus g is monomorphic. As a result, $(l_p)_{p \in X}: A \to \Pi_{p \in X} A_p$ is a co-universal monomorphism.

(vii) It follows from (v) and (vi) that $(l_q): B \to \Pi_{q \in \mathrm{Spec}(B)} B_q$ is a regular and co-universal monomorphism and that it factors through the morphism $g: B \to G$. According to [8], Proposition 1.2, g is a regular monomorphism. As a result, $(l_p)_{p \in X}: A \to \Pi_{p \in X} \mathrm{A}_p$ is a co-universal regular monomorphism. ∎

Proposition (3.5.2). *The family $(l_p: A \to A_p)_{p \in \mathrm{Spec}(A)}$ of localizations of an object A is a co-universal co-union of epimorphisms.*

Proof. Let $f: A \to B$ be any morphism and, for any $p \in \mathrm{Spec}(A)$, let $(m_p: B \to M_p, n_p: A_p \to M_p)$ be the pushout of (f, l_p). Let $e: B \to E$ be an epimorphism through which any morphism m_p factors in the form $m_p = h_p e$. Let $q \in \mathrm{Spec}(B)$ and $p(q) = \mathrm{Spec}(f)(q)$, and let $r_q: M_{p(q)} \to B_q$ be the morphism such that $r_q m_{p(q)} = l_q$ and $r_q n_{p(q)} = f_q$. The morphism $h = (r_q h_{p(q)})_{q \in \mathrm{Spec}(B)}: E \to \Pi_{q \in \mathrm{Spec}(B)} B_q$ is such that $he = (l_q)_{q \in \mathrm{Spec}(B)}$. According to Proposition (3.5.1), (l_q) is a regular monomorphism and thus e is an isomorphism. It follows that $(m_p: B \to M_p)_{p \in \mathrm{Spec}(A)}$ is a co-union of epimorphisms. As a result, $(l_p: A \to A_p)_{p \in \mathrm{Spec}(A)}$ is a co-universal co-union. ∎

Theorem (3.5.3). *In a Zariski category, the class of local objects is a regular (thus proper or strong) cogenerating class.*

Proof. According to Proposition (3.5.1) any object is a regular (and thus extremal) subobject of a product of local objects. ∎

3.6 LOCAL PROPERTIES

Proposition (3.6.1). *An object A is reduced (regular) if and only if each of its local object A_p is reduced (simple).*

Proof. Let A be reduced. According to Proposition (3.1.1), A_p is a cofiltered co-intersection of singular quotients of A. According to Proposition (2.3.3), A_p is reduced. Conversely, let each A_p be reduced. According to Proposition (3.5.1), A is a subobject of $\Pi_{p\in \mathrm{Spec}(A)}A_p$. From Definition (2.3.1), it follows that $\Pi_{p\in \mathrm{Spec}(A)}A_p$ is reduced and A is also reduced. Let A be regular. According to Proposition (2.11.5), A_p is regular. According to Proposition (2.12.7), A_p is simple. Conversely, let each A_p be simple. According to Proposition (3.5.1), A is a regular subobject of $\Pi_{p\in \mathrm{Spec}(A)}A_p$. Thus it follows from Proposition (2.11.5) that $\Pi_{p\in \mathrm{Spec}(A)}A_p$ is regular and A is also regular. ■

Definition (3.6.2). *The* localization functor *at $p\in\mathrm{Spec}(A)$ is the pushout functor $L_p: A/A \to A_p/A$ along the morphism $l_p: A \to A_p$.*

Proposition (3.6.3). *For any object A, the family of localization functors $(L_p: A/A \to A_p/A)_{p\in \mathrm{Spec}(A)}$ is conservative, i.e. reflects isomorphisms.*

Proof. Let (B,f), (C,g) be a pair of objects in A/A and $p\in \mathrm{Spec}(A)$. Let $(m_p: A_p \to M_p, h_p: B \to M_p)$ be the pushout of (l_p,f) and let $(n_p\colon A_p \to N_p,\ k_p: C \to N_p)$ be the pushout of (l_p, g). Then $(M_p, m_p) = L_p(B,f)$ and $(N_p, n_p) = L_p(C,g)$. Let $h = (h_p)_{p\in \mathrm{Spec}(A)}: B \to \Pi_{p\in \mathrm{Spec}(A)}M_p$ and $k = (k_p)_{p\in \mathrm{Spec}(A)}$. According to Proposition (3.5.2), the families (h_p), (k_p) are co-unions of epimorphisms. Thus h and k are extremal monomorphisms. For any morphism $u: (B,f) \to (C,g)$, Let $L(u) = \Pi_{p\in \mathrm{Spec}(A)}L_p(u)$. Then the relation $L(u)h = ku$ holds. Let $u, v: (B,f) \rightrightarrows (C,g)$ be a pair of morphisms such that $L_p(u) = L_p(v)$ for any $p\in \mathrm{Spec}(A)$. Then $L(u) = L(v)$; thus $ku = kv$ and hence $u = v$. It follows that the family of functors $(L_p)_{p\in \mathrm{Spec}(A)}$ is faithful. Thus this family reflects monomorphisms and epimorphisms. Moreover, if u is such that $L(u)$ is an isomorphism, then u is an epimorphism and the relation $L(u)h = ku$ implies that u is an extremal monomorphism. Thus u is an isomorphism. As a result, the family $(L_p)_{p\in \mathrm{Spec}(A)}$ is conservative. ■

Corollary (3.6.4). *A morphism $f: A \to B$ is a monomorphism (an epimorphism, a regular epimorphism) if and only if its pushout along each localization of A is a monomorphism (an epimorphism, a regular epimorphism).*

Proposition (3.6.5). *A morphism is flat if and only if its local morphisms are flat.*

Proof. Let $f: A \to B$ be a morphism. Let us assume that f is flat, $p \in \text{Spec}(B)$, and $q = \text{Spec}(f)\ (p)$. The morphisms l_p, l_q are flat. According to Proposition (1.6.2), the morphism $l_p f = f_p l_q$ is flat, as is the morphism f_p. Conversely, let us suppose that each f_p ($p \in \text{Spec}(B)$) is flat. Let $p \in \text{Spec}(B)$ and $q = \text{Spec}(f)\ (p)$. Let us denote by $F: A/\mathbf{A} \to B/\mathbf{A}$, $F_p: A_q/\mathbf{A} \to B_p/\mathbf{A}$, $L_p: B/\mathbf{A} \to B_p/\mathbf{A}$, and $L_q: A/\mathbf{A} \to A_q/\mathbf{A}$ the pushout functors along f, f_p, l_p, and l_q respectively. Since f_p and l_q are flat, the functor F_p and L_q preserve monomorphisms; thus the functors $L_p F = F_p L_q$ also preserves monomorphisms. According to Proposition (3.6.3), the family of functors $(L_p)_{p \in \text{Spec}(B)}$ is conservative and thus is faithful. Therefore it reflects monomorphisms. Consequently, the functor F preserves monomorphisms. As a result, f is flat. ■

Proposition (3.6.6). *The local morphisms of an epimorphism (regular epimorphism) are epimorphisms (regular epimorphisms).*

Proof. Let $f: A \to B$ be a morphism, let $p \in \text{Spec}(B)$, and let $q = \text{Spec}(f)\ (p)$. Then $l_p f = f_p l_q$. If f is an epimorphism, $l_p f$ is an epimorphism and thus f_p is an epimorphism. Let us assume that f is a regular epimorphism. Let $(g: A_q \to C, h: B \to C)$ be the pushout of (l_q, f) and let $u: C \to B_p$ be the morphism such that $uh = l_p$ and $ug = f_p$. Since the morphisms l_p and h are presingular, the morphism u is presingular (Proposition (1.9.5). Let s be a proper congruence on C. Since g is a regular epimorphism, s is of the form $s = g_{*c}(r)$ where r is a proper congruence on A_q. Since f_p is local, $(f_p)_{*c}(r)$ is proper (Proposition (2.12.4)). Thus $u_{*c}(s) = u_{*c}(g_{*c}(r)) = (f_p)_{*c}(r)$ is proper. According to Proposition (2.12.4), the morphism u is local. Consequently, u is an isomorphism and f_p is a regular epimorphism. ■

Definition (3.6.7). A local isomorphism *is a morphism whose local morphisms are isomorphisms.*

Proposition (3.6.8).

(i) *Isomorphisms are local isomorphisms.*
(ii) *Composites of local isomorphisms are local isomorphisms.*
(iii) *Local isomorphisms are flat.*
(iv) *Presingular epimorphisms are local isomorphisms.*

Proof.

(i) and (ii) are obvious.
(iii) Follows from Proposition (3.6.5).
(iv) Let $f: A \to B$ be a presingular epimorphism, $p \in \text{Spec}(B)$, and $q = \text{Spec}(f)\ (p)$. Since 1_p is presingular, $f_p l_q = l_p f$ is presingular. But f_p is

prelocal. According to Theorem (1.9.6), f_p is an isomorphism. As a result f is a local isomorphism. ∎

Proposition (3.6.9). *If $f: A \to B$ is a finitely presentable local isomorphism the map* $\mathrm{Spec}(f): \mathrm{Spec}(B) \to \mathrm{Spec}(A)$ *is open.*

Proof. Let $D(r)$, where $r \in \mathrm{CongCodis}(B)$, be an arbitrary affine open set in $\mathrm{Spec}(B)$. Let $p \in D(r)$ and $q = \mathrm{Spec}(f)(p)$. Let $u: \Delta(r) \to B_p$ be the morphism such that $u\delta(r) = l_p$ and let $n = f_p^{-1}u$. Then $n\delta(r)f = f_p^{-1}u\delta(r)f = f_p^{-1}l_pf = f_p^{-1}f_pl_q = l_q = l_q$. Since $\mathrm{I}_q = \wedge_{q \in D(s)} \delta(s)$ is a filtered colimit, there exists $s \in \mathrm{CongCodis}(A($ such that $q \in D(s)$ and a morphism $g: \Delta(r) \to \Delta(s)$ such that $g\delta(r)f = \delta(s)$. Let us prove that $D(s) \subset \mathrm{Spec}(f)\ (D(r))$. Let $q' \in D(s)$ and $v: \Delta(s) \to A_{q'}$ such that $v\delta(s) = l_{q'}$. According to Theorem (2.13.3), the morphism $vg\delta(r): B \to A_{q'}$ factors in the form $hl_{p'}$ where $p' \in \mathrm{Spec}(B)$ and $h: B_{p'} \to A_{q'}$ is local. According to Proposition (1.9.2), the morphism $l_{p'}$ factors through $\delta(r)$. Thus $p' \in D(r)$ (Proposition (3.1.1)). Then we have $hl_{p'}f = vg\delta(r)f = v\delta(s) = l_{q'}$. According to Theorem (2.13.3), there exists a local morphism $k: A_{q'} \to B_{p'}$ such that $kl_{q'} = l_{p'}f$ and $hk = 1_{A_{q'}}$. It follows that $q' = \mathrm{Spec}(f)(p')$, and so $q' \in \mathrm{Spec}(f)D(r)$. As a result $D(s) \subset \mathrm{Spec}(f)\,(D(r))$ and the map $\mathrm{Spec}(f)$ is open. ∎

3.7 SEMISINGULAR EPIMORPHISMS

Definition (3.7.1). *A* semisingular epimorphism *is a morphism which is the simultaneous codisjunctor of some family of pairs of morphisms* (cf. Definition (1.1.9)).

Proposition (3.7.2). *For any morphism f the following assertions are equivalent.*

(i) *f is a semisingular epimorphism.*

(ii) *f is the simultaneous codisjunctor of the family of all pairs of parallel morphisms (with a finitely presentable domain) which are codisjointed by f.*

(iii) *f is the simultaneous codisjunctor of the family of all congruences (finitely generated congruences) which are codisjointed by f.*

Proof. The proof is analogous to the proof of Proposition (1.8.2). ∎

Notation. For any morphism $f: A \to B$, the set of congruences r on A which are codisjointed by f is denoted by $\mathcal{R}(f)$. According to Proposition (3.7.2), any semisingular epimorphism f is the simultaneous codisjunctor of $\mathcal{R}(f)$.

Proposition (3.7.3). *An epimorphism is semisingular if and only if it is a local isomorphism.*

Proof. Let $f: A \to B$ be an epimorphism.

(i) Let us assume that f is semisingular. Let $p \in \mathrm{Spec}(B)$ and $q = \mathrm{Spec}(f)(p)$. Let $r \in \mathcal{R}(f)$. Then $f_q l_q = l_p f$ codisjoints r, i.e. $(f_p l_q)_{*c}(r) = 1_{B_p \times B_p}$. According to Proposition (2.12.4), $(f_p)_{*c}$ preserves proper congruences or, equivalently, reflects unit congruences. Therefore the relation $(f_p)_{*c}((l_q)_{*c}(r)) = (f_p l_q)_{*c}(r) = 1_{B_p \times B_p}$ implies $(l_q)_{*c}(r) = 1_{A_q \times A_q}$. Thus l_q codisjoints r. The existence of a morphism $g: B \to A_q$ such that $gf = l_q$ follows from the universal property of f. According to Proposition (1.9.5), the morphism g is presingular since the morphism l_q is. The relation $f_p gf = f_p l_q = l_p f$ implies $f_p g = l_p$. According to the essential uniqueness of the presingular factorization (Theorem (1.9.6)), the local morphism f_p is an isomorphism. As a result f is a local isomorphism.

(ii) Let us assume that f is a local isomorphism. Let us consider a local object L with maximal congruence m and a morphism $g: A \to L$ which simultaneously codisjoints $\mathcal{R}(f)$. Let $p = \mathrm{Spec}(g)(m)$. Then g does not codisjoint p and thus neither does f. Therefore $f_{*c}(p)$ is proper. According to Proposition (3.3.7), there exists $q \in \mathrm{Spec}(B)$ such that $\mathrm{Spec}(f)(q) = p$. Since the morphism $f_q: A_p \to B_q$ is an isomorphism, the morphism l_p factors through f in the form $l_p = uf$. According to Theorem (2.13.3), g factors in the form $g = hl_p$. Consequently, g factors through f in the form $g = huf$.

Now let us consider a morphism $v: A \to C$ which simultaneously codisjoints $\mathcal{R}(f)$. Then for any $p \in \mathrm{Spec}(C)$, the morphism $l_p v: A \to C_p$ simultaneously codisjoints $\mathcal{R}(F)$ and therefore, according to the preceding analysis, it factors through f in the form $l_p v = w_p f$. If we denote by l the morphism $(l_p)_{p \in \mathrm{Spec}(C)}: C \to \Pi_{p \in \mathrm{Spec}(C)} C_p$ and by w the morphism $(w_p)_{p \in \mathrm{Spec}(C)}: B \to \Pi_{p \in \mathrm{Spec}(C)} C_p$, then we have $wf = lv$. Because f is epimorphic and l is a regular monomorphism (Proposition (3.5.1)), there is a morphism $u: B \to C$ such that $uf = v$. As a result, f is the simultaneous codisjunctor of $\mathcal{R}(f)$ and thus is a semisingular epimorphism. ■

Proposition (3.7.4).

(i) *Presingular epimorphisms are semisingular.*

(ii) *Semisingular epimorphisms are flat epimorphisms.*

(iii) *Semisingular epimorphisms are co-universal.*

(iv) *The class of semisingular epimorphisms is closed under compositions, co-intersections, and colimits.*

Proof.

(i) Presingular epimorphisms are local isomorphisms (Proposition (3.6.8)) and thus are semisingular (Proposition (3.7.3)).

(ii) Semisingular epimorphisms are local isomorphisms (Proposition (3.7.3)) and thus are flat (proposition (3.6.8)).

(iii) If $f: A \to B$ is a semisingular epimorphism which is a simultaneous codisjunctor of the family $((g_i, h_i) : C_i \rightrightarrows A)_{i \in I}$, the pushout of f along a morphism $u : A \to C$ is the simultaneous codisjunctor of the family $(ug_i, uh_i)_{i \in I}$.

(iv) The closure under compositions follows from Proposition (3.7.3) and the fact that the classes of epimorphisms and local isomorphisms are closed under compositions (Proposition (3.6.8)). The closure under co-intersections follows from the closure under colimits. Let $(f_j : A_j \to B_j)_{j \in \mathbf{J}}$ be a diagram of semisingular epimorphisms whose colimit is $((\alpha_j, \beta_j) : f_j \to f)_{j \in \mathbf{J}}$ with $f : A \to B$. Let $g : A \to C$ be a morphism which simultaneously codisjoints $\mathcal{R}(f)$. For any $j \in \mathbf{J}$, the morphism $g\alpha_j : A_j \to C$ simultaneously codisjoints $\mathcal{R}(f_j)$; thus it factors uniquely through f_j in the form $g\alpha_j = h_j f_j$. We obtain an inductive cone $(h_j)_{j \in \mathbf{J}}$ which defines a morphism $h : B \to C$. Then $hf = g$. As a result, f is the simultaneous codisjunctor of $\mathcal{R}(f)$ and so is semisingular. ∎

Semisingular extensions

A semisingular bimorphism $f : A \to B$ is also called a *semisingular extension of A*. The full subcategory of $A/\mathbf{A}$ whose objects are the semisingular extensions of A is small and is associated to a preorder. Let it be denoted by **ExtSemSing** (A). It is a complete prelattice whose joins are co-intersections of epimorphisms (cf. § 1.7, p. 37).

Theorem (3.7.5). *Any object A has a semisingular hull, i.e. a semisingular extension $\psi_A : A \to G(A)$ which factors uniquely through any semisingular extension of A.*

Proof. Take the top object in **ExtSemSing** (A). ∎

Definition (3.7.6). *An object is* semisingularly closed *if any of its semisingular extensions is an isomorphism. With flat morphisms between them, they form the category* ***SemSingClA***.

Theorem (3.7.7). *To any object is associated a universal semisingularly closed object, i.e.* ***SemSingClA*** *is a reflective subcategory of* $\mathbf{A}_{\mathrm{flat}}$. *The reflection is precisely the semisingular hull.*

Proof. The proof is analogous to the proof of Theorem (1.7.5). ∎

Semisingular quotients

Semisingular epimorphisms $f : A \to B$ represent semisingular quotient objects of A. Let us denote by QuotSemSing (A) the set of semisingular quotients of A. It is a complete lattice whose meets are the co-intersections of quotients of A.

3.8 SUBSETS CLOSED UNDER GENERIZATIONS AND AFFINE SUBSETS

Notation. For any family $\mathcal{R} = (r_i)_{i\in I}$ of congruences on an object A, let us denote by $D(\mathcal{R})$ the subset of $\mathrm{Spec}(A)$ defined by

$$D(\mathcal{R}) = \bigcap_{i\in I} D(r_i) = \{p\in \mathrm{Spec}(A) : \forall i\in I, r_i \not\leq p\}.$$

Proposition (3.8.1) $D(\mathcal{R}) = \{p\in \mathrm{Spec}(A) : l_p$ *simultaneously codisjoints* $\mathcal{R}\}$.

Proof. The proof follows from Proposition 3.1.1. ■

Proposition (3.8.2). *The subsets of* $\mathrm{Spec}(A)$ *of the form* $D(\mathcal{R})$ *are precisely the subsets closed under generizations.*

Proof. The subsets of the form $D(\mathcal{R})$ are precisely intersections of open subsets of $\mathrm{Spec}(A)$. Thus they are the subsets closed under generizations. ■

Proposition (3.8.3). *A morphism* $f: A \to B$ *codisjoints simultaneously a family* $\mathcal{R}$ *of congruences on* A *if and only if the image* $\mathrm{ImSpec}(f)$ *of the map* $\mathrm{Spec}(f)$ *is included in* $D(\mathcal{R})$.

Proof. According to Proposition (3.3.6), f codisjoints $r\in\mathcal{R}$ if and only if the image of $\mathrm{Spec}(f)$ is included in $D(r)$. ■

Proposition (3.8.4). *For any morphism f, the set* $D(\mathcal{R}(f))$ *is precisely the closure of* $\mathrm{ImSpec}(f)$ *under generizations.*

Proof. Let Y be the closure of $\mathrm{ImSpec}(f)$ under generizations. Then Y is the intersection of all open subsets of $\mathrm{Spec}(A)$ containing $\mathrm{ImSpec}(f)$, i.e. $Y = \bigcap\{D(r) : \mathrm{ImSpec}(f) \subset D(r)\}$. According to proposition (3.3.6), $\mathrm{Imspec}(f) \subset D(r) \Leftrightarrow f$ codisjoints $r \Leftrightarrow r\in\mathcal{R}(f)$. Thus $Y = \bigcap_{r\in\mathcal{R}(f)} D(r) = D(\mathcal{R}(f))$. ■

Proposition (3.8.5). *For a flat morphism f,* $D(\mathcal{R}(f)) = \mathrm{ImSpec}(f)$.

Proof. The proof follows from the fact that the image of $\mathrm{Spec}(f)$ is closed under generizations (Proposition (3.3.7) and from Proposition (3.8.4). ■

Definition (3.8.6). *An* affine subset *of* Spec(A) *is a set of the form* $D(\mathcal{R})$ *where* $\mathcal{R}$ *is a simultaneously codisjunctable family of congruences on* A.

According to Proposition (3.8.2), an affine subset of Spec(A) is closed under generizations.

Proposition (3.8.7). *A family* $\mathcal{R}$ *of congruences on* A *is simultaneously codisjunctable if and only if the set* $D(\mathcal{R})$ *is affine.*

Proof. The condition is obviously necessary. Conversely, let us assume that $D(\mathcal{R})$ is affine. There exists a simultaneous codisjunctable family $\mathcal{S}$ of congruences on A such that $D(\mathcal{R}) = D(\mathcal{S})$. According to Proposition (3.8.3), for any morphism $f: A \to B$, f simultaneously codisjoints $\mathcal{R} \Leftrightarrow$ ImSpec$(f) \subset D(\mathcal{R}) = D(\mathcal{S}) \Leftrightarrow f$ simultaneously codisjoints $\mathcal{S}$. Consequently the simultaneous codisjunctor of $\mathcal{S}$ is also a simultaneous codisjunctor of $\mathcal{R}$. As a result, $\mathcal{S}$ is simultaneously codisjunctable. ■

Proposition (3.8.8). *If* $f: A \to B$ *is a semisingular epimorphism which is a simultaneous codisjunctor of a family of congruences* $\mathcal{R}$, *the map* Spec(f) : Spec $(B) \to$ Spec(A) *induces an homeomorphism between* Spec(B) *and its affine image* $D(\mathcal{R})$ *in* Spec(A).

Proof. According to Proposition (3.7.4), f is a flat epimorphism. Thus according to Proposition (3.3.8), Spec(f) induces a homeomorphism between Spec(B) and its image ImSpec(f). According to Proposition (3.8.3), ImSpec$(f) \subset D(\mathcal{R})$. Let $p \in D(\mathcal{R})$. According to Proposition (3.8.1), the morphism l_p simultaneously codisjoints $\mathcal{R}$. Therefore l_p factors through f in the form $l_p = gf$. Then (Spec(f) $(g^{*c}(m_p)) = f^{*c}(g^{*c}(m_p)) = l_p^{*c}(m_p)$ $= p$. Consequently, $p \in$ ImSpec(f). As a result, ImSpec$(f) = D(\mathcal{R})$. ■

Corollary (3.8.9). *Affine sets are compact.*

Proof. In the preceding notation, the affine set $D(\mathcal{R})$ is the image of the compact space Spec(B) (Theorem (3.1.9)) by the continuous map Spec(f). ■

Proposition (3.8.10). *For a pair* $\mathcal{R}$, $\mathcal{S}$ *of families of congruences on* A, *such that* $\mathcal{R}$ *or* $\mathcal{S}$ *is simultaneous codisjunctable,* $D(\mathcal{R}) = D(\mathcal{S}) \Leftrightarrow \mathcal{R}$ *and* $\mathcal{S}$ *have the same simultaneous codisjunctor.*

Proof. The proof follows from Propositions (3.8.3) and (3.8.8). ■

Proposition (3.8.11). *The assignment* $f \mapsto$ ImSpec(f) *establishes an isomorphism between the ordered set* QuotSemSing(A) *of semisingular quotients* *of A and the ordered set* $S_a(\mathrm{Spec}(A))$ *of affine subsets of* $\mathrm{Spec}(A)$.

Proof. Let $f, g \in \mathrm{QuotSemSing}(A)$. Then $f \leqslant g \Leftrightarrow \mathcal{R}(g) \subset \mathcal{R}(f) \Leftrightarrow D(\mathcal{R}(f)) \subset D(\mathcal{R}(g)) \Leftrightarrow \mathrm{ImSpec}(f) \subset \mathrm{ImSpec}(g)$ (Proposition (3.8.8)). It follows that the map $\mathrm{ImSpec} : \mathrm{QuotSemSing}(A) \to S_a(\mathrm{Spec}(A))$ is well defined, injective, order-preserving, and order-reflecting. It is also surjective since any affine subset $D(\mathcal{R})$ of $\mathrm{Spec}(A)$ is identical with $\mathrm{ImSpec}(f)$ where f is the simultaneous codisjunctor of $\mathcal{R}$. ■

Proposition (3.8.12). *Affine open sets of* $\mathrm{Spec}(A)$ *are precisely the open sets of* $\mathrm{Spec}(A)$ *which are affine.*

Proof. Obviously any affine open set is open and affine. If $D(r)$ is an open set which is also an affine set then, according to Proposition (3.8.7), the congruence r is codisjunctable. Thus $D(r)$ is an affine open set. ■

Proposition (3.8.13). *Singular epimorphisms are precisely finitely presentable semisingular epimorphisms.*

Proof. According to Proposition (1.8.5), singular epimorphisms are finitely presentable semisingular epimorphisms. Let $f : A \to B$ be a finitely presentable semisingular epimorphism. According to Proposition (3.7.3), f is a local isomorphism. According to Proposition (3.6.9), $\mathrm{ImSpec}(f)$ is an open set of $\mathrm{Spec}(A)$. Thus there exists $r \in \mathrm{Cong}(A)$ such that $D(r) = \mathrm{ImSpec}(f)$. According to Proposition (3.8.5), $D(r) = D(\mathcal{R}(f))$. According to Proposition (3.8.10), r and $\mathcal{R}(f)$ have the same simultaneous codisjunctor. Then f is the codisjunctor of r. As a result f is a singular epimorphism. ■

Corollary (3.8.14). *An epimorphism is singular if and only if it is a finitely presentable local isomorphism.*

Proof. The proof follows from Proposition (3.7.3). ■

3.9 DIRECT FACTOR CONGRUENCES AND CLOPEN SUBSETS

Theorem (3.9.1). *Direct factor morphisms are precisely the morphisms which are both regular and singular epimorphisms.*

Proof. We know that direct factor morphisms are regular singular epimorphisms (cf. Definition (2.11.1)). Let $f : A \to B$ be a regular singular epimorphism. As a singular epimorphism it is of the form $f = \delta(r) : A \to \Delta(r)$ with $r \in \mathrm{CongCodis}(A)$. According to Proposition (3.3.10), $\mathrm{ImSpec}(f) = D(r)$. As a regular epimorphism f is of the form $f = q_s : A \to A/s$ with

$s \in \mathrm{Cong}(A)$. According to Proposition (3.3.11), $\mathrm{ImSpec}(f) = V(s)$. It follows that $D(r) \cup D(s) = \mathrm{Spec}(A)$ is a disjoint open covering of $\mathrm{Spec}(A)$. Consequently the restriction morphisms $\rho_{D(r)} : A \to \tilde{A}(D(r))$, $\rho_{D(s)} : A \to \tilde{A}(D(s))$ of the structure sheaf $\tilde{A}$ (cf. § 3.4) are the projections of a product $A \simeq \tilde{A}(D(r)) \times \tilde{A}(D(s))$. Therefore $f = \rho_{D(r)}$ is a direct factor morphism. ■

Definition (3.9.2). *A* direct factor congruence *on A is a congruence r on A such that the quotient $q_r : A \to A/r$ of A by r is a direct factor morphism.*

The ordered set of direct factor congruences on A is denoted by $\mathrm{CongDirFac}(A)$.

In order to characterize the direct factor congruences among congruences, we need the following notions.

Modular complement

Let r, s be two congruences on A. The *meet* of r and s is said to be *modular* if it satisfies the modular identity in $\mathrm{Cong}(A) : \forall t \leqslant r, t \overset{c}{\vee} (r \wedge s) = r \wedge (t \overset{c}{\vee} s)$. A *modular complement* of r is a complement r' of r in $\mathrm{Cong}(A)$ such that the meet of r and r' is modular. The congruence r is *modularly complemented* if it has a modular complement. The congruence r is said to be *radically minimal* if it is a minimal element among congruences s with the same radical as r.

Proposition (3.9.3). *For a congruence r on A the following assertions are equivalent.*

(i) *r is a direct factor congruence.*
(ii) *r is modularly complemented in* $\mathrm{Cong}(A)$.
(iii) *r is radically minimal and the set $D(r)$ is closed in* $\mathrm{Spec}(A)$.

Proof.

(i) ⇒ (ii) Let $q_r : A \to A/r$ be the quotient of A by r. Let $q' : A \to Q'$ be such that (q_r, q') is a product. According to Proposition (1.4.2), q' is a regular epimorphism which is the quotient of a congruence r' on A. Because q' is a complement of q in $\mathrm{QuotReg}(A)$, the congruence r' is a complement of r in $\mathrm{Cong}(A)$ (Proposition (1.5.2). Let $s \in \mathrm{Cong}(A)$ such that $s \leqslant r$, let $q_s : A \to A/s$ be the quotient of A by s, let $\bar{q}_r : A/s \to A/r$ be such that $\bar{q}_r q_s = q_r$, and let $\bar{q}_{r'} : A/s \to A/(s \overset{c}{\vee} r')$ be the pushout of $q_{r'}$ along q_s. Because finite products are co-universal in **A**, the pair $(\bar{q}_r, \bar{q}_{r'})$ is a product in **A**. Therefore $s = r \wedge (s \overset{c}{\vee} r')$ and thus $s \overset{c}{\vee} (r \wedge r') = s = r \wedge (s \overset{c}{\vee} r')$. It follows that the meet $r \wedge r' = \Delta_A$ is modular and that r' is a modular complement of r.

(ii) ⇒ (iii) Let r' be a modular complement of r in Cong(A). Let $s \in \text{Cong}(A)$ be such that $s \leqslant r$ and $\text{rad}(s) = \text{rad}(r)$. Then $D(s) = D(r)$ (Theorem (3.1.5)). Thus $D(r' \overset{c}{\vee} s) = D(r') \cup D(s) = D(r') \cup D(r) = D(r' \overset{c}{\vee} r) = D(1_{A \times A})$ (Proposition (3.1.2)). It follows that $\text{rad}(r' \overset{c}{\vee} s) = 1_{A \times A}$ (Theorem (3.1.5)) and $r' \overset{c}{\vee} s = 1_{A \times A}$ (Proposition (2.4.13)). The modular identity for $r \wedge r' = \Delta_A$ gives $s = s \overset{c}{\vee} (r \wedge r') = r \wedge (s \overset{c}{\vee} r') = r \wedge 1_{A \times A} = r$. As a result, r is radically minimal. Moreover, the relations $D(r) \cup D(r') = D(r \overset{c}{\vee} r') = D(1_{A \times A}) = \text{Spec}(A)$ and $D(r) \cap D(r') = D(r \wedge r') = D(\Delta_A) = \varnothing$ imply that $D(r)$ is closed in Spec(A).

(iii) ⇒ (i) Let $U = D(r)$ and $V = V(r) = \text{Spec}(A) \setminus D(r) = \text{Spec}(A) \setminus U$. According to the definition of sheaves, the pair of restriction morphisms $(\rho_U : A \to \tilde{A}(U), \rho_V : A \to \tilde{A}(V))$ is a product. Let $(h : A/r \to B, g : \tilde{A}(U) \to B)$ be the pushout of (q_r, ρ_U). Let us assume that B is not terminal. According to Corollary (2.1.9), B has a simple quotient $k : B \to K$. According to Theorems (2.12.6) and (2.13.3), the morphism $kg\rho_U = khq_r : A \to K$ factors in the form $khq_r = mk_p l_p$ with $p \in \text{Spec}(A)$. Because ρ_U and q_r are regular epimorphisms and m is a monomorphism, there exist two morphisms $n : \tilde{A}(U) \to K(p)$, $t : A/r \to K(p)$ such that $n\rho_U = k_p l_p = tq_r$. Thus $k_p l_p$ coequalizes r. According to Proposition (3.1.1), $p \in V(r) = V$. Then $k_p l_p$ factors through ρ_V as well as through ρ_U. This is impossible. Consequently B is terminal and the pushout of ρ_U along q_r is null. Because finite products are co-universal, the pushout of ρ_V along q_r is isomorphic, i.e. q_r factors though ρ_V. Let $s \in \text{Cong}(A)$ be the kernel pair of ρ_V. Then ρ_V is the quotient of A by s. Thence $s \leqslant r$ and $D(s) \subset D(r) = U$. Let $p \in U$. The morphism $k_p l_p : A \to K(p)$ factors through ρ_U. Thus it does not factor through ρ_V and hence it does not coequalize s. Consequently, it codisjoints s (Proposition (2.6.2)). Thus $p \in D(s)$ (Proposition (3.1.1)). It follows that $U \subset D(s)$ and so $D(r) = U = D(s)$. Because r is radically minimal, $r = s$. As a result, r is a direct factor congruence. ■

Theorem (3.9.4). *The ordered set* CongDirFac(A) *of direct factor congruences on A is isomorphic to the ordered set* $\mathcal{B}(\text{Spec}(A))$ *of clopen sets of* Spec(A).

Proof. The order-preserving map $D : \text{Cong}(A) \to \Omega(\text{Spec}(A))$ induces an order-preserving map $D_f : \text{CongDirFac}(A) \to \mathcal{B}(\text{Spec}(A))$ (Proposition (3.9.3)). Let $r_1, r_2 \in \text{CongDirFac}(A)$ such that $D_f(r_1) \subset D_f(r_2)$. Then $r_1 \wedge r_2 \leqslant r_1$ and $D(r_1 \wedge r_2) = D(r_1) \cap D(r_2) = D(r_1)$. Because r_1 is radically minimal (Proposition (3.9.3)), $r_1 \wedge r_2 = r_1$, i.e. $r_1 \leqslant r_2$. It follows that D_f is injective and induces an isomorphism of ordered sets between CongDirFac(A) and its image. Let U be a clopen set of Spec(A) and $V = \text{Spec}(A) \setminus U$. Then $(\rho_U : A \to \tilde{A}(U), \rho_V : A \to \tilde{A}(V))$ is a product. Let $r, s \in \text{Cong}(A)$ be the kernel pairs of ρ_U and ρ_V respectively. Then r and s are direct factor congruences on A and are complementary in Cong(A). If

$p \in U$, then $k_p l_p$ factors through ρ_U and thus $k_p l_p$ coequalizes r, and according to Proposition (3.1.1) $p \in V(r) = D(s)$. It follows that $U \subset D(s)$. Similarly, we prove that $V \subset D(r)$. Consequently, $U = D(r)$ where $r \in \mathrm{CongDirFac}(A)$. As a result, D_f is an isomorphism. ■

Corollary (3.9.5). *The set* CongDirFac(A) *of direct factor congruences on* A *is a boolean algebra.*

The direct factor congruence functor

Let $f: A \to B$ be a morphism in **A**, let r be a direct factor congruence on A, and let $s = f_{*c}(r)$ be its direct image by f. The quotient $q_s: B \to B/s$ of B by s is the pushout along f of the quotient $q_r: A \to A/r$ of A by r. Because finite products are co-universal, q_s is a direct factor morphism. Thus s is a direct factor congruence. Therefore the map f_{*c} induces a map CongDirFac$(A) \to$ CongDirFac(B) which preserves joins and complements, and thus is a morphism of boolean algebras. If **Bool** denotes the category of boolean algebras, the *direct factor congruence functor* CongDirFac : **A** $\to$ **Bool** assigns the set CongDirFac(A) of direct factor congruences on A to an object A, and assigns the restriction of the map f_{*c} to a morphism f. We denote by Cl: **TopSpec**$^{\mathrm{op}} \to$ **Bool** the clopen *functor* which assigns the boolean algebra of clopen sets of X to a spectral space X, and assigns the restriction of the inverse image map f^{-1} to a continuous map f.

Proposition (3.9.6). *The direct factor congruence functor* CongDirFac : **A** $\to$ ***Bool*** *is isomorphic to the functor* Cl. Spec^{op} *which is the composite of the opposite of the spectrum functor* Spec : **A**$^{\mathrm{op}} \to$ ***TopSpec*** *by the clopen functor* Cl : ***TopSpec***$^{\mathrm{op}} \to$ ***Bool***.

Proof. The proof follows from Theorem (3.9.4). ■

Proposition (3.9.7). *The direct factor congruence functor* CongDirFac : **A** $\to$ ***Bool*** *preserves filtered colimits.*

Proof.

(i) Let us assume that any object of **A** is reduced, i.e. **A** = **RedA**. Let $(\alpha_i: A_i \to A)_{i \in \mathbf{I}}$ be a filtered colimit in **A**. Let $r \in \mathrm{CongDirFac}(A)$ and let r' be its modular complement (Proposition (3.9.3)). Then r and r' are finitely generated congruences on A. Because the finitely generated congruence functor Cong_0 : **A** $\to \vee$-**SemLat** preserves filtered colimits (Proposition (1.5.5)), there exist $i \in \mathbf{I}$ and r_i, $s_i \in \mathrm{Cong}_0(A_i)$ such that $\alpha_{i*c}(r_i) = r$ and $\alpha_{i*c}(s_i) = r'$. Then $\alpha_{i*c}(r_i \wedge s_i) = \alpha_{i*c}(r_i) \wedge \alpha_{i*c}(s_i) = r \wedge r' = \Delta_A$ because $\alpha_{i*c} = \alpha_{i*cr}$ is a morphism of lattices (cf. § 2.4, p. 57). Similarly,

$\alpha_{i*c}(r_i \overset{c}{\vee} s_i) = \alpha_{i*c}(r_i) \overset{c}{\vee} \alpha_{i*c}(s_i) = r \overset{c}{\vee} r' = 1_{A\times A}$. Then there exists a morphism $u : i \to j$ in **I** such that $A_u(r_i \wedge s_i) = \Delta_{A_j}$ and $A_u(r_i \vee s_i) = 1_{A_j \times A_j}$. Let $r_j = A_{u*c}(r_i)$ and $r_j' = A_{u*c}(s_i)$. Then r_j and r_j' are complementary elements in $\mathrm{Cong}_o(A_j) = \mathrm{CongRad}_o(A_j)$. Because $\mathrm{CongRad}_o(A_j)$ is a bounded distributive lattice (Theorem (2.4.10)), r_j' is a modular complement of r_j. According to Proposition (3.9.3), r_j is a direct factor congruence on A_j and $a_{j*c}(r_j) = r$. It follows easily that the functor CongDirFac : **A** → **Bool** preserves filtered colimits.

(ii) Let us consider any Zariski category **A**. Then the category **RedA** of reduced objects of **A** is also a Zariski category (Theorem (12.3.2) below). Let $\mathrm{Spec}_{\mathbf{A}} : \mathbf{A}^{\mathrm{op}} \to$ **TopSpec** and $\mathrm{Spec}_{\mathbf{RedA}} : (\mathbf{RedA})^{\mathrm{op}} \to$ **TopSpec** be the prime spectrum functors associated to **A** and **RedA** respectively. If R : **A** → **RedA** is the reflector of the inclusion functor **RedA** → **A**, the two functors $\mathrm{Spec}_{\mathbf{RedA}}.R^{\mathrm{op}}$ and $\mathrm{Spec}_{\mathbf{A}}$ are isomorphic (Proposition (3.3.3)). Thence the two functors $\mathrm{Cl.Spec}^{\mathrm{op}}_{\mathbf{RedA}}.R$ and $\mathrm{Cl.Spec}^{\mathrm{op}}_{\mathbf{A}} : \mathbf{A} \to$ **Bool** are isomorphic. According to Proposition (3.9.6), the functor $\mathrm{Cl.Spec}^{\mathrm{p}}{}_{\mathbf{RedA}}$ is isomorphic to the direct factor congruence functor associated to **RedA**. It preserves filtered colimits by (i). As R also preserves filtered colimits, the functor $\mathrm{Cl.Spec}^{\mathrm{op}}_{\mathbf{RedA}}.\ R$ preserves filtered colimits. Thus $\mathrm{Cl.Spec}^{\mathrm{op}}_{\mathbf{A}}$ preserves filtered colimits. But according to Proposition (3.9.6), it is isomorphic to the functor CongDirFac : **A** → **Bool**. ■

Corollary (3.9.8). *The square $Z^2 = Z \times Z$ of the initial object Z of A is a finitely presentable object.*

Proof. The proof follows from [9], § 2, or [10], § 1.3, by noticing that the fact that Z^2 is finitely presentable is not used to prove that it is a coclassifier object for direct factors, and that the functor Δ : **A** → **Bool** is isomorphic to the functor CongDirFac. ■

Proposition (3.9.9). *The prime spectrum functor* $\mathrm{Spec} : A^{\mathrm{op}} \to$ ***TopSpec*** *preserves finite coproducts.*

Proof. Let $(q_1 : A \times B \to A, q_2 : A \times B \to B)$ be the product of a pair of objects A, B. According to Proposition (1.4.2), q_1 and q_2 are regular epimorphisms which are quotients of $A \times B$ by r_1 and r_2 respectively. The relations $r_1 \wedge r_2 = \Delta_{A\times B}$ and $r_1 \vee r_2 = 1_{A\times B, A\times B}$ imply the relation $V(r_1) \cup V(r_2) = \mathrm{Spec}(A\times B)$ and $V(r_1) \cap V(r_2) = \varnothing$. Consequently, $V(r_1)$ and $V(r_2)$ and complementary clopen subsets of $\mathrm{Spec}(A\times B)$, so that $\mathrm{Spec}(A\times B) = V(r_1) \amalg V(r_2)$. According to Proposition (3.3.11), the map $\mathrm{Spec}(q_1)$ (or $\mathrm{Spec}(q_2)$) induces a homeomorphism between $\mathrm{Spec}(A)$ (or $\mathrm{Spec}(B)$) and $V(r_1)$ (or $V(r_2)$). As a result, $\mathrm{Spec}(A\times B) \simeq \mathrm{Spec}(A) \amalg \mathrm{Spec}(B)$. Moreover, $\mathrm{Spec}(1) = \varnothing$. ■

3.10 INDECOMPOSABLE OBJECTS

According to Theorem (3.10.1) below, all the results of [9] or [10], Chapter I, apply to Zariski categories.

Theorem (3.10.1). *A Zariski category is a locally indecomposable category* [9, 10].

Proof. The proof follows from the definition of locally indecomposable categories [9] and from Propositions (1.4.1) and (1.4.3) and Corollary (3.9.8). ■

Definition (3.10.2). *An object of* **A** *is* indecomposable *if it has precisely two direct factors* [9, 10].

Proposition (3.10.3). *The following assertions for an object A are equivalent.*

(i) *A is indecomposable.*

(ii) *There are precisely two direct factor congruences on A.*

(iii) Spec(A) *is non-empty and connected.*

Proof.

(i) ⇔ (ii) follows from Proposition (1.5.2) and Definition (3.9.2).

(ii) ⇔ (iii) According to Theorem (3.9.4) we have the equivalences: A has precisely two direct factor congruences ⇔ Spec(A) has precisely two clopen sets ⇔ Spec(A) is non-empty and connected. ■

Definition (3.10.4). *An* indecomposable quotient of A *is a regular quotient object of A whose codomain is an indecomposable object.* [9, 10].

Definition (3.10.5). *A* congruence *r on A is* indecomposable *if the quotient of A* by *r is indecomposable.*

Proposition (3.10.6). *For a congruence r on A, the following assertions are equivalent.*

(i) *r is indecomposable.*

(ii) *The subset $V(r)$ of* Spec(A) *is non-empty and connected.*

Proof. The prime spectrum of the quotient A/r is homeomorphic to $V(r)$ (Proposition (3.3.11)). Then r is indecomposable ⇔ A/r is indecomposable ⇔ Spec(A/r) is non-empty and connected (Proposition (3.10.3)) ⇔ $V(r)$ is non-empty and connected. ■

Definition (3.10.7). *An* indecomposable component *of an object A is a maximal indecomposable quotient of A* [9, 10].

Proposition (3.10.8). *For a congruence r on A, the following assertions are equivalent.*

(i) $q_r : A \to A/r$ *is an indecomposable component of* A.

(ii) r *is radically minimal and* $V(r)$ *is a connected component of* Spec(A).

Proof.

(i) ⇒ (ii) As A/r is indecomposable, r is indecomposable. Thus $V(r)$ is connected and non-empty (Proposition (3.10.6)). Let F be the connected component of $V(r)$ in Spec(A). As it is a closed subset, it is of the form $F = V(s)$ with $s \in \text{Cong}(A)$. According to Proposition (3.10.6), s is indecomposable. The congruence $r \wedge s$ is such that $V(r \wedge s) = V(r) \cup V(s) = F$ is connected and non-empty. According to Proposition (3.10.6), $r \wedge s$ is indecomposable. Thus the quotient $q_{r\wedge s} : A \to A/r \wedge s$ is indecomposable and such that $q_r \leqslant q_{r\wedge s}$ in QuotReg(A). Because q_r is maximal, $q_r = q_{r\wedge s}$ and thus $r = r \wedge s$, hence $r \leqslant s$ and $V(s) \subset V(r)$. It follows that $F = V(s) = V(r)$. As a result, $V(r)$ is a connected component of Spec(A). Moreover, let $t \in \text{Cong}(A)$ such that $t \leqslant r$ and $V(t) = V(r)$. According to Proposition (3.10.6), t is indecomposable. The relation $q_r \leqslant q_t$ implies $q_r = q_t$ and thus $r = t$. It follows that r is radically minimal.

(ii) ⇒ (i) According to Proposition (3.10.6), q_r is indecomposable. Let $t \leqslant r$ such that q_t is indecomposable. Then $V(t)$ is connected. The relation $V(r) \subset V(t)$ implies the equality $V(r) = V(t)$. Because r is radically minimal, $r = t$. Therefore q_r is an indecomposable component of A. ∎

Proposition (3.10.9). *The set of indecomposable components of an object A is in one-to-one correspondence with the set of connected components of* Spec(A).

Proof. Let us denote by V the map which assigns to an indecomposable component $q_r : A \to A/r$ of A the connected component $V(r)$ of Spec(A) (Proposition (3.10.8)). Let q_s be another indecomposable component of A such that $V(s) = V(r)$. The congruence $r \wedge s$ is such that $V(r \wedge s) = V(r) \cup V(s) = V(r)$. As r is radically minimal, $r \wedge s = r$, i.e. $r \leqslant s$. Similarly, $s \leqslant r$. Thence $r = s$. Therefore the map V is injective. Let F be a connected component of Spec(A) and let $\mathfrak{F}$ be the set of clopen sets of Spec(A) containing F. It is an ultrafilter of $\mathcal{B}(\text{Spec}(A))$. Because Spec$(A)$ is a compact space (Theorem (3.1.9)), $F = \bigcap_{U \in \mathfrak{F}} U$. For any $U \in \mathfrak{F}$, let $q_U : A \to \tilde{A}(U)$ be the restriction morphism of $\tilde{A}$ on U and let r_U be the kernel pair of q_U. Let $r = \overset{c}{\vee}_{U \in \mathfrak{F}} r_U$. According to [10], Proposition 1.6.1, $q_r = \wedge_{U \in \mathfrak{F}} q_U$ is an indecomposable component of A. Moreover,

$$V(r) = V(\overset{c}{\vee}_{U \in \mathfrak{F}} r_U) = \bigcap_{U \in \mathfrak{F}} V(r_U) = \bigcap_{U \in \mathfrak{F}} U = F.$$

As a result, the map V is bijective. ∎

4
SCHEMES

The localization process associates to any object a sheaf of local objects on its prime spectrum, while the globalization process rebuilds any object from the continuous family of its localized objects. By means of these processes, any object turns into a geometrical object: its affine scheme. By making these processes functorial, we obtain an equivalence between the category of affine schemes on a Zariski category **A** and the dual of **A**. Numerous properties of objects and morphisms in **A** have nice geometrical interpretations in the associated category of affine schemes. For example, a morphism in **A** is a regular epimorphism (or singular epimorphism etc.) if and only if its affine spectrum is a closed immersion (or open immersion etc.). Schemes on a Zariski category are built up by the classical method of gluing affine schemes together using modelled spaces and locally modelled spaces instead of ringed spaces and locally ringed spaces. Numerous classical properties and constructions of schemes are valid, finite limits and arbitrary coproducts of schemes exist, and the general gluing construction can be performed. Moreover, separated schemes are disjunctable; hence all disjunctors exist in the category of separated schemes and they are open immersions. In particular, affine schemes are disjunctable, and the open immersions into an affine scheme are precisely its singular subobjects.

4.1 MODELLED SPACES AND LOCALLY MODELLED SPACES

Definition (4.1.1)

(i) *A* modelled space on *A is a topological space X equipped with a sheaf* $O_X : \Omega(X)^{op} \to A$ *based on X with values in A called the structure sheaf on X.*

(ii) *A* morphism of modelled spaces *on A, from a modelled space X to a modelled space Y, is a continuous map* $f: X \to Y$ *equipped with a natural transformation* $f^{\#} : O_Y \to O_X(f^{-1})^{op}$, *where* $f^{-1} : \Omega(Y) \to \Omega(X)$ *denotes the functor which assigns* $f^{-1}(U)$ *to U.*

The modelled spaces on **A** and their morphisms constitute the *category* **SpModA** *of modelled spaces on* **A**, where the composite of two morphisms $f: X \to Y$ and $g: Y \to T$ is defined as the continuous map $g \circ f: X \to T$ equipped with the natural transformation $(g \circ f)^{\#} = (f^{\#}(g^{-1})^{op})g^{\#}$.

The *forgetful functor* **SpModA** → **Esp** assigns the base topological space

X to a modelled space X and the continuous map $f: X \to Y$ to a morphism $f: X \to Y$.

Proposition (4.1.2). *The forgetful functor SpModA → Esp is a fibration and a cofibration* [18].

Proof.

(i) Let X be a topological space, let Y be a modelled space on **A**, and let $f: X \to Y$ be a continuous map. Let $F: \Omega^{op}(X) \to \mathbf{A}$ be the left Kan extension of $O_Y: \Omega^{op}(Y) \to \mathbf{A}$ along the functor $(f^{-1})^{op}: \Omega(Y)^{op} \to \Omega(X)^{op}$ ([31], p. 232), and let $\psi: O_Y \to F(f^{-1})^{op}$ be the universal morphism. Let $O_X: \Omega^{op}(X) \to \mathbf{A}$ be the universal sheaf based on X associated to the presheaf F, and let $\alpha: F \to O_X$ be the universal morphism. Then O_X makes X a space modelled on **A** and the natural transformation $f^{\#} = (\alpha(f^{-1})^{op})\psi: O_Y \to O_X(f^{-1})^{op}$ makes $f: X \to Y$ a morphism of modelled spaces. Using the universal properties of the left Kan extension and the universal sheaf, it is easy to prove that this morphism $f: X \to Y$ is an inverse image of the modelled space Y along the continuous map $f: X \to Y$.

(ii) Let X be a modelled space on **A**, let Y be a topological space, and let $f: X \to Y$ be a continuous map. Equipped with the sheaf $O_Y = O_X(f^{-1})^{op}: \Omega(Y)^{op} \to \mathbf{A}$, Y is modelled on **A**. The continuous map $f: X \to Y$ is a morphism of modelled spaces with $f^{\#} = 1_{O_Y}$. It can immediately be seen that $f: X \to Y$ is the direct image of the modelled space X along the continuous map f. ■

Notation. Let $f: X \to Y$ be a continuous map. The inverse image along f of a modelled space (Y, O_Y) is denoted by $(X, f^*(O_Y))$, while the direct image along f of a modelled space (X, O_X) is denoted by $(Y, f_*(O_X))$. Any morphism $(f, f^{\#}): (X, O_X) \to (Y, O_Y)$ of modelled spaces defines canonically a morphism $(1_Y, f^{\#}): (Y, f_*(O_X)) \to (Y, O_Y)$ as well as a morphism $(X, O_X) \to (X, f^*(O_Y))$ denoted by $(1_X, f_{\#})$. For any $x \in X$, the stalk morphism of f at x is $f^{\#}_x = (f_{\#})_x: O_{Y, f(x)} \to O_{X,x}$ which is the stalk of the morphism of sheaves $f_{\#}$. The morphism $(f, f^{\#})$ is cartesian with respect to the fibration if and only if $(1_X, f_{\#}): (X, O_X) \to (X, f^*(O_Y))$ is an isomorphism, i.e. if $f_{\#}: f^*(O_Y) \to O_X$ is an isomorphism of sheaves on X.

Proposition (4.1.3). *The category* **SpModA** *of modelled spaces on* **A** *is complete and cocomplete.*

Proof. The category **Esp** is complete and cocomplete, and for any topological space X the fibre of the forgetful functor **SpModA** → **Esp** at X is complete and cocomplete as it is the opposite of the category **Sh**$[X, \mathbf{A}]$ of sheaves based on X with values in **A** ([21], II, § 6). Then the result follows from the fact that the forgetful functor **SpModA** $\to$ **Esp** is a bifibration. ■

The *singleton functor* $J: \mathbf{A}^{\mathrm{op}} \to$ **SpModA** assigns to an object A the singleton modelled space $J(A)$ whose stalk is A, and to a morphism $f: A \to B$ the unique continuous map $J(f) : J(B) \to J(A)$ equipped with the stalk morphism f.

The *global section functor* Γ: **SpModA** $\to \mathbf{A}^{\mathrm{op}}$ assigns the object of global sections $\Gamma(X) = 0_X(X)$ of the sheaf O_X to a modelled space X, and the morphism $\Gamma(f) = f^{\#}_Y : O_Y(Y) \to O_X(X)$ induced by $f^{\#}$ to a morphism of modelled spaces $f: X \to Y$.

Proposition (4.1.4). *The singleton functor* $J: A^{\mathrm{op}} \to$ *SpModA is full and faithful and right-adjoint to the global section functor* Γ : *SpModA* $\to A^{\mathrm{op}}$.

Proof. It is obvious that, for any modelled space X on **A** and any object A in **A**, morphisms $X \to JA$ are in one-to-one natural correspondence with morphisms $A \to O_X(X) = \Gamma(X)$. ■

Definition (4.1.5).

(i) *A* locally modelled space on *A is a modelled space on A whose stalks are local objects of A.*

(ii) *A* morphism of locally modelled spaces on *A is a morphism* $f: X \to Y$ *of modelled spaces on A such that the stalk morphism* $f^{\#}_x : O_{Y, f(x)} \to O_{X,x}$ *at each* $x \in X$ *is a local morphism of A.*

Locally modelled spaces on **A** and their morphisms constitute the *category* **SpModLocA** *of locally modelled spaces on* **A**, as a subcategory of **SpModA**.

The prime spectrum functor Σ : $\mathbf{A}^{\mathrm{op}} \to$ SpModLocA

For any object A of **A**, the prime spectrum $\mathrm{Spec}(A)$ equipped with its structure sheaf $\tilde{A}$ is a locally modelled space on **A** (Definition (3.4.3) and Theorem (3.4.4)) denoted by $\Sigma(A)$. Let $f: A \to B$ be a morphism in **A**. Let us consider the continuous map $\mathrm{Spec}(f) : \mathrm{Spec}(B) \to \mathrm{Spec}(A)$ (cf. § 3.1, p. 86). Let $s \in \mathrm{CongRadCodis}_o(A)$ and $t = f_{*\mathrm{cr}}(s)$. Then $(\mathrm{Spec}(f)^{-1})(D(s)) = D(t)$. Let $\delta(s) : A \to \tilde{A}(D(s))$ and $\delta(t) : B \to \tilde{B}(D(t))$ be the codisjunctors of s and t respectively. Then $\delta(t)$ is the pushout of $\delta(s)$ along f (Propositions (1.1.5) and (2.4.7)). Let us denote by $\tilde{f}_{D(s)} : \tilde{A}(D(s)) \to \tilde{B}(D(t))$ the morphism such that $\tilde{f}_{D(s)}\delta(s) = \delta(t)f$. The family of morphisms $(\tilde{f}_{D(s)})$, where s runs over $\mathrm{CongRadCodis}_o(A)$, defines uniquely a natural transformation $\tilde{f}: \tilde{A} \to \tilde{B}((\mathrm{Spec}(f)^{-1})^{\mathrm{op}})$ which gives the continuous map $\mathrm{Spec}(f)$ the structure of a morphism of modelled spaces on **A** denoted by

$\Sigma(f):\Sigma(B)\to\Sigma(A)$. Let $p\in\mathrm{Spec}(B)$ and $q=\mathrm{Spec}(f)(p)$. The stalk morphism $\tilde{f}_p:A_q\to B_p$ of $\tilde{f}$ at p satisfies the relation $\tilde{f}_p l_q = l_p f$. According to Definition (2.13.4), $\tilde{f}_p$ is precisely the local morphism of f at p. Therefore it is a local morphism. As a result, $\Sigma(f):\Sigma(B)\to\Sigma(A)$ is a morphism of locally modelled spaces on **A**. Therefore we obtain the functor $\Sigma:\mathbf{A}^{\mathrm{op}}\to\mathbf{SpModLocA}$ which is called the *prime spectrum functor*. This functor is also denoted by $\mathrm{Spec}:\mathbf{A}^{\mathrm{op}}\to\mathbf{SpModLocA}$.

Theorem (4.1.6). *The prime spectrum functor Σ: $A^{\mathrm{op}}\to$ SpModLocA is full and faithful and right-adjoint to the global section functor Γ: SpModLocA $\to A^{\mathrm{op}}$.*

Proof. Let us prove that $1_A:A\to\Gamma(\Sigma(A))$ is a universal morphism from A to the functor Γ^{op}.

(i) Let X be a locally modelled space with its sheaf denoted by O. For any open set U of X denote the restriction morphism by $\rho_U^X:O(X)\to O(U)$, and for any $x\in U$ denote the restriction morphism at x by $\rho_x^U:O(U)\to O_x$. Let $h:A\to\Gamma^{\mathrm{op}}(X)=O(X)$ be a morphism in **A**. According to Theorem (2.13.3), the morphism $\rho_x^X h:A\to O_x$ factors uniquely in the form $\rho_x^X h=\psi_x l_{f(x)}$, where $f(x)\in\mathrm{Spec}(A)$, $l_{f(x)}:A\to A_{f(x)}$ is the localization of A at $f(x)$, and $\psi_x:A_{f(x)}\to O_x$ is a local morphism. By assigning $f(x)$ to any $x\in X$, we define a map $f:X\to\mathrm{Spec}(A)$.

(ii) We now prove that f is continuous. Let $D(r)$ with $r\in\mathrm{CongRadCodis}_o(A)$ be an affine open set of $\mathrm{Spec}(A)$. According to the construction of the structure sheaf $\tilde{A}$, the codisjunctor of r is the restriction morphism $\rho_{D(r)}:A\to\tilde{A}(D(r))$. Let $y\in X$. Then $f(y)\in D(r)$ if and only if $l_{f(y)}$ codisjoints r (Proposition (3.1.1)). Because ψ_y is local, $l_{f(y)}$ codisjoints r if and only if $\psi_y l_{f(y)}=\rho_y^X h$ codisjoints r, or equivalently if and only if $\rho_y^X h$ factors through $\rho_{D(r)}$. Let $x\in f^{-1}(D(r))$, i.e. $f(x)\in D(r)$. The morphism $\rho_x^X:O(X)\to O_x$ is the filtered colimit of the family of restriction morphisms $\rho_U^X:O(X)\to O(U)$ where U runs over open neighbourhoods of x. Because the morphism $\rho_{D(r)}$ is finitely presentable (Proposition (1.8.5)), there exist an open neighbourhood U of x and a morphism $h_U:\tilde{A}(D(r))\to O(U)$ such that $\rho_U^X h=h_U\rho_{D(r)}$. For any $y\in U$, the restriction morphism $\rho_y^X:O(X)\to O_y$ factors through ρ_U^X. Thus the morphism ρ_y^Xh factors through $\rho_{D(r)}$ and consequently $f(y)\in D(r)$, i.e. $y\in f^{-1}(D(r))$. It follows that $U\subset f^{-1}(D(r))$. As a result, f is continuous.

(iii) Let $D(r)\in\Omega_a(\mathrm{Spec}(A))$ with $r\in\mathrm{CongRadCodis}_o(A)$. Let $U=f^{-1}(D(r))$. For any $x\in U$, let us denote the restriction morphism of $\tilde{A}$ at $f(x)$ by $\rho_{f(x)}^{D(r)}:\tilde{A}(D(r))\to A_{f(x)}$, and the morphism $(\psi_x\rho_{f(x)}^{D(r)})_{x\in U}$: $\tilde{A}(D(r))\to\Pi_{x\in U}O_x$ by h_r. According to the proof of Proposition (3.5.1), $(\rho_x^U)_{x\in U}:O(U)\to\Pi_{x\in U}O_x$ is a regular monomorphism. Because $\rho_{D(r)}$ is epimorphic, the relation $h_r\rho_{D(r)}=(\rho_x^U)_{x\in U}\rho_U^X h$ implies the existence of a morphism $\psi_{D(r)}: \tilde{A}(D(r)) \to O(U)$ such that $\psi_{D(r)}\rho_{D(r)} = \rho_U^X h$. This family of morphisms $(\psi_{D(r)})$, where $D(r)$ runs over the set $\Omega_a(\mathrm{Spec}(A))$, defines uniquely a natural transformation $\psi: \tilde{A} \to O(f^{-1})^{\mathrm{op}}$. In this way we obtain a morphism of modelled spaces $f: X \to \Sigma(A)$ with $f^\# = \psi$. For any $x \in X$, the stalk morphism $f_x^\#: A_{f(x)} \to O_x$ satisfies $f_x^\# l_{f(x)} = \rho_x h$. As $l_{f(x)}$ is epimorphic, $f_x^\# = h_x$ and thus $f_x^\#$ is local. As a result, we obtain a morphism of locally modelled spaces $f: X \to \Sigma(A)$ such that $\Gamma^{\mathrm{op}}(f) 1_A = h$.

(iv) Let us show that any morphism of locally modelled spaces $g: X \to \Sigma(A)$ such that $\Gamma^{\mathrm{op}}(g) 1_A = h$ must be equal to f. Let $x \in X$. The relation $g_x^\# l_{g(x)} = \rho_x^X h$ with g_x^* local implies that $g(x) = f(x)$ (Theorem (2.13.3)). The relation $g_{D(r)}^\# \rho_{D(r)} = \rho_U^X h$ with $U = g^{-1}(D(r))$ and $\rho_{D(r)}$ epimorphic implies $g_{D(r)}^\# = f_{D(r)}^\#$. Therefore $g = f$.

(v) As a result, the functor $\Gamma^{\mathrm{op}}: (\mathbf{SpModLocA})^{\mathrm{op}} \to \mathbf{A}$ has a left-adjoint which is full and faithful ([31], p. 88, Theorem 1). As the functor $\Sigma^{\mathrm{op}}: \mathbf{A} \to (\mathbf{SpModLocA})^{\mathrm{op}}$ satisfies $\Gamma^{\mathrm{op}}\Sigma^{\mathrm{op}} = 1_{\mathbf{A}}$, this left-adjoint is precisely Σ^{op}. It follows that Σ is full and faithful and right-adjoint to Γ. ■

Notation. Let X be a locally modelled space. For any point x of X, the residue simple object of the local object $O_{X,x}$ is denoted by $K_X(x)$ or $K(x)$ and is called the *residue simple object of X at x*. Let $f: X \to Y$ be a morphism of locally modelled spaces. For any point x of X, the morphism $K_Y(f(x)) \to K_X(x)$ canonically associated to the local morphism $f_x^\#: O_{Y,f(x)} \to O_{X,x}$ is denoted by f^x and is called the *residue morphism of f at x*.

4.2 PRIME SPECTRUM OF A MODELLED SPACE

Notation. Let X be a modelled space on $\mathbf{A}$. For any pair of open sets U, U' of X such that $U \subset U'$ the restriction morphism of O_X is denoted by $\rho_U^{U'}: O_X(U') \to O_X(U)$, and for any element x in U the canonical morphism $O_X(U) \to O_{X,x}$ is denoted by ρ_x^U. Let us define the set $\mathrm{Spec}(X)$ and, for any pair (U, r) of an open set U of X and a codisjunctable radical congruence r on $O_X(U)$, let us define the subset $D(U, r)$ as follows:

$$\mathrm{Spec}(X) = \amalg_{x \in X} \mathrm{Spec}(O_{X,x})$$

$$D(U, r) = \{(x, p) \in \mathrm{Spec}(X) : x \in U \quad \text{and} \quad r \nleq \mathrm{Spec}(\rho_x^U)(p)\}$$

According to the notation on p. 81,

$$D(U, r) = \{(x, p) \in \mathrm{Spec}(X) : x \in U \quad \text{and} \quad \mathrm{Spec}(\rho_x^U)(p) \in D(r)\}.$$

According to Proposition (3.1.1),

$$D(U, r) = \{(x, p) \in \mathrm{Spec}(X) : x \in U \quad \text{and} \quad l_p \rho_x^U \text{ codisjoints } r\}.$$

Proposition (4.2.1). *The set of subsets of* Spec(X) *of the form* $D(U, r)$ *is closed under finite intersections.*

Proof. We have

$$D(X, 1_{O_X(X) \times O_X(X)}) = \text{Spec}(X).$$

Let $D(U, r), D(V, s)$ be a pair of subsets of Spec(X). Then $W = U \cap V$ is an open set of X and $t = (\rho^U_W)_{*\text{cr}}(r) \wedge (\rho^V_W)_{*\text{cr}}(s)$ is a codisjunctable radical congruence on $O_X(W)$ such that

$$D(W, t) = \{(x, p) : x \in W \text{ and } \text{Spec}(\rho^W_x)(p) \in D((\rho^U_W)_{*\text{cr}}(r)) \cap D((\rho^V_W)_{*\text{cr}}(s))\} = \{(x, p) : x \in W \text{ and } \text{Spec}(\rho^W_x)(p) \in D((\rho^U_W)_{*\text{cr}}(r)) \text{ and } \text{Spec}(\rho^W_x)(p) \in D((\rho^V_W)_{*\text{cr}}(s))\} = \{(x, p) : x \in U \text{ and } \text{Spec}(\rho^U_x)(p) \in D(r)\} \cap \{(x, p) : x \in V \text{ and } \text{Spec}(\rho^V_x)(p) \in D(s)\} = D(U, r) \cap D(V, s).$$ ■

Definition (4.2.2). *The* prime spectrum *of a modelled space X is the topological space* Spec(X) *whose points are the pairs (x, p) with $x \in X$ and $p \in$* Spec$(O_{X,x})$ *and whose topology, called the* Zariski topology, *has as an open base the set of subsets of the form*

$$D(U, r) = \{(x, p) : x \in U \text{ and } r \not\leq \text{Spec}(\rho^U_x)(p)\}$$

where U runs through the open sets of X and r runs through the radical codisjunctable congruences on $O_X(U)$.

The open sets of Spec(X) of the form $D(U, r)$ are called the *basic open sets.*

The prime spectrum functor for modelled spaces Spec_{sp} : **SpModA** → **Esp**

Let $f: X \to Y$ be a morphism of modelled spaces on **A**. Let us define the map Spec(f) : Spec$(X) \to$ Spec(Y) by Spec$(f)(x, p) = (f(x), \text{Spec}(f^\#_x)(p))$. Let $D(U, r)$ be a basic open set in Spec(Y). According to (2.4.6), $(f^\#_U)_{*\text{cr}}(r)$ is a codisjunctable radical congruence on $O_X(f^{-1}(U))$. Let $U' = f^{-1}(U)$ and $r' = (f^\#_U)_{*\text{cr}}(r)$. Then $D(U', r')$ is a basic open set of Spec(X). For any $(x, p) \in$ Spec(X), we have $(x, p) \in (\text{Spec}(f))^{-1}(D(U, r)) \Leftrightarrow f(x) \in U$ and $\text{Spec}(\rho^U_{f(x)})(\text{Spec}(f^\#_x)(p)) \in D(r) \Leftrightarrow x \in U'$ and $\text{Spec}(f^\#_U)(\text{Spec}(\rho^{U'}_x)(p)) \in D(r) \Leftrightarrow x \in U'$ and $\text{Spec}(\rho^{U'}_x)(p) \in (\text{Spec}(f^\#_U))^{-1}(D(r)) \Leftrightarrow x \in U'$ and $\text{Spec}(\rho^{U'}_x)(p) \in D(r') \Leftrightarrow (x, p) \in D(U', r')$. Therefore $(\text{Spec}(f))^{-1}(D(U, r)) = D(U', r')$. It follows that the map Spec(f) is continuous.

By assigning, the space Spec(X) to a modelled space X and the continuous map Spec(f) to a morphism of modelled spaces $f: X \to Y$, we define the *prime spectrum functor for modelled spaces* $\mathrm{Spec}_{sp} : \mathbf{SpModA} \rightarrow \mathbf{Esp}$.

The structure sheaf on the prime spectrum of a modelled space

Definition (4.2.3). *The* ordered set $C(X)$ *is the set of pairs* (U, r) *where* U *is an open set of* X *and* r *is a codisjunctable radical congruence on* $O_X(U)$, *whose order is defined by* $(U, r) \leqslant (U', r')$ *if and only if* $U \subset U'$ *and* $r \leqslant (\rho_U^{U'})_{*cr}(r')$.

Definition (4.2.4). *The* ordered set $\Omega_o(\mathrm{Spec}(X))$ *is the set of basic open sets of Spec*(X) *ordered by inclusion. It is a site whose Grothendieck topology is induced by the canonical topology on the ordered set* $\Omega(\mathrm{Spec}(X))$ *of open sets of Spec*(X). *We know that it is sufficient to define a sheaf on* $\Omega_o(\mathrm{Spec}(X))$ *in order to have a sheaf on the space* $\mathrm{Spec}(X)$.

Definition (4.2.5) *The* order-preserving map $D : C(X) \rightarrow \Omega_o(\mathrm{Spec}(X))$ *assigns to a pair* (U, r) *the set* $D(U, r)$ *described previously. It is order-preserving because, for* $(U, r) \leqslant (U', r')$ *and any element* $(x, p) \in D(U, r)$, *we have* $x \in U'$ *and*

$$\mathrm{Spec}(\rho_x^{U'})(p) = \mathrm{Spec}(\rho_U^{U'})(\mathrm{Spec}(\rho_x^U)(p)) \in \mathrm{Spec}(\rho_U^{U'})(D(r)) \subset \mathrm{Spec}(\rho_U^{U'})(D(\rho_U^{U'})_{*cr}(r') = \mathrm{Spec}(\rho_U^{U'})(\mathrm{Spec}(\rho_U^{U'})^{-1}(D(r')) \subset D(r').$$

Hence $(x, p) \in D(U', r')$.

Definition (4.2.6). *The* presheaf $\Delta : C(X)^{\mathrm{op}} \rightarrow A$ *defined on* $C(X)$ *with values in* A *assigns to an object* (U, r) *the codomain of a codisjunctor* $\delta_{U,r} : O_X(U) \rightarrow \Delta(U, r)$ *of the congruence* r, *and to a morphism* $(U, r) \leqslant (U', r')$ *the unique morphism* $\Delta((U, r) \leqslant (U', r')) : \Delta(U', r') \rightarrow \Delta(U, r)$ *fulfilling the relation* $\Delta((U, r) \leqslant (U', r'))\, \delta_{U',r'} = \delta_{U,r} \rho_U^{U'}$.

Definition (4.2.7). *The* presheaf $\bar{\Delta} : \Omega_o(\mathrm{Spec}(X))^{\mathrm{op}} \rightarrow A$ *is the left Kan extension* ([31], p. 232) *of the functor* $\Delta : C(X)^{\mathrm{op}} \rightarrow A$ *along the functor* $D^{\mathrm{op}} : C(X)^{\mathrm{op}} \rightarrow \Omega_o(\mathrm{Spec}(X))^{\mathrm{op}}$.

Definition (4.2.8). *The* sheaf $\tilde{\Delta} : \Omega_o(\mathrm{Spec}(X))^{\mathrm{op}} \rightarrow A$ *on the site* $\Omega_o(\mathrm{Spec}(X))$ *with values in* $\mathbf{A}$ *is the universal sheaf associated to the presheaf* $\bar{\Delta}$.

Definition (4.2.9). *The* structure sheaf $\tilde{X}$ *on* $\mathrm{Spec}(X)$ *is the canonical extension of the sheaf* $\tilde{\Delta}$ *defined on the open base* $\Omega_0(\mathrm{Spec}(X))$ *of* $\mathrm{Spec}(X)$.

Proposition (4.2.10). *The stalk of the structure sheaf* $\tilde{X}$ *at the point* (x, p) *is the localized object* $O_{X,x,p}$ *of* $O_{X,x}$ *at* p.

Proof.

(i) Let us consider a filtered colimit $(\alpha_i : A_i \to A)_{i \in \mathbf{I}}$ in $\mathbf{A}$. Let $p \in \mathrm{Spec}(A)$. For any $i \in \mathbf{I}$, let $p_i = \mathrm{Spec}(\alpha_i)(p)$ and $A_{p_i} = (A_i)_{p_i}$, and let β_i be the local morphism such that $\beta_i l_{p_i} = l_p \alpha_i$. We obtain a filtered diagram $(A_{p_i})_{i \in \mathbf{I}}$ and an inductive cone $(\beta_i : A_{p_i} \to A_p)$ based on it. Let $(\gamma_i : A_{p_i} \to B)_{i \in \mathbf{I}}$ be the colimit of this diagram and let $\beta : B \to A_p$ be the morphism defined by $\beta\gamma_i = \beta_i$ for any $i \in \mathbf{I}$. The morphism β is local, as it is a filtered colimit of the local morphisms $(\beta_i)_{i \in \mathbf{I}}$ (Proposition (2.12.8)). On the other hand, the colimit of the diagram of presingular epimorphisms $(l_{p_i} : A_i \to A_{p_i})_{i \in \mathbf{I}}$ is a presingular epimorphism (Proposition (1.9.5)) l: $A \to B$ which satisfies the relation $\beta l = l_p$. According to the essential uniqueness of the presingular factorization, the morphism β is an isomorphism. Consequently, $(\beta_i : A_{p_i} \to A_p)_{i \in \mathbf{I}}$ is a filtered colimit, i.e. $A_p = \xrightarrow{lim}_{i \in \mathbf{I}} A_{p_i}$.

(ii) Now let us consider an element (x, p) of $\mathrm{Spec}(X)$. Let $C(X)_{x,p}$ be the ordered subset of $C(X)$ whose elements are the pairs (U, r) such that $(x, p) \in D(U, r)$. Then

$$\frac{\lim}{(U,r) \in C(X)_{x,p}} \to \Delta(U, r)$$

$$\simeq \frac{\lim}{x \in U \in \Omega(X)} \to \left(\frac{\lim}{\mathrm{Spec}(\rho_x^U)(p) \in D(r)} \to \Delta(U, r) \right)$$

$$= \frac{\lim}{x \in U \in \Omega(X)} \to (O_X(U)_{\mathrm{Spec}(\rho_x^U)(p)}).$$

According to (i), this colimit is $O_{X,x,p}$.

(iii) Let us denote by $\Omega_0(\mathrm{Spec}(X))_{x,p}$ the subset of $\Omega_0(\mathrm{Spec}(X))$ whose members are those basic open sets which contain (x, p). Let $D_{x,p}$: $C(X)_{x,p} \to \Omega_0(\mathrm{Spec}(X))_{x,p}$ be the map induced by the map D and let $\Delta_{x,p}$: $C(X)_{x,p}^{\mathrm{op}} \to \mathbf{A}$ and $\overline{\Delta_{x,p}} : \Omega(\mathrm{Spec}(X))_{x,p}^{\mathrm{op}} \to \mathbf{A}$ be the functors induced by Δ and $\bar{\Delta}$ respectively. The functor $\bar{\Delta}_{x,p}$ is the left Kan extension of $\Delta_{x,p}$ along $D_{x,p}^{\mathrm{op}}$. From this fact follows an isomorphism $\xrightarrow{lim} \Delta_{x,p} \simeq \xrightarrow{lim} \bar{\Delta}_{x,p}$, i.e. an isomorphism

$$\frac{\lim}{(U,r) \in C(X)_{x,p}} \to \Delta(U, r) \simeq \frac{\lim}{(x,p) \in D(U,r)} \to \bar{\Delta}(D(U,r))$$

According to (ii),

$$\frac{\lim}{(x,p) \in D(U,r)} \to \bar{\Delta}(D(U,r)) = O_{X,x,p}.$$

As the stalks of $\bar{\Delta}$, $\tilde{\Delta}$, and $\tilde{X}$ at (x, p) are the same, the stalk of $\tilde{X}$ at (x, p) is $O_{X,x,p}$. ■

The prime spectrum functor for modelled spaces Σ_{sp} : SpModA → SpModLocA.

For any modelled space X on **A**, the prime spectrum $\mathrm{Spec}(X)$ equipped with its structure sheaf $\tilde{X}$ is a locally modelled space (Proposition (4.2.10) denoted by $\Sigma_{sp}(X)$. Let $f: X \to Y$ be a morphism of modelled spaces on **A**. Let us denote by $C(X)$, $\Omega_o(\mathrm{Spec}(X))$, D_X, Δ_X, $\bar{\Delta}_X$, $\tilde{\Delta}_X$, and $\tilde{X}$ the ordered sets and functors described previously for the modelled space X, and by $C(Y)$, $\Omega_o(\mathrm{Spec}(Y))$, D_Y, Δ_Y, $\bar{\Delta}_Y$, $\tilde{\Delta}_Y$, $\tilde{Y}$ the corresponding notions relative to the modelled space Y. Let $C(f): C(Y) \to C(X)$ be the order-preserving map defined by $C(f)(U,r) = (f^{-1}(U), (f^{\#}_U)_{*cr}(r))$. Let $s(f): \Delta_Y \to \Delta_X(C(f)^{op})$ be the natural transformation whose value for an object (U,r) in $C(Y)$ is the unique morphism $s(f)_{(U,r)}: \Delta(U,r) \to \Delta(f^{-1}(U), (f^{\#}_U)_{*cr}(r))$ which satisfies the relation

$$s(f)_{U,r}\delta_{U,r} = \delta_{f^{-1}(U),(f^{\#}_U)_{*cr}(r)} f^{\#}_U .$$

Let us denote by $\Omega_o(\mathrm{Spec}(f)): \Omega_o(\mathrm{Spec}(Y)) \to \Omega_o(\mathrm{Spec}(X))$ the map defined by $\Omega_o(\mathrm{Spec}(f))(D(U,r)) = (\mathrm{Spec}(f))^{-1}(D(U,r))$. Then we have the relation $\Omega_o(\mathrm{Spec}(f))D_Y = D_X C(f)$. The morphism $s(f): \Delta_Y \to \Delta_X(C(f))^{op}$ defines, by the universal property of the left Kan extension, a morphism $\overline{s(f)}: \overline{\Delta_Y} \to \overline{\Delta_X}\Omega_o(\mathrm{Spec}(f))^{op}$. Because $\Omega_o(\mathrm{Spec}(f))$ is a continuous functor for the site structures, $\overline{s(f)}$ defines a morphism $\widetilde{s(f)}: \widetilde{\Delta_Y} \to \widetilde{\Delta_X}\Omega_o\ (\mathrm{Spec}(f))^{op}$. The morphism $\widetilde{s(f)}$ extends canonically to a morphism $\tilde{Y} \to \tilde{X}(\mathrm{Spec}(f)^{-1})^{op}$ denoted by $\mathrm{Spec}(f)^{\#}$. We can check that a functor Σ_{sp} : **SpModA** → **SpModLocA** is defined by $\Sigma_{sp}(X) = (\mathrm{Spec}(X), \tilde{X})$ and $\Sigma_{sp}(f) = (\mathrm{Spec}(f), \mathrm{Spec}(f)^{\#})$.

Theorem (4.2.11). *The prime spectrum functor Σ_{sp} : SpModA → SpModLocA for modelled spaces on* **A** *is right-adjoint to the inclusion functor I_{sp} : SpModLocA → SpModA.*

Proof.

(i) Let X be an object in **SpModA**. Let us define the continuous map ε_X: $\mathrm{Spec}(X) \to X$ by $\varepsilon_X(x,p) = x$. Then we define the natural transformation $\varepsilon^{\#}_X: O_X \to (\varepsilon_X)_*(\tilde{X})$ in the following way: for any open set U in X, $(\varepsilon^{\#}_X)_U$: $O_X(U) \to \tilde{X}(\varepsilon_X^{-1}(U))$ is the morphism $O_X(U) \xrightarrow{=} \Delta_X(U, 1_{O_X(U)\times O_X(U)}) \to \bar{\Delta}_X(\varepsilon_X^{-1}(U)) \to \tilde{\Delta}_X(\varepsilon_X{}^{-1}(U)) \xrightarrow{=} \tilde{X}(\varepsilon_X^{-1}(U))$ where the morphisms denoted by $\xrightarrow{=}$ are identities and the others are defined by the properties of the left Kan extension and the associated sheaf. We obtain a morphism $(\varepsilon_X, \varepsilon^{\#}_X): (\mathrm{Spec}(X), \tilde{X}) \to (X, O_X)$ in **SpMpdA**, which is natural in X and therefore defines a natural transformation $\varepsilon: I_{sp}\Sigma_{sp} \to 1_{\mathbf{SpMpdA}}$.

(ii) Let X be an object in **SpModLocA**. Let us define the map $\eta_X: X \to \mathrm{Spec}(X)$ by $\eta_X(x) = (x, m_x)$, where m_x denotes the maximal congruence on $O_{X,x}$. Let us prove that η_X is continuous. Let $D(U,r)$ be a basic open set of $\mathrm{Spec}(X)$. Then

$\eta_X^{-1}(D(U,r)) = \{x \in X : x \in U$ and ρ_x^U codisjoints $r\} =$
$\{x \in U : \rho_x^U$ factors through $\delta_{U,r}\}$.

Let $x \in \eta_X^{-1}(D(U,r))$. Because the morphism ρ_x^U is a filtered colimit $\rho_x^U = \varinjlim_{x \in U' \in \Omega(X)} \rho_{U'}^U$ and the morphism $\delta_{U,r}$ is finitely presentable, there exists an open neighbourhood U' of x included in U such that the morphism $\rho_{U'}^U$ factors through $\delta_{U,r}$. Then, for any element $x' \in U'$, the morphism $\rho_{x'}^U$ factors through $\delta_{U,r}$ and thus codisjoints r. Hence $x' \in \eta_X^{-1}(D(U,r))$. Consequently, $U' \subset \eta_X^{-1}(D(U,r))$ and $\eta_X^{-1}(D(U,r))$ is an open set of X. As a result, the map $\eta_X : X \to \mathrm{Spec}(X)$ is continuous. Let us denote by V the open set $\eta_X^{-1}(D(U,r))$. For any element $x \in V$, the morphism ρ_x^U factors through $\delta_{U,r}$ in the form $\rho_x^U = \bar{\rho}_x^U \delta_{U,r}$. Then we obtain a morphism $(\bar{\rho}_x^U) : \Delta_{(U,r)} \to \Pi_{x \in V} O_{X,x}$. On the other hand, we obtain a morphism ρ_x^V: $O_X(V) \to \Pi_{x \in V} O_{X,x}$ which is easily proved to be a prelocal morphism. The relation $\rho_x^V \rho_V^U = (\rho_x^U)\delta_{U,r}$ implies the existence of a unique morphism $\theta_{U,r}$: $\Delta(U,r) \to O_X(V)$ which satisfies the relation $\theta_{U,r}\delta_{U,r} = \rho_V^U$. In this way we obtain a natural transformation $\theta_X : \Delta_X \to D_X(\eta_X)_*(O_X)$. By the universal property of the left Kan extension, the morphism θ provides a morphism $\bar{\theta}_X : \bar{\Delta}_X \to (\eta_X)_*(O_X)$. By the universal property of the associated sheaf, $\bar{\theta}_X$ provides a morphism $\tilde{\theta}_X : \tilde{\Delta}_X \to (\eta_X)_*(O_X)$. By canonical extension, we obtain a morphism $\tilde{X} \to (\eta_X)_*(O_X)$ denoted by $\eta_X^\#$. It follows a morphism $(\eta_X, \eta_X^\#) : (X, O_X) \to (\mathrm{Spec}(X), \tilde{X})$ in **SpModLocA** which is natural in X. It defines a natural transformation $\eta : 1_{\mathbf{SpModLocA}} \to \Sigma_{sp} I_{sp}$.

(iii) It is straightforward to prove that the morphisms ε, η provide an adjunction between the functors Σ_{sp} and I_{sp}. ■

4.3 SCHEMES

Definition (4.3.1). *An* affine scheme on **A** *is a modelled space on* **A** *isomorphic to the modelled space* Spec(A) *for some object A of* **A**.

The *category* **AffA** *of affine schemes on* **A** is the full subcategory of **SpModA** whose objects are the affine schemes on **A**.

Theorem (4.3.2). *The category* **AffA** *of affine schemes on* **A** *is a full reflective subcategory of the category* **SpModLocA** *of locally modelled spaces on* **A** *and is equivalent to the dual of the category* **A**.

Proof. The proof follows from Theorem (4.1.6). ■

Definition (4.3.3). *An* open set *U of a modelled space X on* **A** *is* affine *if the induced modelled space on U is an affine scheme.*

Definition (4.3.4). *A* scheme *on* **A** *is a modelled space* X *on* **A** *such that any element of* X *has an affine open neighbourhood.*

A scheme on **A** is locally modelled on **A**. The *category* **SchA** of *schemes on* **A** is the full subcategory of **SpModLocA** whose objects are the schemes on **A**.

Theorem (4.3.5). *The category* **AffA** *of affine schemes on* **A** *is a full reflective subcategory of the category* **SchA** *of schemes on* **A**.

Proof. The proof follows from Theorem (4.3.2). ■

Proposition (4.3.6). *The topological space of a scheme is sober.*

Proof. The proof of [22], I, Proposition 2.1.5, holds. ■

Definition (4.3.7). *A scheme is* simple (pseudo-simple, local) *if it is isomorphic to a scheme of the form* $\mathrm{Spec}(A)$ *where* A *is a simple (pseudo-simple, local) object. A scheme is* integral (reduced) *if the value of its structure sheaf on any non-empty open set is integral (reduced). It is* irreducible (connected, compact) *if its underlying space is irreducible (connected, compact).*

Proposition (4.3.8). *A scheme is integral if and only it is both reduced and irreducible.*

Proof. Let X be an integral scheme. For any non-empty open set U of X, $O_X(U)$ is integral and hence reduced. Therefore X is reduced. If U, V were two disjoint non-empty open sets in X, then $O_X(U \cup V) \simeq O_X(U) \times O_X(V)$ would not be integral. Thus X is irreducible. Conversely, let us assume that X is a reduced and irreducible scheme. Let U be a non-empty affine open set in X. Then $O_X(U)$ is reduced and $\mathrm{Spec}(O_X(U)) \simeq U$ is irreducible. According to Proposition (3.3.4), $O_X(U)$ is quasi-primary and, according to Proposition (2.7.5), $O_X(U)$ is integral. Let V be a non-empty affine open set contained in U and let $x \in V$. The restriction morphism $\rho^U_V : O_X(U) \to O_X(V)$ with the localization morphisms $\rho^U_x : O_X(U) \to O_{X,x}$ and $\rho^V_x : O_X(V) \to O_{X,x}$ satisfy the relation $\rho^U_x = \rho^V_x \rho^U_V$. Because $O_X(U)$ is integral, the morphism ρ^U_x is monomorphic (Proposition (2.2.7)). Thus ρ^U_V is monomorphic. Let W be an arbitrary non-empty open set in X. According to the definition of sheaves, if W is covered with non-empty affine open subsets, the value $O_X(W)$ is a connected limit of a diagram of integral objects and monomorphisms, i.e. a diagram of **IntA**. Because **IntA** is closed in **A** under non-empty connected limits (Corollary (2.2.3)), $O_X(W)$ is integral. As a result, X is integral. ■

Proposition (4.3.9). *A scheme is reduced if and only if its stalks are reduced.*

Proof. The stalks of a reduced scheme are reduced because filtered colimits of reduced objects are reduced. Conversely, if a scheme X has reduced stalks, the value $O_X(U)$ of its structure sheaf on an open set U is a subobject of $\Pi_{x \in U} O_{X,x}$ and hence is reduced. ■

Proposition (4.3.10). *An affine scheme* Spec(A) *is integral (reduced, irreducible, non-empty connected) if and only if the object A is integral (reduced, quasi-primary, indecomposable).*

Proof. If Spec(A) is reduced, then $A \simeq \tilde{A}(\mathrm{Spec}(A))$ is reduced. Conversely, if A is reduced, its localized objects are reduced (Proposition (3.6.1)) and therefore Spec(A) is reduced (Proposition (4.3.9)). According to Proposition (3.3.4), the topological space Spec(A) is irreducible if and only if A is quasi-primary. According to Proposition (4.3.8), Spec(A) is integral if and only if Spec(A) is reduced and irreducible, thus if and only if A is reduced and quasi-primary, and thus if and only if A is integral (Proposition (2.7.5)). The equivalence Spec(A) is non-empty connected $\Leftrightarrow$ A is indecomposable, is proved in Proposition (3.10.3). ■

The categories SchSimA, SchPsdSimA, SchIntA, SchRedA, SchIrrA, and SchLocA

SchSimA, **SchPsdSimA**, and **SchRedA** are the full subcategories of **SchA** whose objects are the simple, pseudo-simple, and reduced schemes respectively. **SchIntA** and **SchIrrA** are the subcategories of **SchA** whose objects are the integral and irreducible schemes respectively and whose morphisms are the dominant ones (cf. (3.3.13)). **SchLocA** is the subcategory of **SchA** whose objects are local schemes and whose morphisms are those preserving the closed point. The prime spectrum functor $\Sigma_{\mathbf{A}} : \mathbf{A}^{\mathrm{op}} \to$ **SchA** induces equivalences of categories $(\mathbf{SimA})^{\mathrm{op}} \to$ **SchSimA**, $(\mathbf{PsdSimA})^{\mathrm{op}} \to$ **SchPsdSimA**, and $(\mathbf{LocA})^{\mathrm{op}} \to$ **SchLocA**, and full and faithful functors $(\mathbf{IntA})^{\mathrm{op}} \to$ **SchIntA**, $(\mathbf{RedA})^{\mathrm{op}} \to$ **SchRedA**, and $(\mathbf{QuaPrimA})^{\mathrm{op}} \to$ **SchIrrA** (Propositions (3.3.13) and (3.3.4) below).

Theorem (4.3.11). *To any irreducible scheme is associated a co-universal pseudo-simple scheme, i.e.* ***SchPsdSimA*** *is a coreflective subcategory of* ***SchIrrA***.

Proof. Let X be an irreducible scheme. According to Proposition (4.3.6), X has a generic point x. Let U be an affine open neighbourhood of x. Then U is irreducible with a generic point x. Let $U = \mathrm{Spec}(A)$. Then A is quasi-primary (Proposition (4.3.10)), $x = \text{rad}(A)$ (Proposition (2.7.4)), and $O_{X,x} = O_{U,x} = A_{\text{rad}(A)}$ is the universal pseudo-simple object associated to A (Proposition 2.13.5)). Let us consider the pseudo-simple scheme $\text{Spec}(O_{X,x})$ and the morphism of schemes $\varepsilon_X : \text{Spec}(O_{X,x}) \to X$ defined by $\varepsilon_X(\text{rad}(O_{X,x})) = x$ and $(\varepsilon_X^{\#})_{\text{rad}(O_{X,x})} = 1_{O_{X,x}}$. Let $f: \text{Spec}(B) \to X$ be a morphism of schemes with a pseudo-simple domain. Then $f(\text{rad}(B)) = x$ and $f^{\#}_{\text{rad}(B)} : O_{X,x} \to B$. The morphism of schemes $\text{Spec}(f^{\#}_{\text{rad}(B)}) : \text{Spec}(B) \to \text{Spec}(O_{X,x})$ is such that $\varepsilon_X \text{Spec}(f^{\#}_{\text{rad}(B)}) = f$ and it is the only one. Consequently, $\varepsilon_X : \text{Spec}(O_{X,x}) \to X$ is a co-universal morphism from **SchPsdSimA** to X. *As a result,* **SchPsdSimA** is a coreflective subcategory of **SchIrrA.** ■

Theorem (4.3.12). *To any integral scheme is associated a co-universal simple scheme, i.e.* ***SchSimA*** *is a coreflective subcategory of* ***SchIntA***.

Proof. The previous proof holds if we replace irreducible, quasi-primary, and pseudo-simple by integral, integral, and simple respectively. ■

Corollary (4.3.13). *The functors* $\Sigma_A : (\textbf{\textit{PsdSimA}})^{\text{op}} \to \textbf{\textit{SchIrrA}}$ *and* $\Sigma_{\mathbf{A}} : (\textbf{\textit{SimA}})^{\text{op}} \to \textbf{\textit{SchIntA}}$ *have right-adjoints.*

Notation. We denote by $K_{\mathbf{A}} : (\textbf{SchIrrA})^{\text{op}} \to \textbf{PsdSimA}$ and $K_{\mathbf{A}} : (\textbf{SchIntA})^{\text{op}} \to \textbf{SimA}$ the left-adjoints to the functors $(\Sigma_{\mathbf{A}})^{\text{op}} : \textbf{PsdSimA} \to (\textbf{SchIrrA})^{\text{op}}$ and $(\Sigma_{\mathbf{A}})^{\text{op}} : \textbf{SimA} \to (\textbf{SchIntA})^{\text{op}}$ respectively. For $X \in \textbf{SchIrrA}$ (or **SchIntA**), $K_{\mathbf{A}}(X)$ is called the *pseudo-simple* (or *simple*) *object of X*.

Proposition (4.3.14). *For an integral scheme X and any non-empty open set U of X, the canonical morphism* $\rho^U : O_X(U) \to K_{\mathbf{A}}(X)$ *is monomorphic.*

Proof. Let x be the generic point of X. First, let us assume that U is affine, i.e. $U = \text{Spec}(A)$. According to Proposition (4.3.10), A is integral. According to Proposition (2.13.6), $\rho_x^U : O_X(U) \to O_{U,x} = O_{X,x}$ is monomorphic. For an arbitrary non-empty open set U of X, $\rho_x^U : O_X(U) \to O_{X,x} = K_{\mathbf{A}}(X)$ is monomorphic as it is a limit of monomorphisms. ■

Theorem (4.3.15). *To any scheme is associated a co-universal reduced scheme, i.e.* ***SchRedA*** *is a coreflective subcategory of* ***SchA***.

Proof. Let $I : \textbf{RedA} \to \mathbf{A}$ be the inclusion functor and let $R : \mathbf{A} \to \textbf{RedA}$ be the reflector (Theorem (2.3.2)). Let X be a scheme. Let us consider the presheaf P on X with values in $\mathbf{A}$ defined by $P = IRO_X : \Omega^{\text{op}}(X) \to \mathbf{A}$. Let U, V be affine open sets of X such that $V \subset U$. According to Definition (3.1.10) and Proposition (3.3.10), the inclusion morphism $V \to U$ is of the form $\text{Spec}(f)$ where $f: A \to B$ is a singular epimorphism in $\mathbf{A}$ which is the codisjunctor of a congruence r on A such that $V = D(r)$. Then $R(f)$: $R(A) \to R(B)$ is a singular epimorphism in **RedA** (Proposition (2.3.4)) and also in **A**. It is the codisjunctor in **A** of a congruence $\bar{r}$ on $R(A)$ such that $D(\bar{r}) = D(r) = V$ (Proposition (3.3.3)). Then $P(V) = RO_X(V) \simeq R(B) \simeq \widetilde{R(A)}(D(r)) = \widetilde{R(A)}(V)$. Consequently, the restriction P/U of the presheaf P to U is isomorphic to the sheaf $\widetilde{R(A)}$ and the modelled space $(U, P/U)$ is a reduced affine scheme (Proposition (4.3.10)). Let $O_{R(X)}$ be the sheaf on X associated to the presheaf P. Then, for any affine open set U of X, $(U, O_{R(X)}/U) \simeq (U, P/U)$ is a reduced affine scheme. Consequently, $(X, O_{R(X)})$ is a reduced scheme. Let us define the morphism of schemes $\Psi : O_X \to O_{R(X)}$ by its value at an open set U which is the composite of the universal morphism $r_{O_X(U)} : O_X(U) \to P(U)$ with the canonical morphism $a_U : P(U) \to O_{R(X)}(U)$. Then we obtain a morphism of schemes $(1_X, \Psi) : (X, O_{R(X)}) \to (X, O_X)$ which is indeed a co-universal morphism from **SchRedA** to (X, O_X). ■

Examples

Schemes on the category **CRng** are usual schemes. On **RedCRng**, they are usual reduced schemes. Schemes on the category **CAlg**(A) of commutative algebras over a ring A are usual schemes over the affine scheme Spec(A). Schemes on **Bool** are locally boolean spaces. Schemes on **Mod** are precisely quasi-coherent modules. Schemes on **CAlg** are the quasi-coherent algebras.

4.4 SUBSCHEMES, IMMERSIONS, AND EMBEDDINGS OF SCHEMES

A morphism of schemes which is cartesian with respect to the fibration **SpModA** $\to$ **Esp** (Proposition (4.1.2)) is simply called cartesian.

Proposition (4.4.1). *A morphism of schemes $f: X \to Y$ is cartesian if and only if the morphisms $f^{\#}_x : O_{Y,f(x)} \to O_{X,x}$ are isomorphisms for all $x \in X$.*

Proof. According to the notation on p. 113, f is cartesian if and only if the morphism $f_{\#} : f^*(O_y) \to O_X$ is an isomorphism of sheaves on X. This is equivalent to the fact that all the morphisms $f^{\#}_x = (f_{\#})_x : O_{Y,f(x)} \to O_{X,x}$ are isomorphisms (cf. proof of Proposition (3.5.1). ■

Proposition (4.4.2). *A morphism $f: A \to B$ in* **A** *is a local isomorphism if and only if* Spec(f) : Spec$(B) \to$ Spec(A) *is a cartesian morphism of schemes.*

Proof. The proof follows from Proposition (4.4.1) and from the fact that, for any $p \in$ Spec(B), we have $(\text{Spec}(f))^{\#}_p = f_p$. ■

Proposition (4.4.3). *A morphism of schemes $f: X \to Y$ is an isomorphism if and only if f is a homeomorphism and the morphisms $f_x^\#$ are isomorphisms for $x \in X$.*

Proof. The proof follows from Proposition (4.4.1) and the fact that f is an isomorphism if and only if f is cartesian and the map f is a homeomorphism. ∎

Definition (4.4.4).

(i) *A* scheme X *is* induced *by a scheme Y if X is a subspace of Y and the inclusion map $X \to Y$ underlies a cartesian morphism of schemes.*

(ii) *An* embedding of schemes *is a morphism of schemes $f: X \to Y$ which induces an isomorphism between X and a scheme induced by Y.*

Proposition (4.4.5). *For a morphism of schemes $f: X \to Y$ the following assertions are equivalent.*

(i) *f is an embedding of schemes.*

(ii) *f is a topological embedding and a cartesian morphism.*

(iii) *f is a topological embedding and the morphisms $f_x^\#$ are isomorphisms for all $x \in X$.*

Proof. The proof follows immediately from Proposition (4.4.1). ∎

Proposition (4.4.6). *A morphism $f: A \to B$ in* **A** *is a semisingular epimorphism if and only if* $\mathrm{Spec}(f) : \mathrm{Spec}(B) \to \mathrm{Spec}(A)$ *is an embedding of schemes.*

Proof. Let f be a semisingular epimorphism. According to Proposition (3.8.8), the map $\mathrm{Spec}(f)$ is a topological embedding. According to Propositions (3.7.3) and (4.4.2), $\mathrm{Spec}(f)$ is a cartesian morphism. Thus $\mathrm{Spec}(f)$ is an embedding of schemes. Conversely, let us assume that $\mathrm{Spec}(f)$ is an embedding of schemes. Then $\mathrm{Spec}(f)$ is a monomorphism in the category **SchA**, and thus is also a monomorphism in the category **AffA**. It follows that f is an epimorphism in **A** (theorem (4.3.2)). According to Proposition (4.4.5), the morphisms $f_p = (\mathrm{Spec}(f))_p^\#$ are isomorphisms for all $p \in \mathrm{Spec}(B)$. It follows that f is a local isomorphism and therefore that f is a semisingular epimorphism (Proposition (3.7.3)). ∎

Definition (4.4.7). *A* subset *of a scheme X is* affine *if it is the underlying set of an affine scheme induced by X.*

Proposition (4.4.8). *A subset of* $\mathrm{Spec}(A)$ *is an affine subset of the topological space* $\mathrm{Spec}(A)$ *if and only if it is an affine subset of the scheme* $\mathrm{Spec}(A)$.

Proof. According to Proposition 3.8.11, the affine subsets of the space Spec(A) are precisely the images of the maps Spec(f) where $f: A \to B$ are semisingular epimorphisms. According to Proposition (4.4.6), these images are precisely the images of affine embedding of schemes, i.e. the affine subsets of the scheme Spec(A). ■

Definition (4.4.9).

(i) *An* open subscheme *of a scheme X is a scheme U whose topological space is an open subset of X and whose structure sheaf O_U is the restriction O_X/U of the structure sheaf O_X of X.*

(ii) *An* open immersion of schemes *is a morphism of schemes $f: X \to Y$ which induces an isomorphism of X with an open subscheme of Y.*

Proposition (4.4.10). *A morphism of schemes $f: X \to Y$ is an open immersion if and only if f induces a homeomorphism between the topological space X and an open subset of Y and, for any $x \in X$, the morphism $f_x^\#: O_{Y,f(x)} \to O_{X,x}$ is an isomorphism. It is an embedding of schemes.*

Proof. If U is an open set of a scheme X and $f: U \to X$ is the continuous inclusion map, then the sheaf $f^*(O_X)$ is isomorphic to the restriction O_X/U of O_X on U. Then the result follows from Proposition (4.4.5). ■

Proposition (4.4.11). *A morphism $f: A \to B$ in* **A** *is a singular epimorphism if and only if* Spec(f) : Spec$(B) \to$ Spec(A) *is an open immersion of schemes.*

Proof. Let f be a singular epimorphism. According to Proposition (3.3.10), the map Spec(f) induces a homeomorphism between Spec(B) and an open set of Spec(A). According to Proposition (4.4.6), Spec(f) is an embedding of schemes. It follows that Spec(f) is an open immersion of schemes. Conversely, let us assume that Spec(f) is an open immersion of schemes. According to Proposition (4.4.6), f is a semisingular epimorphism. Then f is the simultaneous codisjunctor of $\mathcal{R}(f)$ (notation on p. 100). The image of Spec(f) is an open set of Spec(A) of the form $D(r)$ with $r \in$ Cong(A). Then $D(r) =$ ImSpec$(f) = D(\mathcal{R}(f))$ (Proposition (3.8.5)). According to Proposition (3.8.3), for an arbitrary morphism $g: A \to C$, g simultaneously codisjoints $\mathcal{R}(f) \Leftrightarrow$ ImSpec$(g) \subset D(\mathcal{R}(f)) \Leftrightarrow$ ImSpec$(g) \subset D(r) \Leftrightarrow g$ codisjoints r. It follows that the simultaneous codisjunctor of $\mathcal{R}(f)$ is the codisjunctor of r, and so f is a singular epimorphism. ■

Proposition (4.4.12). *A subset of* Spec(A) *is an affine open set of the space* Spec(A) *if and only if it is an affine open set of the scheme* Spec(A).

Proof. In any case, affine open subsets are subsets which are both affine and open (Proposition (3.8.12)). Thus the result follows from Proposition (4.4.8). ■

Definition (4.4.13).

(i) *A* closed subscheme *of a scheme X is a scheme Y together with a morphism $i: Y \to X$ called the inclusion morphism such that*

(a) *Y is a closed subset of X and i is the inclusion map, and*

(b) *the morphism $i_y^{\#}: O_{X,y} \to O_{Y,y}$ is a regular epimorphism for any $y \in Y$.*

(ii) *A* closed immersion of schemes *is a morphism of schemes $f: X \to Y$ which induces an isomorphism of X onto a closed subscheme of Y.*

Proposition (4.4.14). *A morphism of schemes $f: X \to Y$ is a closed immersion if and only if f induces a homeomorphism between the topological space X and a closed subset of Y and, for any $x \in X$, the morphism $f_x^{\#}: O_{Y,f(x)} \to O_{X,x}$ is a regular epimorphism.*

Proof. The proof is immediate. ■

Proposition (4.4.15). *A morphism $f: A \to B$ in* **A** *is a regular epimorphism if and only if* $\mathrm{Spec}(f): \mathit{Spec}(B) \to \mathrm{Spec}(A)$ *is a closed immersion of schemes.*

Proof.

(i) Let f be a regular epimorphism. According to Proposition (3.3.11), the map $\mathrm{Spec}(f)$ induces a homeomorphism between $\mathrm{Spec}(B)$ and its closed image in $\mathrm{Spec}(A)$. According to Proposition (3.6.6), the local morphisms of f are regular epimorphisms. According to Proposition (4.4.14), $\mathrm{Spec}(f)$ is a closed immersion of schemes.

(ii) Let us assume that $\mathrm{Spec}(f)$ is a closed immersion. Let $f = mg$ be the regular factorization of f, where $g: A \to C$ is a regular epimorphism and $m: C \to B$ is a monomorphism. According to Proposition (3.3.13), the map $\mathrm{Spec}(m)$ is dominant, i.e. $\mathrm{ImSpec}(m)$ is dense in $\mathrm{Spec}(C)$. Thus $\mathrm{Spec}(g)\ (\mathrm{ImSpec}(m))$ is dense in $\mathrm{Spec}(g)\ (\mathrm{Spec}(C))$, i.e. $\mathrm{ImSpec}(f)$ is dense in $\mathrm{ImSpec}(g)$. But $\mathrm{ImSpec}(f)$ is closed by assumption. Thus $\mathrm{Spec}(f)$ and $\mathrm{Spec}(g)$ have the same image. Because both maps $\mathrm{Spec}(f)$ and $\mathrm{Spec}(g)$ induce homeomorphisms between their domains and images, it follows that the map $\mathrm{Spec}(m)$ is a homeomorphism. Let $p \in \mathrm{Spec}(B)$ and $q = \mathrm{Spec}(m)\ (p)$. Then $m_p g_q = f_p$. Because f_p and g_q are regular epimorphisms, m_p is a regular epimorphism. But the morphism m_p is also monomorphic as a filtered colimit of the diagram of monomorphisms $(\tilde{m}_U: \tilde{C}(V) \to \tilde{B}(U))$, where U runs over the set of affine open sets of $\mathrm{Spec}(B) \simeq \mathrm{Spec}(C)$ containing p. Therefore m_p is an isomorphism. It follows that $\mathrm{Spec}(m)$:

$\mathrm{Spec}(B) \to \mathrm{Spec}(C)$ is an isomorphism of affine schemes. Thus m is an isomorphism in $\mathbf{A}$ and hence f is a regular epimorphism. ■

Definition (4.4.16).

(i) *A* subscheme *of a scheme X is a scheme Y together with a morphism $i : Y \to X$ called the inclusion morphism such that*

(a) *Y is a locally closed subset of X and i is the inclusion map, and*

(b) *the morphism $i^{\#}_y : O_{X,y} \to O_{Y,y}$ is a regular epimorphism for any $y \in Y$.*

(ii) *An* immersion of schemes *is a morphism of schemes $f : X \to Y$ which induces an isomorphism of X onto a subscheme of Y.*

Proposition (4.4.17). *A morphism of schemes $f : X \to Y$ is an immersion if and only if f induces a homeomorphism between the topological space X and a locally closed subset of Y and, for any $x \in X$, the morphism $f^{\#}_x$: $O_{Y,f(x)} \to O_{X,x}$ is a regular epimorphism.*

Proof. The proof is immediate. ■

Proposition (4.4.18).

(i) *Any subscheme is a closed subscheme of an open subscheme.*

(ii) *Any immersion is the composite of a closed immersion followed by an open immersion.*

Proof. The proof is immediate. ■

Proposition (4.4.19). *For any point x of a scheme X, there exists a canonical embedding of schemes $l_x : \mathrm{Spec}(O_{X,x}) \to X$ whose image is the set of generizations of x.*

Proof. Let us denote by $\eta : 1_{\mathbf{SchA}} \to \Sigma_{\mathbf{A}}\Gamma_{\mathbf{A}}$ the unit of the adjunction $(\Gamma_{\mathbf{A}}, \Sigma_{\mathbf{A}}) : \mathbf{A}^{\mathrm{op}} \to \mathbf{SchA}$ (cf. Theorem (4.1.6)). Let us consider an affine open subscheme U of X containing x and let $i_U : U \to X$ be the inclusion morphism. The morphism $\rho^U_x : O_X(U) \to O_{X,x}$ is the localization of $O_X(U)$ at the prime congruence $(\rho^U_x)^{*c}(m_x)$. Thus the morphism $\mathrm{Spec}(\rho^U_x)$: $\mathrm{Spec}(O_{X,x}) \to \mathrm{Spec}(O_X(U))$ is an embedding of schemes (Proposition (4.4.6)) whose image is the set of generizations of x in $\mathrm{Spec}(O_X(U))$ (Proposition (3.3.9)). Therefore the morphism $l^U_x = i_U \eta_U^{-1}\,\mathrm{Spec}(\rho^U_x)$: $\mathrm{Spec}(O_{X,x}) \to X$ is an embedding of schemes whose image is the set of generizations of x in X. Let V be an affine open subscheme of X containing x and included in U, and let $i^U_V : V \to U$, $i_V : V \to X$ be the inclusion morphisms. We have $\mathrm{Spec}(\rho^U_V)\eta_V = \eta_U i^U_V$ and $\rho^U_x = \rho^V_x \rho^U_V$, and thus $l^U_x = i_U \eta_U^{-1}\mathrm{Spec}(\rho^U_V)\,\mathrm{Spec}(\rho^V_x) = i_U i^U_V \eta_V^{-1}\,\mathrm{Spec}(\rho^V_x) = i_V \eta_V^{-1}\mathrm{Spec}(\rho^V_x) = l^V_x$. It follows that, for any affine open subscheme W of X containing x, we have

$l_x^U = l_x^{U \cap W} = l_x^W$. Consequently the morphism l_x^U does not depend on U and can be denoted by $l_x : \mathrm{Spec}(O_{X,x}) \to X$. ■

Definition (4.4.20). *For any point x of a scheme X, the scheme* $\mathrm{Spec}(O_{X,x})$ *is called the* local scheme of X *at x and the morphism* $l_x : \mathrm{Spec}(O_{X,x}) \to X$ *is called the* localization of X *at x.*

Theorem (4.4.21). *The category SchLocA of local schemes on A is a multicoreflective subcategory of* ***SchA****, and the family of localizations of a scheme X is a co-universal family of morphisms from* ***SchLocA*** *to X.*

Proof. Let $x, y \in X$ and $g : \mathrm{Spec}(A) \to \mathrm{Spec}(O_{X,x})$, $h : \mathrm{Spec}(A) \to \mathrm{Spec}(O_{X,y})$ be local morphisms of schemes such that $l_x g = l_y h$. Then $x = l_x(m_x) = l_x(g(m_A)) = l_y(h(m_A)) = l_y(m_y) = y$. Let $f : \mathrm{Spec}(A) \to X$ be a morphism of schemes with a local domain. Let $x = f(m_A)$. For any $y \in \mathrm{Spec}(A)$, we have $y \leqslant m_A$. Thus y is a generization of m_A (Proposition (3.1.8)) and hence $f(y)$ is a generization of $f(m_A) = x$. Therefore $f(y)$ belongs to the image of l_x (Proposition (4.4.19). Since l_x is a topological embedding (Proposition (4.4.5)), there exists a unique continuous map $g : \mathrm{Spec}(A) \to \mathrm{Spec}(O_{X,x})$ such that $l_x g = f$. Since l_x is cartesian (Proposition (4.4.5)), there exists a unique morphism of schemes $g : \mathrm{Spec}(A) \to \mathrm{Spec}(O_{X,x})$ such that $l_x g = f$ and this morphism preserves the closed point. As a result, the family $(l_x)_{x \in X}$ of localizations of X is a co-universal family of morphisms from **SchLocA** to X, and **SchLocA** is a multicoreflective subcategory of **SchA** [7]. ■

Proposition (4.4.22). *The class* $\mathfrak{M}$ *of morphisms of* **SchIrrA** (or **SchIntA**) *which are non-empty open subschemes admits a calculus of right fractions* ([35], Chapter 19).

Proof. Let us be in the category **SchIrrA**. The class $\mathfrak{M}$ is contained in the class of monomorphisms. It contains the identities and is closed under compositions. Let $i : U \to X$ in $\mathfrak{M}$ and let $g : Y \to X$. Let $V = g^{-1}(U)$. As U is non-empty and g is dominant, V is a non-empty open set of Y. As Y is irreducible, V is dense in Y. Then the open subscheme $V \to Y$ belongs to $\mathfrak{M}$ and is the inverse image of i along g. As a result, $\mathfrak{M}$ admits a calculus of right fractions in **SchIrrA**. The proof in **SchIntA** is similar. ■

Notation. The category **SchIrrA** $[\mathfrak{M}^{-1}]$ (or **SchInt** $[\mathfrak{M}^{-1}]$) of right fractions of **SchIrrA** (or **SchIntA**) with respect to $\mathfrak{M}$ is denoted by $\mathbf{SchIrr}_{\mathrm{rat}}\mathbf{A}$ (or $\mathbf{SchInt}_{\mathrm{rat}}\mathbf{A}$). Its morphisms are called *rational morphisms* of irreducible (or integral) schemes. Its isomorphisms are called *birational equivalences* of irreducible (or integral) schemes, and isomorphic objects are said to be *birationally equivalent*. The restriction of the canonical functor **SchIrrA** →

SchIrr$_{\text{rat}}$**A** (or **SchIntA** → **SchInt**$_{\text{rat}}$**A**) to the category **SchPsdSimA** (or **SchSimA**) is a full embedding which enables us to identify **SchPsdSimA** (or **SchSimA**) with a subcategory of **SchIrr**$_{\text{rat}}$**A** (or **SchInt**$_{\text{rat}}$**A**). The functor $K_{\mathbf{A}}: (\mathbf{SchIrrA})^{\text{op}} \to \mathbf{PsdSimA}$ (or $K_{\mathbf{A}}: (\mathbf{SchIntA})^{\text{op}} \to \mathbf{SimA}$) (notation on p. 124) sends morphisms of $\mathfrak{M}$ onto isomorphisms. Thus it extends canonically to a functor $(\mathbf{SchIrr}_{\text{rat}}\mathbf{A})^{\text{op}} \to \mathbf{PsdSimA}$ (or $(\mathbf{SchInt}_{\text{rat}}\mathbf{A})^{\text{op}} \to \mathbf{SimA}$), which is also denoted by $K_{\mathbf{A}}$.

Proposition (4.4.23). *The category* **SchPsdSimA** (or **SchSimA**) *is a full coreflective subcategory of* **SchIrr**$_{\text{rat}}$**A** *(or* **SchInt**$_{\text{rat}}$**A**).

Proof. The proof follows from Propositions (4.3.11) and (4.3.12). ■

4.5 LIMITS, COLIMITS, AND DISJUNCTORS OF SCHEMES

Definition (4.5.1). *A* gluing diagram *of SchA is a family* $(X_i)_{i\in I}$ *of objects of SchA together with, for any distinct pair* $(i,j) \in I \times I$, *an open subscheme* X_{ij} *of* X_i *and an isomorphism* $\varphi_{ij}: X_{ij} \to X_{ji}$ *in SchA, such that*

(i) $\varphi_{ji} = \varphi_{ij}^{-1}$

(ii) $\varphi_{ij}(X_{ij} \cap X_{ik}) = X_{ji} \cap X_{jk}$

(iii) $\varphi_{ik} = \varphi_{jk}\varphi_{ij}$ on $X_{ij} \cap X_{ik}$.

It is, in fact, a diagram of **SchA** described as follows. Let $\mathbf{G}(I)$ be the diagram type whose set of objects is I^2, which has one morphism α_i: $(i,j) \to (i,i)$ and one morphism $\beta_{ij}: (i,j) \to (j,i)$ for any distinct pair $(i,j) \in I^2$. Let $\varphi: \mathbf{G}(I) \to \mathbf{SchA}$ be the diagram of **SchA** defined by $\varphi(i,j) = X_i$, $\varphi(i,j) = X_{ij}$ for $i \neq j$, let $\varphi(\alpha_i)$ equal the inclusion morphism $X_{ij} \to X_i$, and $\varphi(\beta_{ij}) = \varphi_{ij}$. This diagram will be denoted by $(X_{ij})_{(i,j)\in\mathbf{G}(I)}$. A colimit of such a diagram will be called a *gluing colimit*.

Proposition (4.5.2). *The category* **SchA** *has gluing colimits.*

Proof. The proof is, in fact, exactly, the classical gluing procedure (see for example, [22], §4.1.7, or [24], §2.12). Let us give a brief description of this construction. Let $(X_{ij})_{(i,j)\in\mathbf{G}(I)}$ be a gluing diagram. Let $(\gamma_i: X_i \to E)_{i\in I}$ be the disjoint union of the family of sets $(X_i)_{i\in I}$. We define an equivalence relation $\sim$ on E by $x \sim y$ if and only if $x = y$ or there exist $(i,j) \in I^2$ with $i \neq j$ and $x_i \in X_{ij}$ such that $x = \gamma_i(x_i)$ and $y = \gamma_j(\varphi_{ij}(x_i))$. This equivalence relation induces the equality on each subset $\gamma_i(X_i)$. Let X be the quotient set $E/\sim$, let $\Psi: E \to X$ be the canonical projection, and, for any $i \in I$, let $\Psi_i = \Psi\gamma_i: X_i \to X$. The family of injective maps $(\Psi_i: X_i \to X)_{i\in I}$ provides a colimit in **Set** of the diagram of sets $(X_{ij})_{(i,j)\in\mathbf{G}(I)}$. We obtain the colimit in **Esp** of the diagram of topological spaces $(X_{ij})_{(i,j)\in\mathbf{G}(I)}$ by taking on X the final topology defined by the set of maps $(\Psi_i: X_i \to X)_{i\in I}$ and the topology on each X_i. Then the family of continuous and open maps $(\Psi_i : X_i \to X)_{i\in I}$ provides a colimit for $(X_{ij})_{(i,j)\in\mathbf{G}(I)}$ in **Esp** such that $(\Psi_i(X_i))_{i\in I}$ is an open covering of X and, for any distinct pair $(i,j) \in I^2$, $\Psi_i(X_i) \cap \Psi_j(X_j) = \Psi_i(X_{ij}) = \Psi_j(X_{ji})$. Let $\mathcal{B}$ be the open base of X whose members are the open sets included in some $\psi_i(X_i)$. Let us define the structure sheaf O_X on $\mathcal{B}$ in the following way. For any $U \in \mathcal{B}$, let us choose one $i \in I$ such that $U \subset \psi_i(X_i)$ and let $O_X(U) = O_{X_i}(\psi_i^{-1}(U))$. Let $V \in \mathcal{B}$ such that $V \subset U$. Let $j \in I$ be the chosen index such that $V \subset U_j$ and let $O_X(V) = O_{X_j}(\psi_j^{-1}(V))$. Let us define the restriction morphism $O_X(U) \to O_X(V)$ as the composite of the restriction morphism $O_{X_i}(\psi_i^{-1}(U)) \to O_{X_i}(\psi_i^{-1}(V))$ followed by the isomorphism $(\psi_{ji}^{\#})_{\psi_i^{-1}(V)} : O_{X_i}(\psi_i^{-1}(V)) \to O_{X_j}(\psi_j^{-1}(V))$. For any $i \in I$, the natural transformation $\psi_i^{\#} : O_X \to (\psi_i)_*(O_{X_i})$ has, as its value at $V \in \mathcal{B}$ with chosen index j, the morphism $(\psi_i^{\#})_V = (\psi_{ij}^{\#})_{\psi_j^{-1}(V)} : O_{X_j}(\psi_j^{-1}(V)) \to O_{X_i}(\psi_i^{-1}(V))$. Then we obtain a scheme (X, O_X) and a family of open immersions of schemes $(\psi_i, \psi_i^{\#}) : (X_i, O_{X_i}) \to (X, O_X)$ which provides a colimit in **SchA** for the diagram $(X_{ij})_{(i,j)\in\mathbf{G}(I)}$. ∎

Corollary (4.5.3). *The category SchA has coproducts.*

Proof. Any family of schemes $(X_i)_{i\in I}$ defines a gluing diagram $(X_{ij})_{(i,j)\in\mathbf{G}(I)}$ where $X_{ij} = \varnothing$ for $i \neq j$. ∎

Proposition (4.5.4). *The category SchA has finite products.*

Proof. The proof is, indeed, the classical construction ([24], Theorem 3.3). We give a brief description of it here. Because the category **A** is cocomplete, the category **AffA** is complete. Because **AffA** is a reflective subcategory of **SchA**, the inclusion functor **AffA** $\to$ **SchA** preserves limits. Thus **SchA** has as its terminal object $(\mathbf{Spec}(Z), \tilde{Z})$, a pair of affine schemes $(\mathrm{Spec}(A), \tilde{A})$, $(\mathrm{Spec}(B), \tilde{B})$ has as its product in **SchA** the scheme $(\mathrm{Spec}(A \amalg B), \widetilde{A \amalg B})$, and a pair of morphisms of affine schemes $Spec(f) : \mathrm{Spec}(B) \to \mathrm{Spec}(A)$, $\mathrm{Spec}(f') : \mathrm{Spec}(B') \to \mathrm{Spec}(A')$ have as their product the morphism $\mathrm{Spec}(f \amalg f') : \mathrm{Spec}(B \amalg B') \to \mathrm{Spec}(A \amalg A')$. According to Propositions (1.8.4) and (4.4.11), the product of a pair of affine open immersions is an affine open immersion. Let X, Y be a pair of schemes. Let $(X_i)_{i\in I}$ be an affine open covering of X and let $(Y_j)_{j\in J}$ be an affine open covering of Y. For any pair $(i,j) \in I \times J$, let us consider the product of affine schemes $T_{ij} = X_i \times Y_j$. For any pair $((i,j),(i',j')) \in (I \times J)^2$ such that $(i,j) \neq (i',j')$, let us consider the product of affine schemes $(X_i \cap X_{i'}) \times (Y_j \cap Y_{j'})$ which is isomorphic to an open subscheme of T_{ij} denoted by $T_{(ij,\, i'j')}$ and isomorphic to an open subscheme of $T_{i'j'}$ denoted by $T_{(i'j',\, ij)}$, and let $\varphi_{(ij,i'j')} : T_{(ij,i'j')} \to T_{i'j',ij)}$ be the isomorphism deduced from the two preceding ones. In this way we obtain a gluing diagram

$(T_{(ij,i'j')})_{(ij,i'j')\in \mathbf{G}(I\times J)}$ whose colimit provides a product of the pair of schemes (X, Y). ■

Proposition (4.5.5). *The category* **SchA** *has equalizers and they are immersions.*

Proof. The proof is classical ([22], §§ I.5.1.5, I.5.1.6). As the category **AffA** has equalizers preserved by the inclusion functor **AffA** $\to$ **SchA**, any pair $(f, g) : X \rightrightarrows Y$ of parallel morphisms between affine schemes has an equalizer $k : K \to X$ which is a closed immersion with $K = \{x \in X : f(x) = g(x) \text{ and } f_x^{\#} = g_x^{\#}\}$. Let $(f, g) : X \rightrightarrows Y$ be an arbitrary parallel pair of morphisms in **SchA**. Let $K = \{x \in X : f(x) = g(x) \text{ and } f_x^{\#} = g_x^{\#}\}$. Let $x \in K$, and let V_x be an affine open set of Y containing $f(x)$. There exists an affine open neighbourhood U_x of x such that $f(U_x) \subset V_x$ and $g(U_x) \subset V_x$. Let us equip U_x, V_x with the affine scheme structure induced by U, V, and let us denote the morphisms of schemes induced by f, g by $f_x, g_x : U_x \rightrightarrows V_x$. Let $k_x : K_x \to U_x$ be the equalizer of (f_x, g_x). Then $K_x = K \cap U_x$. If K is equipped with the topology induced by that of X, then $(K_x)_{x \in K}$ is an open covering of K. Moreover, K_x is an affine subscheme of X. By gluing the family of schemes $(K_x)_{x \in K}$ together, we obtain the equalizer of (f, g) in **SchA**. According to Definition (4.4.16), K is a subscheme of X. ■

Definition (4.5.6). *A scheme X is* separated *if its diagonal morphism $\Delta_x : X \to X \times X$ is a closed immersion.*

Proposition (4.5.7). *Affine schemes are separated.*

Proof. For any object A in **A**, the codiagonal morphism $\nabla_A : A \amalg A \to A$ is a regular epimorphism. According to Proposition (4.4.15), the diagonal $\Delta_{\mathrm{Spec}(A)} = \mathrm{Spec}(\nabla_A)$ is a closed immersion. ■

Proposition (4.5.8). *Separated schemes are disjunctable objects in* **SchA** *and the corresponding disjunctors are open immersions.*

Proof. Let us consider a separated scheme X and the product $(p_1 : X \times X \to X,\ p_2 : X \times X \to X)$ of X with itself. The diagonal morphism $\Delta_X : X \to X \times X$ is the equalizer of (p_1, p_2). Let U be the open subset of $X \times X$ which is the complement of the closed subset $\Delta_X(X)$. If U is equipped with the structure of the scheme induced by $X \times X$, the open immersion $U \to X \times X$ is the disjunctor of (p_1, p_2). ■

Corollary (4.5.9). *Affine schemes are disjunctable objects in* **SchA** *and, for any object A in* **A**, *the singular monomorphisms $X \to \mathrm{Spec}(A)$ in* **SchA** *are precisely the open immersions $X \to \mathrm{Spec}(A)$.*

Proof. The first part follows from Propositions (4.5.7) and (4.5.8). The second part follows from the fact that, for any congruence, $r = (r_1, r_2) : R \rightrightarrows A$ on an object A in **A**, the disjunctor of the pair $(\text{Spec}(r_1), \text{Spec}(r_2)) : \text{Spec}(A) \rightrightarrows \text{Spec}(R)$ is the open subscheme $D(r)$ of $\text{Spec}(A)$. ∎

Proposition (4.5.10). *The category* ***SchRedA*** *of reduced schemes has gluing colimits and finite limits.*

Proof. The proof follows from Propositions (4.5.2), (4.5.4), and (4.5.5), and Theorem (4.3.15). ∎

4.6 LOCALLY FINITELY PRESENTABLE SCHEMES

Definition (4.6.1). *A scheme X on* **A** *is* locally finitely presentable *if each of its points x has an affine open neighbourhood U such that $O_X(U)$ is a finitely presentable object in* **A**.

Proposition (4.6.2). *An object A in* **A** *is finitely presentable if and only if* $\text{Spec}(A)$ *is a locally finitely presentable scheme.*

Proof. If A is a finitely presentable object, $\text{Spec}(A)$ is an affine open neighbourhood of each of its points such that $\tilde{A}(\text{Spec}(A)) \simeq A$ is finitely presentable and so $\text{Spec}(A)$ is a locally finitely presentable scheme. Let $\text{Spec}(A)$ be a locally finitely presentable scheme. Let $p \in \text{Spec}(A)$. There exists an affine open neighbourhood of p of the form $D(r)$ with $r = (r_1, r_2) : R \rightrightarrows A \in \text{CongCodis}(A)$ such that $\tilde{A}(D(r))$ is finitely presentable. Then $q_p : A \to A/p$ does not coequalize (r_1, r_2). According to Axiom (1.2.2) of Zariski categories, there exists a finitely presentable codisjunctable object X and a morphism $m : X \to R$ such that $q_p r_1 m \neq q_p r_2 m$. Let $q : A \to Q$ be the coequalizer of $(r_1 m, r_2 m)$ and let s be the kernel pair of q. Then q_p does not coequalize s and so $p \in D(s)$. The quotient $q_r : A \to A/r$ satisfies $q_r r_1 m = q_r r_2 m$. Thus it factors through q and so $q_r \leqslant q$ in $\text{QuotReg}(A)$. It follows that $s \leqslant r$ in $\text{Cong}(A)$ and $D(s) \subset D(r)$. The codisjunctor of $(r_1 m, r_2 m)$ is obviously a codisjunctor for s and thus $s \in \text{CongCodis}(A)$. According to the construction of the structure sheaf $\tilde{A}$ (cf. § 3.4), the restriction morphisms $\tilde{A}(\text{Spec}(A)) \to \tilde{A}(D(r))$ and $\tilde{A}(\text{Spec}(A)) \to \tilde{A}(D(s))$ are the codisjunctors of r and s respectively. Therefore $\tilde{A}(D(s))$ is a singular quotient of $\tilde{A}(D(r))$. According to Corollary (1.8.6), $\tilde{A}(D(s))$ is finitely presentable. When p runs over $\text{Spec}(A)$, the open sets $D(s)$ form an open covering of $\text{Spec}(A)$ from which a finite covering can be extracted since $\text{Spec}(A)$ is compact. As a result, there exists a finite family of finitely presentable codisjunctable objects $(X_i)_{i \in I}$ and pairs of morphisms $((g_i, h_i) : X_i \rightrightarrows A)_{i \in I}$ with coequalizers $q_i : A \to Q_i$ and codisjunctors $\delta_i : A \to \Delta_i$ such that $\wedge_{i \in I} q_i = O_A$ in QuotReg(A) and $(\delta_i : A \to \Delta_i)_{i \in I}$ is an effective monomorphic family with finitely presentable codomains. Because the object A is a filtered colimit of finitely presentable objects and the objects X_i are finitely presentable, there exist a finitely presentable object B, a morphism $f : B \to A$, and a family of pairs of morphisms $(u_i, v_i) : X_i \rightrightarrows B$ with coequalizers k_i and codisjunctors $d_i : B \to D_i$ such that $(fu_i, fv_i) = (g_i, h_i)$ and $\wedge_{i \in I} k_i = O_B$ in QuotReg(B). According to the construction of the structure sheaf $\tilde{B}$, $(d_i : B \to D_i)_{i \in I}$ is an effective monomorphic family of morphisms. Moreover, there are morphisms $m_i : D_i \to \Delta_i$ such that (m_i, δ_i) is the pushout of (d_i, f). For any $(i, i') \in I^2$, let $(\bar{d}_i : D_i \to D_{i,i'}, \bar{d}_{i'} : D_{i'} \to D_{ii'})$ be the pushout of $(d_i, d_{i'})$, let $\bar{\delta}_i : \Delta_i \to \Delta_{ii'}, \bar{\delta}_{i'} : \Delta_{i'} \to \Delta_{ii'})$ be the pushout of $(\delta_i, \delta_{i'})$, and let $m_{ii'} : D_{ii'} \to \Delta_{ii'}$ be the morphism such that $(m_{ii'}, \bar{\delta}_i)$ is the pushout of $(\bar{d}_i, m_i)$. Because the objects $\Delta_i, \Delta_{ii'}$ are finitely presentable, we can choose the object B and the morphism $f : B \to A$ such that the morphisms m_i and $m_{ii'}$ are split epimorphisms, and furthermore that there exist sections s_i and $s_{ii'}$ of m_i and $m_{ii'}$ respectively such that $\bar{d}_i s_i = s_{ii'} \bar{\delta}_i$ and $\bar{d}_{i'} s_{i'} = s_{ii'} \bar{\delta}_{i'}$. Then the family of morphisms $(s_i \delta_i : A \to D_i)_{i \in I}$ is such that, for any $(i, i') \in I^2$, we have $\bar{d}_i s_i \delta_i = s_{ii'} \bar{\delta}_i \delta_i = s_{ii'} \bar{\delta}_{i'} \delta_{i'} = \bar{d}_{i'} s_{i'} \delta_{i'}$. Because $(d_i : B \to D_i)_{i \in I}$ is an effective monomorphic family, there exists a unique morphism $s : A \to B$ such that $d_i s = s_i \delta_i$ for any $i \in I$. Then $\delta_i fs = m_i d_i s = m_i s_i \delta_i = \delta_i$ for any $i \in I$. Consequently, $fs = 1_A$ and thus A is a split quotient of B. Because B is finitely presentable, A is also. ■

Proposition (4.6.3). *Finite limits and gluing colimits of locally finitely presentable schemes are locally finitely presentable.*

Proof. It is obvious that gluing colimits of locally finitely presentable schemes are locally finitely presentable. The functor $\Sigma : \mathbf{A}^{\mathrm{op}} \to \mathbf{SchA}$ preserves finite limits and sends finitely presentable objects on locally finitely presentable affine schemes (Proposition (4.6.2)). Since the class of finitely presentable objects of $\mathbf{A}$ is stable under finite colimits, the class of locally finitely presentable affine schemes is stable under finite limits. According to the construction of finite limits of schemes (cf. Propositions (4.5.4) and (4.5.5)), finite limits of locally finitely schemes are built up by gluing together locally finitely presentable affine schemes. Therefore they are locally finitely presentable. ■

Notation. Let us denote by $\mathbf{A}_0$, $\mathbf{IntA}_0$, $\mathbf{RedA}_0$, $\mathbf{QuaPrimA}_0$, $\mathbf{PrimA}_0$, $\mathbf{IrrA}_0$, $\mathbf{RegA}_0$, and $\mathbf{JacA}_0$ the full subcategories of $\mathbf{A}$, $\mathbf{IntA}$, $\mathbf{RedA}$, $\mathbf{QuaPrimA}$, $\mathbf{PrimA}$, $\mathbf{IrrA}$, $\mathbf{RegA}$, and $\mathbf{JacA}$ respectively whose objects are the finitely presentable ones. Let us denote by $\mathbf{SchA}_0$, $\mathbf{AffA}_0$, $\mathbf{SchIntA}_0$, $\mathbf{SchInt}_{\mathrm{rat}}\mathbf{A}_0$, $\mathbf{SchIrrA}_0$, $\mathbf{SchIrr}_{\mathrm{rat}}\mathbf{A}_0$, and $\mathbf{SchRedA}_0$ the full subcategories of $\mathbf{SchA}$, $\mathbf{AffA}$, $\mathbf{SchIntA}$, $\mathbf{SchInt}_{\mathrm{rat}}\mathbf{A}$, $\mathbf{SchIrrA}$, $\mathbf{SchIrr}_{\mathrm{rat}}\mathbf{A}$, and $\mathbf{SchRedA}$ respectively whose objects are the locally finitely presentable ones. Let us denote by $J_{\mathbf{A}_0} : \mathbf{A}_0 \to \mathbf{A}$ the inclusion functor, by $\Sigma_{\mathbf{A}_0} : \mathbf{A}_0^{\mathrm{op}} \to \mathbf{SchA}_0$ the functor induced by $\Sigma_{\mathbf{A}} : \mathbf{A}^{\mathrm{op}} \to \mathbf{SchA}$, and by $K_{\mathbf{A}_0} : (\mathbf{SchIrr}_{\mathrm{rat}}\mathbf{A}_0)^{\mathrm{op}} \to \mathbf{PsdSimA}$ and $K_{\mathbf{A}_0} : (\mathbf{SchInt}_{\mathrm{rat}}\mathbf{A}_0)^{op} \to \mathbf{SimA}$ the restrictions of the functors $K_{\mathbf{A}} : (\mathbf{SchIrr}_{\mathrm{rat}}\mathbf{A})^{\mathrm{op}} \to \mathbf{PsdSimA}$ and $K_{\mathbf{A}} : (\mathbf{SchInt}_{\mathrm{rat}}\mathbf{A})^{\mathrm{op}} \to \mathbf{SimA}$ respectively. Proposition (4.6.2) asserts that $\Sigma_{\mathbf{A}_0}$ induces an equivalence of categories between $\mathbf{A}_0^{\mathrm{op}}$ and $\mathbf{AffA}_0$, and Proposition (4.6.3) asserts that $\mathbf{SchA}_0$ is finitely complete.

Theorem (4.6.4). *The functor $K_{\mathbf{A}_0} : (\mathbf{SchIrr}_{\mathrm{rat}}\mathbf{A}_0)^{op} \to \mathbf{PsdSimA}$ is full and faithful.*

Proof. Let X and Y be a pair of objects in $\mathbf{SchIrr}_{\mathrm{rat}}\mathbf{A}_0$ with generic points x and y respectively, and let $k : K_{\mathbf{A}_0}(Y) = 0_{Y,y} \to K_{\mathbf{A}_0}(X) = O_{X,x}$ be a morphism in $\mathbf{PsdSimA}$. Let V be a non-empty affine open set of Y. Because the object $O_Y(V)$ is finitely presentable and $O_{X,x} = \xrightarrow{lim}_{\varnothing \neq U \in \Omega(X)} O_X(U)$ is a filtered colimit, there exists a non-empty affine open set U of X and a morphism $m : O_Y(V) \to O_X(U)$ such that $\rho_x^U m = k\rho_y^V$. Let $i : U \to X$ and $j : V \to Y$ be the inclusion morphism of schemes, let $g : U \to V$ be the morphism of schemes $\mathrm{Spec}(m)$, and let $\overline{(i, jg)} : X \to Y$ be the morphism in the category $\mathbf{SchIrr}_{\mathrm{rat}}\mathbf{A}_0$. We have $K_{\mathbf{A}_0}\overline{((i, jg))} = m_x = k$. Let us assume that the morphism $k : K_{\mathbf{A}_0}(Y) \to K_{\mathbf{A}_0}(X)$ is of the form $k = K_{\mathbf{A}_0}\overline{((i', f))}$ for some morphism $\overline{(i' : U' \to X, f : U' \to Y)}$ in $\mathbf{SchIrr}_{\mathrm{rat}}(\mathbf{A}_0)$. Then we can choose U, U', and V such that $U' = U \subset f^{-1}(V)$. The pair of morphisms $(m, \rho_U^{f^{-1}(V)} f_V^{\#}) : O_Y(V) \rightrightarrows O_X(U)$ is such that

$$\rho_x^U \rho_U^{f^{-1}(V)} f_V^{\#} = \rho_x^{f^{-1}(V)} f_V^{\#} = f_x \rho_y^V = k\rho_y^V = \rho_x^U m.$$

Since $O_Y(V)$ is finitely presentable, we can choose U such that $m = \rho_U^{f^{-1}(V)} f_V^{\#}$. Then $f = gj$ and $\overline{(i', f)} = \overline{(i, jg)}$. It follows that the functor $K_{\mathbf{A}_0}$ is full and faithful. ■

Corollary (4.6.5). *The functor $K_{A_0} : (\mathbf{SchInt}_{\mathrm{rat}}A_0)^{\mathrm{op}} \to \mathbf{SimA}$ is full and faithful.*

Corollary (4.6.6). *For a pair of locally finitely presentable irreducible schemes, X, Y, the following assertions are equivalent.*

(i) *X and Y are birationally equivalent.*

(ii) $K_{\mathbf{A}_0}(X) \simeq K_{\mathbf{A}_0}(Y)$.

Proposition (4.6.7). *For a locally noetherian [16] Zariski category A, $\mathbf{SchRed}A_0$ is a coreflective subcategory of $\mathbf{SchA}_0$.*

Proof. Let (X, O_X) be an object in $\mathbf{SchA}_0$ and let $(X, O_{R(X)})$ be the co-universal reduced scheme associated to it (Proposition (4.3.15)). Let U be an affine open set of X. Then $(U, O_{R(X)}/U)$ is an affine scheme isomorphic to $\mathrm{Spec}(R(O_X(U)))$ (cf. proof of Proposition (4.3.15)). Since $O_X(U)$ is finitely presentable, $R(O_X(U)) \simeq O_X(U)/\mathrm{rad}(O_X(U))$ is also. Therefore $(X, O_{R(X)})$ is locally finitely presentable. It follows easily that $\mathbf{SchRedA}_0$ is a coreflective subcategory of $\mathbf{SchA}_0$. ■

Corollary (4.6.8). *For a locally noetherian Zariski category A, the category $SchRedA_0$ has finite limits and gluing colimits.*

Proof. The proof follows from Proposition (4.6.3). ■

Definition (4.6.9). *A* scheme *on A is* finitely presentable *if it is locally finitely presentable and compact.*

The *category* **SchFinPrA** of *finitely presentable schemes on* **A** is the full subcategory of $\mathbf{SchA}_0$ whose objects are the finitely presentable ones.

5
JACOBSON ULTRASCHEMES

An object A in a Zariski category is a Jacobson object if any prime congruence on A is the meet of maximal congruences. The prime spectrum of a Jacobson object is a Jacobson space, i.e. a space in which any closed set is the closure of the set of its closed points. Thus the maximal spectrum of a Jacobson object is very dense in its prime spectrum, and for many purposes it can replace the prime spectrum. Jacobson schemes are obtained by gluing together affine schemes of Jacobson objects, while Jacobson ultraschemes are obtained by gluing together their affine ultraschemes. The ultraschemes are introduced in order to remove the generic points and deal only with the closed points.

5.1 JACOBSON OBJECTS

Definition (5.1.1).

(i) *An object A is a* Jacobson object *if any prime congruence on A is the meet of maximal congruences.*

(ii) *A morphism $f: A \to B$ is a* Jacobson morphism *if A and B are Jacobson objects and the inverse image by f of any maximal congruence on B is a maximal congruence on A.*

The *category* **JacA** *of Jacobson objects of* **A** is the subcategory of **A** whose objects and morphisms are the Jacobson ones.

Proposition (5.1.2). *The category of Jacobson objects of A contains the category of regular objects of A as a full subcategory.*

Proof. The proof follows from Propositions (2.11.4) and (2.11.5). ■

Proposition (5.1.3). *The intersection of the category of Jacobson objects of A with the category of local objects of A is the category of pseudo-simple objects of A.*

Proof. The proof follows from Proposition (2.6.2). ■

Let us recall that a topological space is a *Jacobson space* if any of its closed subsets is the closure of the set of its closed points ([21], 0, 2.8). Let us define a *Jacobson map* $f: X \to Y$ as a continuous map which preserves closed points between two Jacobson spaces. The Jacobson spaces and maps are the objects and morphisms of the *category of Jacobson spaces* **TopJac**.

Proposition (5.1.4). *An object (morphism) of A is a Jacobson object (morphism) if and only if its prime spectrum is a Jacobson space (map).*

Proof. Let A be an object in **A**. According to proposition (3.2.1) and Definition (3.2.2), the set of closed points of $\mathrm{Spec}(A)$ is the maximal spectrum of A, i.e. $\mathrm{Spec}_{\max}(A)$. Thus $\mathrm{Spec}(A)$ is a Jacobson space if and only if $\mathrm{Spec}_{\max}(A)$ is very dense in $\mathrm{Spec}(A)$ ([21], 0, 2.6).

Let us assume that A is a Jacobson object. Then any radical congruence on A is the meet of maximal congruences. Let $V(r)$, where r is a radical congruence on A, be a closed set in $\mathrm{Spec}(A)$. Let $D(s)$, where s is a radical congruence on A, be an open set in $\mathrm{Spec}(A)$ such that $D(s) \cap V(r) \neq \emptyset$. Then $D(s) \not\subset D(r)$; hence $s \not\leqslant r$ and there exists $m \in \mathrm{Spec}_{\max}(A)$ such that $r \leqslant m$ and $s \not\leqslant m$. Then $m \in V(r) \cap D(s)$. Thus $V(r) \cap D(s) \cap \mathrm{Spec}_{\max}(A) \neq \emptyset$. It follows that $V(r) \cap \mathrm{Spec}_{\max}(A)$ is dense in $V(r)$. As a result, $\mathrm{Spec}_{\max}(A)$ is very dense in $\mathrm{Spec}(A)$ and hence $\mathrm{Spec}(A)$ is a Jacobson space.

Conversely, let us assume that $\mathrm{Spec}(A)$ is a Jacobson space. Let $p \in \mathrm{Spec}(A)$. Let $r \in \mathrm{Cong}(A)$ such that $r \not\leqslant p$. Then $p \in V(p) \cap D(r)$. Thus $V(p) \cap D(r) \neq \emptyset$ and hence $V(p) \cap D(r) \cap \mathrm{Spec}_{\max}(A) \neq \emptyset$. Let $m \in V(p) \cap D(r) \cap \mathrm{Spec}_{\max}(A)$. Then $r \not\leqslant m$ and $p \leqslant m$. It follows that $p = \bigwedge_{psm \in \mathrm{Spec}_{\max}(A)} m$. As a result, A is a Jacobson object. On the other hand, it is obvious that a morphism $f: A \to B$ between Jacobson objects is a Jacobson morphism if and only if $\mathrm{Spec}(f)$ is a Jacobson map. ■

Proposition (5.1.5). *The category of Jacobson objects of A is closed in A under finite products, regular quotients, and singular quotients.*

Proof. The terminal object is obviously a Jacobson object. Let A, B be a pair of Jacobson objects. According to Proposition (3.9.9), $\mathrm{Spec}(A \times B) = \mathrm{Spec}(A) \amalg \mathrm{Spec}(B)$. According to Proposition (5.1.4), $\mathrm{Spec}(A)$ and $\mathrm{Spec}(B)$ are Jacobson spaces. Thus $\mathrm{Spec}(A \times B)$ is a Jacobson space ([21], 0,2.8) and hence $A \times B$ is a Jacobson object. Let A be a Jacobson object and let $f: A \to B$ be a regular epimorphism. According to Proposition (3.3.11), $\mathrm{Spec}(B)$ is homeomorphic to a closed subspace of $\mathrm{Spec}(A)$. Because $\mathrm{Spec}(A)$ is a Jacobson space, $\mathrm{Spec}(B)$ is also ([21], 0, 2.8). Therefore B is a Jacobson object. If $d: A \to D$ is a singular epimorphism, then $\mathrm{Spec}(D)$ is homeomorphic to an open subspace of $\mathrm{Spec}(A)$ (Proposition (3.3.10)). Thus $\mathrm{Spec}(D)$ is a Jacobson space ([21], 0, 2.8) and hence D is a Jacobson object. ■

The maximal spectrum functor $\mathbf{Spec}_{\max}: (\mathbf{JacA})^{op} \to \mathbf{TopJac}$

The *maximal spectrum functor* $\mathrm{Spec}_{\max}: (\mathbf{JacA})^{op} \to \mathbf{TopJac}$ is the functor induced by the prime spectrum functor $\mathbf{Spec}: \mathbf{A}^{op} \to \mathbf{TopSpec}$ (cf. § 3.1, p. 86).

Let A be a Jacobson object in $\mathbf{A}$.

Definition (5.1.6). *An* open set of $\mathrm{Spec}_{\max}(A)$ *is* ultra-affine *if it is the intersection with* $\mathrm{Spec}_{\max}(A)$ *of an affine open set of* $\mathrm{Spec}(A)$.

According to Definition (3.1.10), the ultra-affine open sets of $\mathrm{Spec}_{\max}(A)$ are of the form

$$D_{\max}(r) = \{m \in \mathrm{Spec}_{\max}(A) : r \not\leq m\}$$

where r runs through the set of codisjunctable congruences on A.

Proposition (5.1.7). *Ultra-affine open sets of* $\mathrm{Spec}_{\max}(A)$ *form an open basis for the topology on* $\mathrm{Spec}_{\max}(A)$.

Proof. The proof follows from Proposition (3.1.14). ■

Let us denote the *set of ultra-affine open sets of* $\mathrm{Spec}_{\max}(A)$ by $\Omega_{\mathrm{Ult}}(\mathrm{Spec}_{\max}(A))$.

Proposition (5.1.8). *For any Jacobson object A, the correspondence* $U \to U \cap \mathrm{Spec}_{\max}(A)$ *provides an isomorphism of meet semilattices* $\Omega_a(\mathrm{Spec}(A)) \to \Omega_{\mathrm{Ult}}(\mathrm{Spec}_{\max}(A))$.

Proof. The proof follows from the fact that the inclusion map i_A: $\mathrm{Spec}_{\max}(A) \mapsto \mathrm{Spec}(A)$ is a quasi-homeomorphism ([21], 0, 2.8). ■

According to the preceding discussion and to the construction of the structure sheaf on $\mathrm{Spec}(A)$, we can state the following definition.

Definition (5.1.9). *The* structure sheaf *on* $\mathrm{Spec}_{\max}(A)$ *is the sheaf* $\tilde{A}_{max}$*:* $\Omega(\mathrm{Spec}_{\max}(A))^{op} \to \mathbf{A}$ *based on* $\mathrm{Spec}_{\max}(A)$ *with values in* $\mathbf{A}$, *whose value for any ultra-affine open set* $D_{\max}(r)$ *of* $\mathrm{Spec}_{\max}(A)$ *is* $\tilde{A}_{\max}(D_{max}(r)) = \Delta(r)$ *where* $\delta(r): A \to \Delta(r)$ *is the codisjunctor of* r, *and whose restriction morphism along* $D_{\max}(r) \to D_{\max}(s)$ *is the morphism* $\tilde{A}_{\max}(D_{\max}(r) \to D_{\max}(s)): \tilde{A}_{\max}(D_{\max}(s)) \to \tilde{A}_{\max}(D_{\max}(r))$ *whose composite with* $\delta(s)$ *is* $\delta(r)$.

Theorem (5.1.10). *The structure sheaf* $\tilde{A}_{\max}$ *on the maximal spectrum* $\mathrm{Spec}_{\max}(A)$ *of a Jacobson object A has as its stalks the localized objects* *of A at maximal congruences on A and as its object of global sections the object A.*

Proof. The proof follows from Proposition (3.4.4) and from the fact that $i_A : \mathrm{Spec}_{\max}(A) \to \mathrm{Spec}(A)$ is a quasi-homeomorphism. ∎

Definition (5.1.11). *The* Jacobson radical *of an object A is the meet of maximal congruences on A, denoted by* $\mathrm{Jac}(A)$.

5.2 JACOBSON SCHEMES

Definition (5.2.1). *A scheme (morphism of schemes) on A is a* Jacobson scheme (morphism) *if its underlying space (continuous map) is a Jacobson space (map).*

The *category* **SchJacA** *of Jacobson schemes on* **A** is the subcategory of **SchA** whose objects and morphisms are the Jacobson ones. The *category* $\mathbf{SchJacA}_0$ *of locally finitely presentable Jacobson schemes on* **A** is the full subcategory of **SchJacA** whose objects are locally finitely presentable (Definition (4.6.1)). The *category* **AffJacA** (or $\mathbf{AffJacA}_0$) of *affine* (locally finitely presentable affine) *Jacobson schemes* on **A** is the full subcategory of **SchJacA** (or $\mathbf{SchJacA}_0$) whose objects are affine.

Proposition (5.2.2). *The prime spectrum functor $\Sigma : A^{\mathrm{op}} \to SchA$ induces a prime spectrum functor $\Sigma_{\mathrm{Jac}} : (\boldsymbol{JacA})^{\mathrm{op}} \to \boldsymbol{SchJacA}$.*

Proof. The proof follows from Proposition (5.1.4). ∎

Theorem (5.2.3). *The functor Σ_{Jac} induces an equivalence of categories between the dual of the category **JacA** of Jacobson objects of A and the category **AffJacA** of affine Jacobson schemes on A.*

Proof. The proof follows from Theorem (4.3.2). ∎

The maximal spectrum functor $\Sigma_{\max} : (\mathbf{JacA})^{\mathrm{op}} \to \mathbf{SpModLocA}$

For any Jacobson object A, let us denote by $\Sigma_{\max}(A)$ the locally modelled space $\mathrm{Spec}_{\max}(A)$ equipped with the structure sheaf $\tilde{A}_{\max}$ (Definition (5.1.9)). Let $f : A \to B$ be a Jacobson morphism. Let $D_{\max}(r)$, where $r \in \mathrm{Congcodis}(A)$, be an ultra-affine open set of $\mathrm{Spec}_{\max}(A)$ (Definition (5.1.6)). Let $s = f_{*c}(r) \in \mathrm{CongCodis}(B)$ and let $(\delta(s) : B \to \Delta(s), \tilde{f}_r : \Delta(r) \to \Delta(s))$ be the pushout of $(f : A \to B, \delta(r) : A \to \Delta(r))$. According to the construction of $\tilde{A}_{\max}$ and $\tilde{B}_{\max}$ (Definition (5.1.9)), $\tilde{A}_{\max}(D_{\max}(r)) = \Delta(r)$ and $\tilde{B}_{\max}(D_{\max}(s)) = \Delta(s)$. Since $(\mathrm{Spec}_{\max}(f))^{-1}(D_{\max}(r)) = D_{\max}(s)$

(see p. 140), we can define a morphism $\tilde{f}_{\max} : \tilde{A}_{\max} \to (\mathrm{Spec}_{\max}(f))_*(\tilde{B}_{\max})$ whose value at $D_{\max}(r)$ is $\tilde{f}_r$. Indeed, $\tilde{f}_{\max} = \tilde{f}((i_A^{-1})^{\mathrm{op}})^{-1}$ and the stalks of $\tilde{f}_{\max}$ are local morphisms. Then we define the morphism of locally modelled spaces $\Sigma_{\max}(f) : \Sigma_{\max}(B) \to \Sigma_{\max}(A)$ by $\Sigma_{\max}(f) = (\mathrm{Spec}_{\max}(f), \tilde{f}_{\max})$. In this way we obtain a functor $\Sigma_{\max} : (\mathbf{JacA})^{\mathrm{op}} \to \mathbf{SpModLocA}$ called the *maximal spectrum functor* and also denoted by $\mathrm{Spec}_{\max}$.

Proposition (5.2.4). *The maximal spectrum functor* $\Sigma_{\max} : (\boldsymbol{JacA})^{op} \to$ ***SpModLocA*** *is full and faithful.*

Proof. Let A, B be a pair of Jacobson objects in $\mathbf{A}$. Let $(f, g) : A \rightrightarrows B$ be a pair of Jacobson morphisms in $\mathbf{A}$ such that $\Sigma_{\max}(f) = \Sigma_{\max}(g)$, i.e. $(\mathrm{Spec}_{\max}(f), \tilde{f}_{\max}) = (\mathrm{Spec}_{\max}(g), \tilde{g}_{\max})$. Let $i_A : \mathrm{Spec}_{\max}(A) \to \mathrm{Spec}(A)$ and $i_B : \mathrm{Spec}_{\max}(B) \to \mathrm{Spec}(B)$ be the inclusion maps. Then $\mathrm{Spec}(f) i_B = i_A \mathrm{Spec}_{\max}(f) = i_A \mathrm{Spec}_{\max}(g) = \mathrm{Spec}(g) i_B$. Because $\mathrm{Spec}_{\max}(B)$ is dense in $\mathrm{Spec}(B)$ and $\mathrm{Spec}(A)$ is a T_o-space, the equality $\mathrm{Spec}(f) = \mathrm{Spec}(g)$ follows. Moreover, $\tilde{f} = \tilde{f}_{\max}(i_A^{-1})^{\mathrm{op}} = \tilde{g}_{\max}(i_A^{-1})^{\mathrm{op}} = \tilde{g}$. Consequently, $f = g$ (Theorem (4.1.6)). As a result, the functor $\Delta_{\max}$ is faithful. Let $\varphi : \Sigma_{\max}(B) \to \Sigma_{\max}(A)$ be a morphism of locally modelled spaces. Let

$$f = \varphi^{\#}_{\mathrm{Spec}_{\max}(A)} : \tilde{A}_{\max}(\mathrm{Spec}_{\max}(A)) = A \to \tilde{B}_{\max}(\mathrm{Spec}_{\max}(B)) = B.$$

We now prove that $\Sigma_{\max}(f) = \varphi$. Let $p \in \mathrm{Spec}_{\max}(B)$ and $q = \varphi(p)$. We have the relation

$$\varphi^{\#}_p \rho_q^{\mathrm{Spec}_{\max}(B)} = \rho_p^{\mathrm{Spec}_{\max}(A)} \varphi^{\#}_{\mathrm{Spec}_{\max}(A)}$$

which can be written $\varphi^{\#}_p l_q = l_p f$. Because l_q is a presingular epimorphism and $\varphi^{\#}_p$ is local, $\varphi^{\#}_p l_q$ is the presingular factorization of the morphism $l_p f$ (Theorem (1.9.6)). Because of the essential uniqueness of this factorization, $q = f^{*c}(p) = \mathrm{Spec}(f)(p)$ and $\varphi^{\#}_q = f_p$ (cf. Theorem (2.13.3)). Consequently, $\varphi^{\#}_p = (\tilde{f}_{\max})_p$. Thus $\varphi^{\#} = \tilde{f}_{\max}$ and $\varphi = \Sigma_{\max}(f)$. As a result, the functor $\Sigma_{\max}$ is full. ■

5.3 JACOBSON ULTRASCHEMES

Definition (5.3.1).

(i) *An* affine ultrascheme on $\mathbf{A}$ *is a modelled space on* $\mathbf{A}$ *isomorphic to the modelled space* $\mathrm{Spec}_{\max}(A)$ *for some Jacobson object* A *in* $\mathbf{A}$.

(ii) *The* category ***AffUltA*** *of* affine ultraschemes on $\mathbf{A}$ *is the full subcategory of* ***SpModLocA*** *whose objects are the affine ultraschemes on* $\mathbf{A}$.

Theorem (5.3.2). *The category* ***AffUltA*** *of affine ultraschemes on* $\mathbf{A}$ *is equivalent to the dual of the category* ***JacA*** *of Jacobson objects of* $\mathbf{A}$.

Proof. The proof follows from Proposition (5.2.4). ■

Definition (5.3.3).

(i) *An open set U of a modelled space X is* ultra-affine *if the induced modelled space U is an affine ultrascheme.*

(ii) *An* ultrascheme on A *is a modelled space X on A such that any element of X has an ultra-affine open neighbourhood.*

(iii) *The* category **UltSchA** of ultraschemes on A *is the full subcategory of* **SpModLocA** *whose objects are the ultraschemes on A.*

(iv) *The* maximal spectrum functor $\Sigma_{\max} : (\mathbf{JacA})^{\mathrm{op}} \to \mathbf{UltSchA}$ *is induced by the functor $\Sigma_{\max}$ (Proposition (5.2.4)).*

Theorem (5.3.4). *The category* **UltSchA** *of ultraschemes on A is equivalent to the category* **SchJacA** *of Jacobson schemes on A.*

Proof. Let X be a Jacobson scheme. Let us denote the set of closed points of X by $S(X)$ and the inclusion map by $i_X : S(X) \to X$. If we equip $S(X)$ with the topology induced by that of X, the map i_X becomes a quasi-homeomorphism ([21], 0, 2.7), i.e. the order-preserving map $i_X^{-1} : \Omega(X) \to \Omega(S(X))$ is an isomorphism of lattices. If we denote by $O_{S(X)}$ the sheaf on $S(X)$ which is the inverse image of the sheaf O_X along i_X, then the pair $(S(X), O_{S(X)})$ is a locally modelled space which is indeed an ultrascheme on **A**. Let us denote this ultrascheme simply by $S(X)$. Let $f: X \to Y$ be a morphism of Jacobson schemes. The continuous map f induces a continuous map $S(f) : S(X) \to S(Y)$. Let us denote the natural transformation $f^{\#}((i_Y^{-1})^{\mathrm{op}})^{-1}$ by $S(f)^{\#} : O_{S(Y)} \to S(f)_*(O_{S(X)})$. Then the pair $(S(f), S(f)^{\#})$ is a morphism of locally modelled spaces $(S(X), O_{S(X)}) \to (S(Y), O_{S(Y)})$ simply denoted by $S(f)$. In this way we obtain a functor S: **SchJacA** $\to$ **UltSchA**. Let us prove that this functor is an equivalence of categories.

Let X, Y be a pair of objects in **SchJacA**. Let $\varphi : S(X) \to S(Y)$ be a continuous map and let $\varphi^{-1} : \Omega(S(Y)) \to \Omega(S(X))$ be the complete Heyting algebras homomorphism defined by φ. If $i_X : S(X) \to X$ and $i_Y : S(Y) \to Y$ denote the inclusion maps, we obtain a complete Heyting algebras homomorphism $\alpha = (i_X^{-1})^{-1}\varphi^{-1}i_Y^{-1} : \Omega(Y) \to \Omega(X)$. Because X and Y are sober spaces, α defines uniquely a continuous map $f: X \to Y$ such that $f^{-1} = \alpha : \Omega(Y) \to \Omega(X)$. The relation $i_X^{-1}f^{-1} = i_X^{-1}\alpha = \varphi^{-1}i_Y^{-1} : \Omega(Y) \to \Omega(S(X))$ implies $fi_X = i_Y\varphi$, i.e. φ is induced by f. The assignment $\varphi \to f$ defines a map $\mathrm{Hom}_{\mathbf{Top}}(S(X), S(Y)) \to \mathrm{Hom}_{\mathbf{TopJac}}(X, Y)$ which is indeed the inverse of the map $\mathrm{Hom}_{\mathbf{TopJac}}(X, Y) \to \mathrm{Hom}_{\mathbf{Top}}(S(X), S(Y))$ given by restriction. As a result the restriction process yields a bijection $\mathrm{Hom}_{\mathbf{TopJac}}(X, Y) \to \mathrm{Hom}_{\mathbf{Top}}(S(X), S(Y))$.

Furthermore, for any pair of objects X, Y in **SchJacA**, where the map $i_Y : S(Y) \to Y$ is a quasi-homeomorphism, the assignment $\theta \mapsto \theta((i_Y^{-1})^{\mathrm{op}})^{-1}$ provides a bijection between the set of natural transformations $O_Y \to f_*(O_X)$ and the set of natural transformations $O_{S(Y)} \to S(f)_*(O_{S(X)})$. It follows that two morphisms $f, g : X \rightrightarrows Y$ in **SchJacA** such that $S(f) = S(g)$ in **UltSchA** are equal. Thus the functor S is faithful.

Let $(\varphi, \varphi^{\#}) : (S(X), O_{S(X)}) \to (S(Y), O_{S(Y)})$ be a morphism in **UltSchA**. According to the preceding discussion, there exist a unique Jacobson continuous map $f : X \to Y$ which induces φ and a unique natural transformation $f^{\#} : O_Y \to f_*(O_X)$ such that $f^{\#}((i_Y^{-1})^{\mathrm{op}})^{-1} = \varphi^{\#}$. Let us prove that $(f, f^{\#}) : (X, O_X) \to (Y, O_Y)$ is a morphism of Jacobson schemes, i.e. a morphism of locally modelled spaces. Let $x \in X$. There exists an affine open neighbourhood U of x and an affine open neighbourhood V of $f(x)$ such that $f(U) \subset V$. Let us denote the map induced by f by $g : U \to V$. Then U, V are Jacobson spaces and g is a Jacobson map. Let us denote the open subscheme of (X, O_X) induced on U by (U, O_U), and the open subscheme of (Y, O_Y) induced on V by (V, O_V). Then (U, O_U) and (V, O_V) are affine Jacobson schemes. The morphism $(f, f^{\#})$ induces a morphism of modelled spaces $(g, g^{\#}) : (U, O_U) \to (V, O_V)$. This morphism induces a morphism of modelled spaces $(\psi, \psi^{\#}) : (S(U), O_{S(U)}) \to (S(V), O_{S(V)})$. But this latter morphism is induced by the morphism $(\phi, \phi^{\#}) : (S(X), O_{S(X)}) \to (S(Y), O_{S(Y)})$. Therefore it is a morphism of locally modelled spaces and hence a morphism of affine ultraschemes. Because the functor $\Sigma_{\max} : (\mathbf{JacA})^{\mathrm{op}} \to \mathbf{SpModLocA}$ is full, $(g, g^{\#}) : (U, O_U) \to (V, O_V)$ is necessarily a morphism of affine Jacobson schemes and thus is a morphism of locally modelled spaces. Consequently, the morphism $g_x^{\#} : O_{V, g(x)} \to O_{U,x}$ is local, i.e. the morphism $f_x^{\#} : O_{Y, f(x)} \to O_{X,x}$ is local. It follows that $(f, f^{\#})$ is a morphism of locally modelled spaces, i.e. a morphism of Jacobson schemes. As a result, the functor S is full. This functor S is essentially surjective by the definition of ultraschemes (Definition (5.3.3)). ■

Definition (5.3.5). *An* ultrascheme X *on* A *is* locally finitely presentable *if each of its points* x *has an ultra-affine open neighbourhood* U *such that* $O_X(U)$ *is a finitely presentable object in* A.

The *category* $\mathbf{UltSchA}_0$ of *locally finitely presentable ultraschemes* on **A** is the full subcategory of **UltSchA** whose objects are the locally finitely presentable ones.

Proposition (5.3.6). *The category* $\mathbf{UltSchA}_0$ *of locally finitely presentable ultraschemes on* A *is equivalent to the category* $\mathbf{SchJacA}_0$ *of locally finitely presentable Jacobson schemes on* A.

Proof. Let X be a Jacobson scheme and let $(X_i)_{i \in I}$ be a family of affine open sets of X. Then $S(X)$ is an ultrascheme and $(S(X_i))_{i \in I}$ is a family of ultra-affine open sets of $S(X)$. Because the inclusion map $i_X : S(X) \to X$ is a quarsi-homeomorphism, $(X_i)_{i \in I}$ is a covering of X if and only if

$(S(X_i))_{i \in I}$ is a covering of $S(X)$. Moreover, for any $i \in I$, $O_{S(X)}(S(X_i)) = O_X(X_i)$, so that $O_X(X_i)$ is a finitely presentable object if and only if $O_{S(X)}(S(X_i))$ is a finitely presentable object. It follows that X is a locally finitely presentable scheme if and only if $S(X)$ is a locally finitely presentable ultrascheme. Then the proposition follows from Theorem (5.3.4). ■

Corollary (5.3.7). *A Jacobson object A in $\mathbf{A}$ is finitely presentable if and only if* $\mathrm{Spec}_{\max}(A)$ *is a locally finitely presentable ultrascheme. Thus we obtain the maximal spectrum functor* $\Sigma_{\max}: (\mathbf{JacA}_o)^{\mathrm{op}} \to \mathbf{UltSchA}_0$.

Proof. The proof follows from Propositions (4.6.2) and (5.3.6). ■

Definition (5.3.8). *An* ultrascheme *on $\mathbf{A}$ is* integral (*reduced*) *if the value of its structure sheaf on any non-empty open set is integral* (*reduced*). *It is* irreducible (connected, compact) *if its underlying space is irreducible* (*connected, compact*).

Proposition (5.3.9). *For a Jacobson scheme X on $\mathbf{A}$, the ultrascheme $S(X)$ of closed points of X is integral* (*reduced, irreducible, connected, compact*) *if and only if the scheme X is integral* (*reduced, irreducible, connected, compact*).

Proof. The proof follows immediately from the fact that $S(X)$ is very dense in X. ■

Proposition (5.3.10). *An ultrascheme on $\mathbf{A}$ is reduced if and only if its stalks are reduced.*

Proof. The proof is similar to the proof of Proposition (4.3.9). ■

Proposition (5.3.11). *If X is an irreducible* (*integral*) *ultrascheme, the object* $\underrightarrow{\lim}_{\varnothing \neq U \in \Omega(X)} O_X(U)$ *is pseudo-simple* (*simple*) *and is precisely the pseudo-simple* (*simple*) *object of the scheme associated to X.* (cf. notation on p. 124).

Proof. The ultrascheme X is of the form $S(Y)$ where Y is a Jacobson scheme. Then Y is irreducible (Proposition (5.3.9)). Let y be the generic point of Y. For any open set U of Y, U is a neighbourhood of $y \Leftrightarrow U \neq \varnothing \Leftrightarrow U \cap X \neq \varnothing$. Thus (Theorem (4.3.11) and notation on p. 124).

$$\underset{\varnothing \neq U \in \Omega(X)}{\underrightarrow{\lim}} O_X(U) = \underset{y \in V \in \Omega(Y)}{\underrightarrow{\lim}} O_Y(V) = K_{\mathbf{A}}(Y).$$ ■

Definition (5.3.12). *If X is an irreducible (integral) ultrascheme, the object* $\varinjlim_{\emptyset \neq U \in \Omega(X)} O_X(U)$ *is called the* pseudo-simple (simple) object of X *and is denoted by* $K_A(X)$ *or* $K(X)$.

Definition (5.3.13). *An ultrascheme on A is* finitely presentable *if it is locally finitely presentable and compact.*

Proposition (5.3.14). *For a Jacobson scheme X on A, the ultrascheme $S(X)$ of closed points of X is finitely presentable if and only if the scheme X is finitely presentable.*

Proof. The proof follows from Propositions (5.3.6) and (5.3.9). ■

6
ALGEBRAIC VARIETIES

The Zariski category **RedCAlg**(k) of reduced commutative algebras over an algebraically closed field k is the most appropriate for the study of algebraic varieties over k. What are the special features of this Zariski category that make classical algebraic geometry work? It is a locally neotherian Zariski category whose initial object is simple and algebraically closed, and in which any object is reduced. We use here the notion of locally neotherian categories introduced by Gabriel and Ulmer [16] and that of algebraically closed objects introduced by Fakir [14] in any locally finitely presentable category. We prove that classical algebraic geometry can be performed on any Zariski category satisfying these properties.

The rational Zariski categories are the locally neotherian Zariski categories whose initial object K is simple and algebraically closed. The finitely presentable objects and locally finitely presentable schemes on such a category are Jacobson, and so it is sufficient to deal with closed points. But these points turn out to be rational, i.e. their residue simple object is isomorphic to K. They are the usual points. Then the structure sheaf of a locally finitely presentable ultrascheme X can be canonically represented as a sheaf of K-valued functions on X, and this representation is faithful if and only if X is reduced.

The reduced rational Zariski categories are the rational Zariski categories in which any object is reduced. For such a category, the category of locally finitely presentable ultraschemes becomes a concrete category, i.e. a category in which any object is an actual set equipped with some structure and any morphism is an actual map preserving the structure. Then we can develop the analogue of the classical sketch of algebraic geometry on an algebraically closed field. Algebraic spaces are defined as being separated finitely presentable ultraschemes. They are the objects of a concrete category and can be defined as sets equipped with a neotherian topology and a sheaf of regular functions on it. Any algebraic space is a union of finitely many irreducible algebraic spaces called algebraic varieties. Any algebraic variety X has its simple object $K(X)$ of rational functions on it, in such a way that the values of its sheaf of regular functions are subobjects of $K(X)$. As well as the morphisms and isomorphisms of algebraic varieties, we have the rational morphisms and birational equivalences of algebraic varieties. It can be proved that two algebraic varieties are birationally equivalent if and only if their objects of rational functions are isomorphic. The category of affine algebraic spaces is equivalent to the dual of the category of finitely presentable objects, while the category of affine algebraic varieties is equivalent to the dual of the category of finitely presentable integral objects.

6.1 ALGEBRAICALLY CLOSED SIMPLE OBJECTS

Simple extensions

A monomorphism $f: A \to B$ is also called an *extension* of A. A monomorphism between two simple objects is called a *simple extension.* Notice that an extension of a simple object need not be simple. Simple extensions *satisfy the amalgamation property* if, for any pair of simple extensions $f: K \to L, g: K \to M$ of a simple object K, there exists a pair of simple extensions $u: L \to N, v: M \to N$ such that $uf = vg$. A Zariski category is *amalgamative* if its simple extensions satisfy the amalgamation property. A morphism $f: A \to B$ is said to be *terminal* if its codomain B is the terminal object, and is said to be *interminable* if its pushout along any non-terminal morphism $g: A \to C$ is not terminal.

Proposition (6.1.1). *For a Zariski category A the following assertions are equivalent.*

(i) *A is amalgamative.*

(ii) *Simple extensions are interminable.*

(iii) *Extensions of simple objects are interminable.*

Proof.

(i) ⇒ (ii) Let $f: K \to L$ be a simple extension. Let $g: K \to C$ be a non-terminal morphism. Then C has a simple quotient $q: C \to M$, and $qg: K \to M$ is a simple extension. There exist simple extensions $m: M \to N$, $n: L \to N$ such that $nf = mqg$. Let $(u: L \to P, v: C \to P)$ be the pushout of (f, g). There exists a morphism $w: P \to N$ such that $wu = n$ and $wv = mq$. Therefore P is not terminal, i.e. v is not terminal. As a result, f is interminable.

(ii) ⇒ (iii) Let $f: K \to B$ be an extension of a simple object K. Then B is not terminal and so it has a simple quotient $q: B \to L$. Then the simple extension $qf: K \to L$ is interminable and therefore f is interminable.

(iii) ⇒ (i) Let $f: K \to L, g: K \to M$ be a pair of simple extensions of K. Let $(u: M \to N, v: L \to N)$ be the pushout of (g, f). The non-terminal object N has a simple quotient $s: N \to S$. Then $su: M \to S, sv: L \to S$ are simple extensions such that $svf = sug$. ■

Examples. The Zariski categories **CRng, RedCRng, RegCRng, CAlg**(k), **RedCAlg**(k), **RegCAlg**(k), **Mod, CAlg, RlLatRng**, and **Bool** are amalgamative.

Algebraically closed objects

Following Fakir ([14], Definition 5.5), a *monomorphism* $m: M \to N$ is *algebraically closed* if, for any monomorphism $f: A \to B$ whose domain is finitely generated and whose codomain is finitely presentable, and any pair of morphisms $p: A \to M, q: B \to N$ such that $qf = mp$, there exists a morphism $d: B \to M$ such that $df = p$. Fakir ([14], Proposition 5.3) has proved that, in categories of algebraic structures, this notion coincides with the notion of algebraically closed subalgebras defined by means of compatible systems of equations. Following [14] Definition 6.1, an object A is *algebraically closed* if any monomorphism with domain A is algebraically closed. We shall use this notion only for simple objects. An *algebraically closed simple extension* of a simple object K is a simple extension $K \to L$ whose codomain is algebraically closed.

Proposition (6.1.2). *For a simple object K, the following assertions are equivalent.*

(i) *K is algebraically closed.*

(ii) *For any pair of morphisms $f: A \to B, g: A \to K$ such that A and B are finitely presentable and the pushout of (f, g) is not terminal, there exists a morphism $h: B \to K$ such that $hf = g$.*

(iii) *For any pair of morphisms $f: A \to B, g: A \to K$ such that f is finitely presentable and the pushout of (f, g) is not terminal, there exists a morphism $h: B \to K$ such that $hf = g$.*

(iv) *Any finitely presentable extension of K splits, i.e. has a retraction.*

Proof.

(i) $\Rightarrow$ (ii) Let $(u: B \to C, v: K \to C)$ be the pushout of (f, g). Then $C \neq 1$. According to Corollary (2.1.9), C has a simple quotient $s: C \to S$. Then $sv: K \to S$ is monomorphic (Proposition (2.1.5)). Let $f = me$ be the regular factorization of f. The relation $svg = suf = sume$, where sv is a monomorphism and e is a regular epimorphism, implies the existence of a morphism $w: E \to K$ such that $we = g$ and $svw = sum$. The object E is finitely generated as a regular quotient of a finitely presentable object [16]. The monomorphism sv is algebraically closed since K is an algebraically closed object. Therefore there exists a morphism $h: B \to K$ such that $hm = w$. Then $hf = hme = we = g$.

(ii) $\Rightarrow$ (iii) First let us prove that any finitely presentable morphism $f: A \to B$ is the pushout along some morphism $\alpha: X \to A$ of a morphism $g: X \to Y$ with a finitely presentable domain and codomain. The category $A/\mathbf{A}$ of objects of $\mathbf{A}$ under A is locally finitely presentable and regular, and the set of objects of the form $(A \amalg X, i_X)$, where X runs over the finitely presentable objects of $\mathbf{A}$ and $i_X: A \to A \amalg X$ is the canonical induction, is a proper generating set in $A/\mathbf{A}$ [16]. Because the object (B, f) is finitely presentable in $A/\mathbf{A}$, there exists a pair of finitely presentable objects X, Y in $\mathbf{A}$ and a pair of morphisms $(u, v) : (A \amalg X, i_X) \rightrightarrows (A \amalg Y, i_Y)$ whose coequalizer in $A/\mathbf{A}$ is $q : (A \amalg Y, i_Y) \to (B, f)$. The morphisms $u, v : A \amalg X \rightrightarrows A \amalg Y$ in $\mathbf{A}$ are of the form $u = \langle i_Y, s\rangle$ and $v = \langle i_Y, t\rangle$ respectively, where $(s, t) : X \rightrightarrows A \amalg Y$ is a pair of morphisms whose coequalizer is $q : A \amalg Y \to B$. Because $\mathbf{A}$ is a locally finitely presentable category, the object $\mathbf{A}$ is the filtered colimit of the finitely presentable objects above it, i.e. $A = \xrightarrow{lim}_{(A_0, \alpha_0) \in (\mathbf{A}_0, A)} A_0$ where $\mathbf{A}_0$ denotes the category of finitely presentable objects in $\mathbf{A}$. Then $A \amalg Y = \xrightarrow{lim}_{(A_0, \alpha_0) \in (A_0, A)} (A_0 \amalg Y)$. Because X is finitely presentable, there exists $(A_0, \alpha_0) \in (\mathbf{A}_0, A)$ and a pair of morphisms $(s_0, t_0) : X \rightrightarrows A_0 \amalg Y$ such that $s = (\alpha_0 \amalg 1_Y)s_0$ and $t = (\alpha_0 \amalg 1_Y)t_0$. Let $q_0 : A_0 \amalg Y \to B_0$ be the coequalizer of (s_0, t_0). Then q is the pushout of q_0 along $\alpha_0 \amalg 1_Y$. Thence if $i_0 : A_0 \to A_0 \amalg Y$ denotes the canonical induction, $f = qi_Y$ is the pushout of $q_0 i_0$ along α_0. Moreover, the objects A_0 and B_0 are finitely presentable in $\mathbf{A}$. As a result, there exists a morphism $f_0 : A_0 \to B_0$ between finitely presentable objects and morphisms $\alpha_0 : A_0 \to A$, $\beta : B_0 \to B$ such that (f, β) is the pushout of (α_0, f_0).

Now let us consider a morphism $g : A \to K$ such that the pushout of (f, g) is not terminal. Then the pushout of $(f_0, g\alpha_0)$ is not terminal. Consequently, there exists a morphism $h_0 : B_0 \to K$ such that $h_0 f_0 = g\alpha_0$. Thus there exists a morphism $h : B \to K$ such that $h\beta = h_0$ and $hf = g$.

(iii) ⇒ (iv) Let $f : K \to B$ be a finitely presentable extension of K. Then B is not terminal; otherwise f would be an isomorphism and K would be terminal. Since the pushout of $(f, 1_K)$ is not terminal, there exists a morphism $h : B \to K$ such that $hf = 1_K$. Therefore the monomorphism f splits.

(iv) ⇒ (i) Let $m : K \to M$ be a monomorphism, let $f : A \to B$ be a monomorphism with a finitely generated domain and a finitely presentable codomain, and let $g : A \to K, h : B \to M$ be such that $hf = mg$. Let $(u : B \to C,$ $v : K \to C)$ be the pushout of (f, g). Then the morphism v is finitely presentable. The object C is not terminal because M is not and there exists a morphism $C \to M$. Since the object K is simple, v is monomorphic. Therefore v is a finitely presentable extension of K. Thus v splits. If $s : C \to K$ is such that $sv = 1_K$, then the morphism $su : B \to K$ is such that $suf = svg = g$. It follows that the monomorphism m is algebraically closed. As a result, the object K is algebraically closed. ■

Proposition (6.1.3). *For any simple object K in an amalgamative Zariski category, the following assertions are equivalent.*

(i) *K is algebraically closed.*

(ii) *K is injective with respect to the class of finitely presentable interminable morphisms.*

(iii) *K is injective with respect to the class of finitely presentable interminable monomorphisms.*

Proof.

(i) ⇒ (ii) Let $f : A \to B$ be a finitely presentable interminable morphism and let $g : A \to K$ be any morphism. Let $(\bar{g} : B \to \bar{B}, \bar{f} : K \to \bar{B})$ be the pushout of (f, g). Then $\bar{f}$ is not terminal and thus is a finitely presentable monomorphism. According to Proposition (6.1.2), there exists a morphism $h : \bar{B} \to K$ such that $h\bar{f} = 1_K$. Then the morphism $h\bar{g} : B \to K$ is such that $h\bar{g}f = g$.

(ii) ⇒ (iii) The proof is immediate.

(iii) ⇒ (i) Let $f : K \to A$ be a finitely presentable monomorphism. According to Proposition (6.1.1), f is interminable. Therefore there exists a morphism $g : A \to K$ such that $gf = 1_K$. According to Proposition (6.1.2), K is algebraically closed. ■

Definition (6.1.4).

(i) *An object A is* transnoetherian *if any finitely presentable morphism $f : A \to B$ has a noetherian codomain.*

(ii) *A Zariski category* **A** *is* simply noetherian *if its simple objects are transnoetherian.*

Examples. The Zariski categories **CRng**, **RedCRng**, **CAlg**(k), **RedCAlg**(k), **RlRng**, **RlOrdRng**, **GradCRng**, **Mod**, **CAlg**, and **Bool** are simply noetherian.

Proposition (6.1.5). *For a transnoetherian simple object K, the following assertions are equivalent.*

(i) *K is algebraically closed.*

(ii) *Any finitely presentable simple extension of K is an isomorphism.*

Proof.

(i) ⇒ (ii) According to Proposition (6.1.2), any finitely presentable simple extension of K has a retraction which must be a monomorphism. Thus it is an isomorphism.

(ii) ⇒ (i) Let $f : K \to A$ be a finitely presentable extension of K. The non-terminal object A has a simple quotient $s : A \to S$. Since K is transnoetherian, A is noetherian and hence s is a finitely presentable morphism. Then sf is a finitely presentable simple extension of K. It must be an isomorphism and hence f has a retraction. Thus according to Proposition (6.1.2), K is algebraically closed. ■

Nullstellensatz

Let us be in a Zariski category **A** with a simple initial object denoted by K. For any congruence r on an object A, let us denote by

$$\Sigma(r) = \{x : A \to K : x \text{ coequalizes } r\}$$

the set of 'solutions of r over K'. For any subset Y of $\mathrm{Hom}_{\mathbf{A}}(A, K)$, let us denote by

$$I(Y) = \wedge \{\ker(x) : x \in Y\}$$

the 'relation on A fulfilled by the elements of Y'. We say that the *Nullstellensatz holds with respect to* K if, for any finitely generated congruence r on any finitely presentable object A, $I(\Sigma(r)) = \mathrm{rad}(r)$.

Theorem (6.1.6). *For a simple initial object K in a Zariski category, the following assertions are equivalent.*

(i) *K is algebraically closed.*

(ii) *For any non-terminal finitely presentable object A, there exists at least one morphism $A \to K$.*

(iii) *The Nullstellensatz holds with respect to K.*

Proof.

(i) $\Leftrightarrow$ (ii) Follows from Proposition (6.1.2) since assertion (ii) is equivalent to the assertion (iv) of Proposition (6.1.2).

(ii) $\Rightarrow$ (iii) Let r be a finitely generated congruence on a finitely presentable object A. As any $x \in \Sigma(r)$ coequalizes r, we have $r \leqslant \ker(x)$ and thus $r \leqslant \wedge_{x \in \Sigma(r)} \ker(x) = I(\Sigma(r))$. As K is a reduced object, $\ker(x)$ is radical and thus $I(\Sigma(r))$ is radical. It follows that $\mathrm{rad}(r) \leqslant I(\Sigma(r))$. Let $t = (t_1, t_2) : T \rightrightarrows A$ be a codisjunctable radical congruence on A such that $t \nleqslant \mathrm{rad}(r)$. Let us prove that $t \nleqslant I(\Sigma(r))$. Let $q_r : A \to A/r$ be the quotient of A by r, and let $q_{\mathrm{rad}(r)} : A \to A/\mathrm{rad}(r)$ be the quotient of A by $\mathrm{rad}(r)$. Because t is codisjunctable, the pair of morphisms $(q_r t_1, q_r t_2)$ is codisjunctable. Let $d : A/r \to D$ be its codisjunctor. The relation $t \nleqslant \mathrm{rad}(r)$ implies that $q_{\mathrm{rad}(r)}$ does not coequalize t, i.e. $q_{\mathrm{rad}(r)} t_1 \neq q_{\mathrm{rad}(r)} t_2$. Because $A/\mathrm{rad}(r)$ is the universal reduced object associated to A/r, the pair $(q_r t_1, q_r t_2)$ is not conjoint (Proposition (2.5.2)), i.e. D is not terminal. The object D is finitely presentable as a singular quotient of the finitely presentable object A/r (Corollary (1.8.6)). Therefore there exists a morphism $n : D \to K$. Let $x = ndq_r : A \to K$. Then $x \in \Sigma(r)$. But x codisjoints t. Hence x does not coequalize t and so $t \nleqslant \ker(x)$. It follows that $t \nleqslant I(\Sigma(r))$. Because any radical congruence is the join of codisjunctable radical congruences (Proposition (2.4.7)), the equality $\mathrm{rad}(r) = I(\Sigma(r))$ follows.

(iii) $\Rightarrow$ (ii) Let A be a non-terminal finitely presentable object. The congruence Δ_A on A is proper and finitely generated. According to Proposition (2.4.13), its radical $\mathrm{rad}(A)$ is proper. According to the *Nullstellensatz*, $I(\Sigma(\Delta_A))$ is proper. Therefore $\Sigma(\Delta_A) \neq \varnothing$, i.e. there exists a morphism $A \to K$. ■

Examples. In **CRng**, **RedCRng**, and **RegCRng** algebraically closed simple objects are precisely algebraically closed fields. In **CAlg**(k), **RedCAlg**(k), and **RegCAlg**(k) they are the algebraically closed field extensions of k. If k is an algebraically closed field, the *Nullstellensatz* holds with respect to k in the categories **CAlg**(k), **RedCAlg**(k), **RegCAlg**(k). In **RlRng**, **RedRlRng**, **RlOrdRng**, and **RedRlOrdRng** algebraically closed simple objects are real closed fields. In **RlAlg**(R), **RedRlAlg**(R), **RlOrdAlg**(R), and **RedRlOrdAlg**(R) they are the real closed field extensions of R. If R is a real closed field, the *Nullstellensatz* holds with respect to R in the categories **RlAlg**(R), **RedRlAlg**(R), **RlOrdAlg**(R), and **RedRlOrdAlg**(R). In **GradCRng**, algebraically closed objects are precisely of the form K or $K[X, X^{-7}]$ where K is an algebraically closed field of degree zero, and X is an homogeneous element of degree 1.

Theorem (6.1.7). *In an amalgamative Zariski category, any simple object has an algebraically closed simple extension.*

Proof. Let K be a simple object. According to Proposition (6.1.1), the pushout $(f: K \to A, g: K \to B)$ of a pair of extensions is not terminal. By induction, we prove that the generalized pushout $(f_i: K \to A_i)_{i \in [1,n]}$ of finitely many extensions is not terminal. Because filtered colimits of non-terminal morphisms are not terminal, generalized pushouts of arbitrary families of extensions $(f_i: K \to A_i)_{i \in I}$ are not terminal. Let $\mathfrak{F}$ be the class of finitely presentable extensions $f: K \to A$. Up to isomorphisms, $\mathfrak{F}$ is a set. Let $g: K \to B$ be the generalized pushout of $\mathfrak{F}$. Then B is not terminal. Thus B has a simple quotient $q: B \to L_1$. Let $l_1 = qg$. Then $l_1: K \to L_1$ is a simple extension which factors through any morphism in $\mathfrak{F}$. Let us build up a sequence of morphisms $l_n: L_{n-1} \to L_n$ by induction on $n \in \mathbb{N}^*$ in the following way. Let $L_0 = K$ and $l_1: L_0 \to L_1$ as constructed above. If $(l_n)_{n<p}$ is built up, the morphism $l_p: L_{p-1} \to L_p$ is the morphism built up as l_1, but starting with L_{p-1} instead of K. Let $(\alpha_n: L_n \to L)_{n \in \mathbb{N}}$ be the colimit of the diagram $(L_n)_{n \in \mathbb{N}}$. The object L is simple, for it is a filtered colimit of simple objects (Proposition (2.1.3)). Let us show that L is algebraically closed. Let $u: X \to Y$ be a morphism with a finitely presentable domain and codomain and let $g: X \to L$ be a morphism such that the pushout of (u, g) is not terminal. There exists $n \in \mathbb{N}$ and a morphism $h: X \to L_n$ such that $\alpha_n h = g$. Let $(v: L_n \to T, w: Y \to T)$ be the pushout of (h, u). It is not terminal because the pushout of (g, u) is not. Since the morphism v is finitely presentable, according to the construction of $l_{n+1}: L_n \to L_{n+1}$ there exists a morphism $t: T \to L_{n+1}$ such that $tv = l_{n+1}$. Then the morphism $\alpha_{n+1} tw: Y \to L$ is such that $\alpha_{n+1} twu = \alpha_{n+1} tvh = \alpha_{n+1} l_{n+1} h = \alpha_n h = g$. Consequently, the object L is algebraically closed (Proposition (6.1.2)). As a result, $\alpha_0: K \to L$ is an algebraically closed simple extension of K. ∎

6.2 RATIONAL SPECTRA

Let us consider a Zariski category **A** whose initial object is simple and is denoted by K.

Definition (6.2.1). *A congruence r on an object A is* rational *if the quotient object of A by r is isomorphic to K.*

Proposition (6.2.2). *A rational congruence is maximal.*

Proof. The proof follows from Proposition (2.1.7).

Definition (6.2.3). *The* rational spectrum *of an object A is the topological space* $\mathrm{Spec}_{\mathrm{rat}}(A)$ *whose elements are the rational congruences on A and whose topology is induced by the Zariski topology on* $\mathrm{Spec}(A)$.

The rational spectrum functor $\mathrm{Spec}_{\mathrm{rat}} : \mathbf{A}^{\mathrm{op}} \to \mathbf{Esp}$

Let $f: A \to B$ be a morphism in **A**. Let $i: K \to A$ be the unique morphism. Let $r \in \mathrm{Spec}_{\mathrm{rat}}(B)$, let $q_r: B \to K$ be the quotient of B by r, and let $s = f^{*c}(r)$. The relation $q_r f i = 1_K$ implies that $q_r f: A \to K$ is a regular epimorphism. As the kernel pair of $q_r f$ is s, it follows that s is rational. As a result, the map $\mathrm{Spec}(f): \mathrm{Spec}(B) \to \mathrm{Spec}(A)$ preserves rational congruences and therefore induces a map $\mathrm{Spec}_{\mathrm{rat}}(f): \mathrm{Spec}_{\mathrm{rat}}(B) \to \mathrm{Spec}_{\mathrm{rat}}(A)$. By assigning $\mathrm{Spec}_{\mathrm{rat}}(A)$ to an object A and $\mathrm{Spec}_{\mathrm{rat}}(f)$ to a morphism f, we define the rational spectrum functor $\mathrm{Spec}_{\mathrm{rat}} : \mathbf{A}^{\mathrm{op}} \to \mathbf{Esp}$.

Proposition (6.2.4). *The initial object K is a classifying object for rational congruences, i.e. for any rational congruence r on an object A, there exists a unique morphism $f: A \to K$ such that the inverse image of Δ_K by f is r.*

Proof. The quotient morphism $q_r: A \to A/r \simeq K$ provides a regular epimorphism $f: A \to K$. Then $f^{*c}(\Delta_K) = r$. Let $g: A \to K$ be another morphism such that $g^{*c}(\Delta_K) = r$. Because g is a regular epimorphism, g is a quotient morphism of A by r. Thus there is an isomorphism $u: K \to K$ such that $g = uf$. But $u = 1_K$; hence $g = f$. ■

Corollary (6.2.5). *The functor* $\mathrm{Spec}_{\mathrm{rat}} : A^{\mathrm{op}} \to \mathit{Set}$ *is representable and isomorphic to the functor* $\mathrm{Hom}_{\mathbf{A}}(-, K)$.

Definition (6.2.6). *A* point *x of a locally modelled space X on A is* rational *if the residue simple object $K(x)$ of X at x is isomorphic to K.*

Let us denote by X_{rat} the set of rational points of X.

Proposition (6.2.7). *If X is the affine scheme of an object A in $\mathbf{A}$, then* $X_{\mathrm{rat}} = \mathrm{Spec}_{\mathrm{rat}}(A)$.

Proof. Let $p \in X_{\mathrm{rat}}$. Then p is a prime congruence on A such that $K(p)$ is isomorphic to K. Thus the morphism $j_p q_p : A \to K(p)$ is a split epimorphism. Hence the monomorphism $j_p : A/p \to K(p)$ is an isomorphism. Therefore $p \in \mathrm{Spec}_{\mathrm{rat}}(A)$. Conversely, if $p \in \mathrm{Spec}_{\mathrm{rat}}(A)$, then $K(p) \simeq A/p \simeq K$ and thus $p \in X_{\mathrm{rat}}$. ■

Proposition (6.2.8). *Any rational point of a scheme is closed.*

Proof. Let x be a rational point of a scheme X. Let $(X_i)_{i \in I}$ be an affine open covering of X. If $i \in I$ is an index such that $x \in X_i$, then x is a rational point in the affine scheme X_i and thus x is closed in X_i (Propositions (3.2.1) and (6.2.2)). It follows that $X - \{x\} = \cup_{i \in I}(X_i - \{x\})$ is an open set of X and hence x is closed in X. ■

The rational point functor $(\text{—})_{\mathrm{rat}} : \mathbf{SchA} \to \mathbf{Esp}$

Let $f : X \to Y$ be a morphism in **SchA**. Let $x \in X$ be a rational point. The morphism $f^x : K(f(x)) \to K(x) \simeq K$ is an isomorphism because it is simultaneously a monomorphism and a split epimorphism. It follows that $K(f(x)) \simeq K$ and thus $f(x)$ is a rational point. Consequently, the map f induces a map $f_{\mathrm{rat}} : X_{\mathrm{rat}} \to Y_{\mathrm{rat}}$. By assigning to any object X the set X_{rat} equipped with the topology induced by the topology on X, and to any morphism $f : X \to Y$ the continuous map $f_{\mathrm{rat}} : X_{\mathrm{rat}} \to Y_{\mathrm{rat}}$ induced by f, we obtain the rational point functor $(\text{—})_{\mathrm{rat}} : \mathbf{SchA} \to \mathbf{Esp}$.

Proposition (6.2.9). *The terminal object* $\mathrm{Spec}(K)$ *is a coclassifying object for rational points of schemes, i.e. for any rational point x of a scheme X, there exists a unique morphism* $f : \mathrm{Spec}(K) \to X$ *such that* $f(\Delta_K) = x$.

Proof. Let $x \in X_{\mathrm{rat}}$ and let U be an open affine subscheme of X such that $x \in U$. Then $x \in U_{\mathrm{rat}}$. According to Proposition (6.2.4), there exists a unique morphism $g : \mathrm{Spec}(K) \to U$ such that $g(\Delta_K) = x$. Thus there exists a morphism $f : \mathrm{Spec}(K) \to X$ such that $f(\Delta_K) = x$. If $h : \mathrm{Spec}(K) \to X$ is another morphism such that $h(\Delta_K) = x$, then h factors through the inclusion morphism $U \to X$ in a morphism $\mathrm{Spec}(K) \to U$ which needs to be g, and therefore $h = f$. ■

Corollary (6.2.10). *The functor* $(\text{—})_{\mathrm{rat}} : \mathbf{SchA} \to \mathbf{Set}$ *is representable and isomorphic to the functor* $\mathrm{Hom}_{\mathbf{SchA}}(\mathit{Spec}(K), \text{—})$.

6.3 RATIONAL ZARISKI CATEGORIES

Recall that, following Gabriel and Ulmer [16], an object A of a category $\mathbf{A}$ is said to be *noetherian* if any decreasing sequence of extremal quotient objects of A is stationary, and that the category $\mathbf{A}$ is said to be *locally noetherian* if it is cocomplete and has a proper generating set formed of finitely presentable noetherian objects whose finite sums are still noetherian objects, or, equivalently, if it is a locally finitely presentable category in which any finitely generated object is finitely presentable.

Definition (6.3.1). *A* Zariski category *is* rational *if it is locally noetherian and its initial object K is simple and algebraically closed.*

Proposition (6.3.2). *For a locally Noetherian Zariski category $\mathbf{A}$ not equivalent to $\mathbf{1}$, the following assertions are equivalent.*

(i) *$\mathbf{A}$ is rational.*

(ii) *Any finitely presentable simple object of $\mathbf{A}$ is initial.*

Proof.

(i) $\Rightarrow$ (ii) Let L be a finitely presentable simple object. The morphism $f: K \to L$ is a finitely presentable extension of K. According to Proposition (6.1.2), f splits, i.e. there exists $r: L \to K$ such that $rf = 1_K$. Since K and L are simple, r is monomorphic. Consequently, r is an isomorphism and L is initial.

(ii) $\Rightarrow$ (i) Since the category is not equivalent to $\mathbf{1}$, the initial object Z is not terminal and thus it has a simple quotient $k: Z \to K$. Since Z is finitely presentable, K is a finitely presentable simple object and thus is initial. Consequently, the category has a simple initial object K. Let $f: K \to A$ be a finitely presentable extension of K. Then A is not terminal and hence it has a simple quotient $q: A \to L$. Since A is finitely presentable, L is also. It follows that L is initial. Then $qf: K \to L$ is an isomorphism and f is a split monomorphism. Consequently, K is algebraically closed (Proposition (6.1.2)). ∎

Proposition (6.3.3). *In a rational Zariski category $\mathbf{A}$ any finitely presentable object (any morphism between finitely presentable objects) is a Jacobson object (morphism) and its rational spectrum coincides with its maximal spectrum, i.e. $\mathbf{A}_0 = \mathbf{Jac}\mathbf{A}_0$ and* $\mathrm{Spec}_{\mathrm{rat}} = \mathrm{Spec}_{\mathrm{max}} : \mathbf{A}_0^{\mathrm{op}} \to \mathbf{Esp}$.

Proof. Let $A \in \mathbf{A}_0$. According to the *Nullstellensatz* (Theorem (6.1.6)), any $p \in \mathrm{Spec}(A)$ is such that $p = \mathrm{rad}(p) = I\Sigma(p) = \wedge_{x \in \Sigma(p)} \ker(x)$ is a meet of maximal congruences. Consequently, A is a Jacobson object. According to Proposition (6.2.2), if p is rational, then p is maximal. Conversely, if p is maximal, A/p is a finitely presentable simple object and thus is isomorphic to K (Proposition (6.3.2)). Therefore p is rational. Consequently, $\mathrm{Spec}_{rat}(A) = \mathrm{Spec}_{max}(A)$. Let $f: A \to B$ be a morphism in $\mathbf{A}_0$. Since $\mathrm{Spec}(f)$ preserves rational congruences (cf. § 6.2, p. 154), it preserves maximal congruences also. Therefore f is a Jacobson morphism and $\mathrm{Spec}_{rat}(f) = \mathrm{Spec}_{max}(f)$. ∎

Proposition (6.3.4). *On a rational Zariski category A, any locally finitely presentable scheme (any morphism between locally finitely presentable schemes) is a Jacobson scheme (morphism) and its closed points are precisely its rational points, i.e.* ***SchA***$_0$ = ***SchJacA***$_0$ *and* $S(X) = X_{rat}$.

Proof. Let X be a locally finitely presentable scheme. There exists an affine open covering $(X_i)_{i \in I}$ of X such that each $O_X(X_i) = O_{X_i}(X_i)$ is a finitely presentable object. According to Proposition (6.3.3), $O_{X_i}(X_i)$ is a Jacobson object. Thus X_i is a Jacobson scheme (Theorem (5.2.3)). It follows that X is a Jacobson space ([21], Proposition 2.8.3), i.e. a Jacobson scheme. According to Proposition (6.2.8), any rational point of X is closed. Conversely, any closed point of X is a closed point of some X_i and thus it belongs to $\mathrm{Spec}_{max}(X_i) = \mathrm{Spec}_{rat}(X_i) \subset X_{rat}$ (Proposition (6.3.3)). Any morphism $f: X \to Y$ of locally finitely presentable schemes preserves rational points (cf. § 6.2, p. 154). Thus it preserves closed points and so is a Jacobson morphism. ∎

Proposition (6.3.5). *For a rational Zariski category A, the category* ***UltSchA***$_0$ *of locally finitely presentable ultraschemes on A is a full subcategory of* ***SpModA*** *and is equivalent to* ***SchA***$_0$.

Proof. As **UltSchA**$_0$ is a full subcategory of **SpModLocA**, we just have to prove that, for a pair X, Y of objects in **UltSchA**$_0$, any morphism of modelled space $f: X \to Y$ is such that, for any $x \in X$, the morphism $f_x^{\#}: O_{Y, f(x)} \to O_{X,x}$ is local. But this morphism is such that $k_x f_x^{\#}: O_{Y, f(x)} \to K(x) \simeq K$ is a split epimorphism and thus is a regular epimorphism. Therefore $(f_x^{\#})^{*c}(m_x)$, the kernel pair of $k_x f_x^{\#}$, is a maximal congruence. Consequently, $f_x^{\#}$ is local. As a result $f: X \to Y$ is a morphism of locally modelled spaces and thus is a morphism of ultraschemes. The equivalences **UltSchA**$_0$ ~ **SchJacA**$_0$ ~ **SchA**$_0$ follow from Propositions (5.3.6) and (6.3.4). ∎

Corollary (6.3.6). *The category* ***UltSchA***$_0$ *is finitely complete and has gluing colimits.*

Proof. The proof follows from Proposition (4.6.3). ∎

Definition (6.3.7). *A locally finitely presentable ultrascheme X on a rational Zariski category is* separated *if its diagonal $\Delta_X : X \to X \times X$ has a closed image.*

Proposition (6.3.8). *For a locally finitely presentable scheme X on a rational Zariski category, the following assertions are equivalent.*

(i) *X is separated.*

(ii) *The ultrascheme $S(X)$ of rational points of X is separated.*

Proof. Let us denote by $S : \mathbf{SchA}_0 \to \mathbf{UltSchA}_0$ the equivalence of categories given by Proposition (5.3.6). This functor S preserves products and diagonals. Thus if $\Delta_X : X \to X \times X$ is the diagonal of X, $S(\Delta_X) : S(X) \to S(X \times X) = S(X) \times S(X)$ is the diagonal $\Delta_{S(X)}$ of $S(X)$. Let us identify the maps $\Delta_X, \Delta_{S(X)}$ with inclusion maps. Let us denote the inclusion maps by $i_X : S(X) \to X$, $i_{X \times X} : S(X) \times S(X) \to X \times X$. Because X and $X \times X$ are Jacobson spaces, i_X and $i_{X \times X}$ are quasi-homeomorphisms. Moreover, since Δ_X is a Jacobson map, the pair of morphisms $(i_X, \Delta_{S(X)})$ is the pullback of the pair $(\Delta_X, i_{X \times X})$ in **Esp**. It follows that, if Δ_X is closed in $X \times X$, then $\Delta_{S(X)}$ is closed in $S(X) \times S(X)$. Consequently, if X is separated, $S(X)$ is also. Conversely, let us assume that $S(X)$ is separated. Let $\bar{X}$ be the closure of Δ_X in $X \times X$. Then $S(\bar{X})$ is a closed set in $S(X) \times S(X)$ and it contains $S(X)$. Let F be a closed set in $S(X) \times S(X)$ containing $S(X)$. Then F is of the form $F = S(G)$, where G is a closed set in $X \times X$. The relation $S(X) \subset S(G)$ implies $X \subset G$ because X and G are locally closed sets in the Jacobson space $X \times X$ ([21], 0, Proposition 2.7.7). Then $\bar{X} \subset G$ and hence $S(\bar{X}) \subset S(G) = F$. Consequently $S(\bar{X})$ is the closure of $S(X)$ in $S(X) \times S(X)$. As $S(X)$ is closed in $S(X) \times S(X)$, we have $S(\bar{X}) = S(X)$. Because $\bar{X}$ and X are locally closed sets in $X \times X$, the equality $\bar{X} = X$ follows, i.e. X is separated. ∎

The sheaf of K-valued functions on a topological space

Let X be a topological space. For any $U \in \Omega(X)$, let us denote by $\Phi_X(U)$ the object K^U which is the product of K by itself U times. For any $V \in \Omega(K)$ such that $V \subset U$, let us denote by $\Phi_X(V \to U) : \Phi_X(U) \to \Phi_X(V)$ the morphism whose composite with the projection $\Pi_x : K^V \to K$ is the projection $\Pi_x : K^U \to K$. We obtain a functor $\Phi_X : \Omega(X)^{\text{op}} \to \mathbf{A}$, which is a sheaf on X with values in $\mathbf{A}$ called the *sheaf of K-valued functions* on X. Let $f : X \to Y$ be a continuous map. Let us define the morphism of sheaves $\Phi_f : \Phi_Y \to f_*(\Phi_X)$ by its value at $V \in \Omega(Y)$ which is the morphism $\Phi_{f,V} : K^V \to K^{f^{-1}(V)}$ defined by $\Pi_x \Phi_{f,V} = \Pi_{f(x)}$. We obtain a morphism of modelled spaces $(f, \Phi_f) : (X, \Phi_X) \to (Y, \Phi_Y)$. It follows a functor $\Phi : \mathbf{Esp} \to \mathbf{SpModA}$ defined by $\Phi(X) = (X, \Phi_X)$ and $\Phi(f) = (f, \Phi_f)$.

The K-evaluation of a locally finitely presentable ultrascheme

Let X be a locally finitely presentable ultrascheme on a rational Zariski category $\mathbf{A}$. According to Proposition (6.3.4), we shall assume that the residue simple object at any point of X is K. Let Φ_X be the sheaf of K-valued functions on the topological space X. For any $U \in \Omega(X)$, let us denote by $\varphi_{X,U} : O_X(U) \to \Phi_X(U)$ the morphism in $\mathbf{A}$ whose composite with the projection $\Pi_x : \Phi_X(U) = K^U \to K$ is the morphism $k_x \rho_x^U : O_X(U) \to K$, where $\rho_x^U : O_X(U) \to O_{X,x}$ is the restriction morphism and $k_x : O_{X,x} \to K$ is the residue simple quotient of $O_{X,x}$. The family of morphisms $(\varphi_{X,U})_{U \in \Omega(X)}$ defines a morphism of sheaves $\varphi_X : O_X \to \Phi_X$ called the *K-evaluation of the structure sheaf of X*. At any point $x \in X$, the stalk of φ_X at x is the morphism $\varphi_{X,x} : O_{X,x} \to \Phi_{X,x}$ naturally defined by the usual colimit process. Let $f : X \to Y$ be a morphism of locally finitely presentable ultraschemes on $\mathbf{A}$. Let $V \in \Omega(Y)$, $U = f^{-1}(V)$, $x \in U$, and $y = f(x)$. Then

$$\Pi_x \varphi_{X,U} f_V^{\#} = k_x \rho_x^U f_V^{\#} = k_x f_x^{\#} \rho_y^V = k_y \rho_y^V = \Pi_y \varphi_{Y,V} = \Pi_x \Phi_{f,V} \varphi_{Y,V}.$$

Consequently, we have the equalities $\varphi_{X,U} f_V^{\#} = \Phi_{f,V} \varphi_{Y,V}$, $(f_*(\varphi_X)) f^{\#} = \Phi_f \varphi_Y$, and $(f, f^{\#})(1_X, \varphi_X) = (1_Y, \varphi_Y)(f, \Phi_f)$. If we denote the inclusion functor $\mathbf{UltSchA}_0 \to \mathbf{SpModA}$ by I_0, and the composite of the forgetful functor $\mathbf{UltSchA}_0 \to \mathbf{Esp}$ and the functor $\Phi : \mathbf{Esp} \to \mathbf{SpModA}$ by Φ_0, we obtain a natural transformation $\varphi : \Phi_0 \to I_0$ given by $\varphi_{(X, O_X)} = (1_X, \varphi_X)$ which is called the *K-evaluation of locally finitely presentable ultraschemes on* $\mathbf{A}$.

Proposition (6.3.9). *For any $x \in X$, the kernel pair of the K-evaluation morphism $\varphi_{X,x} : O_{X,x} \to \Phi_{X,x}$ is the radical of $O_{X,x}$.*

Proof. First, let us assume that X is affine, i.e. X is of the form $X = \mathrm{Spec}_{\mathrm{rat}}(A) = \mathrm{Spec}_{\max}(A)$ where A is a finitely presentable object. The morphism $\varphi_{X,X} : A \to K^X$ is such that its composite with the projection $\Pi_m : K^X \to K$ is the morphism $k_m l_m : A \to K$. Let us prove that the kernel pair of $\varphi_{X,X}$ is the radical of A. On the one hand, $\mathrm{rad}(A)$ is the intersection of the set of maximal congruences on A because A is a Jacobson object. On the other hand, for any maximal congruence m on A, $\ker(\Pi_m \varphi_{X,X}) = \ker(k_m l_m) = (k_m l_m)^{*c}(\Delta_k) = m$. Therefore $\ker(\varphi_{X,X}) = \bigcap_{m \in X} \ker(\Pi_m J_{X,X}) = \bigcap_{m \in X} m = \mathrm{rad}(A)$. Now let us consider any locally finitely presentable ultrascheme X. It follows from the preceding discussion that, for any ultra-affine open set U of X, $\ker(\varphi_{X,U}) = \mathrm{rad}(O_X(U))$. Because kernel pairs and radicals commute with filtered colimits, it follows that, for any $x \in X$, $\ker(\varphi_{X,x}) = \mathrm{rad}(O_{X,x})$. ■

Corollary (6.3.10). *For a reduced locally finitely presentable ultrascheme X, the K-evaluation $\varphi_X : O_X \to \Phi_X$ is monomorphic.*

Proof. Since the stalks of X are reduced (Proposition (5.3.10)), the morphisms $\varphi_{X,x} : O_{X,x} \to \Phi_{X,x}$ are monomorphic and thus φ_X is monomorphic. ■

Examples

If k is an algebraically closed field, the Zariski categories **CAlg**(k), **RedCAlg**(k), **AlgCAlg**(k), and **RedAlgCAlg**(k) are rational. If R is a real closed field, the categories **RlAlg**(R), **RedRlAlg**(R), **RlOrdAlg**(R), and **RedRlOrdAlg**(R) are rational.

6.4 REDUCED RATIONAL ZARISKI CATEGORIES

Definition (6.4.1). *A* Zariski category *is* reduced *if its objects are reduced.*

Proposition (6.4.2). *For a Zariski category the following assertions are equivalent.*

(i) *The category is reduced.*
(ii) *Any local object is reduced.*
(iii) *Any indecomposable object is reduced.*
(iv) *Any irreducible object is integral.*
(v) *The class of simple objects is a cogenerating class.*
(vi) *The class of integral objects is a cogenerating class.*
(vii) *Conjoint morphisms are equal.*
(viii) *Any regular epimorphism with a bijective spectrum is an isomorphism.*
(ix) *Any bijective closed immersion of schemes is an isomorphism.*
(x) *Any bijective immersion of schemes is an isomorphism.*

Proof.

(i) ⇔ (ii) According to Proposition (3.5.1) any object is a subobject of a product of local objects.

(i) ⇔ (iii) According to Theorem (3.10.1) and [9], Proposition 5.2, any object is a subobject of a product of indecomposable objects.

(i) ⇔ (iv) According to Proposition (2.10.3) any reduced irreducible object is integral, and according to Theorem (2.10.4) any object is a subobject of a product of irreducible objects.

(i) ⇔ (v) follows from the definition of a cogenerating class of objects in a category ([31], p. 123) and the definition of reduced objects.

(v) ⇔ (vi) follows from the fact that any integral object is a subobject of a simple object.

(i) ⇔ (vii) follows from Corollary (2.5.5).

(vii) ⇒ (viii) Let $f : A \to B$ be a regular epimorphism, which is a coequalizer of a congruence r, such that the map Spec(f) is bijective. According to

Proposition (3.3.11), the image of $\text{Spec}(f)$ is $V(r)$. Thus $V(r) = \text{Spec}(A)$ so that $D(r) = \varnothing$. According to Proposition (3.1.16), r is conjoint. Therefore $r = \Delta_A$ and f is an isomorphism.

(viii) ⇒ (ix) Let $f: X \to Y$ be a closed immersion of schemes such that the map f is bijective. Let V be an affine open subset of Y and let $U = f^{-1}(V)$. Then f induces a closed immersion of schemes $g: U \to V$ such that the map g is bijective. Since V is affine, U is affine and g is isomorphic to a morphism $\text{Spec}(q)$ where q is a regular epimorphism (Proposition (4.4.14)). Then q is an isomorphism and so is g. It follows that f is an isomorphism (Proposition (4.4.3)).

(ix) ⇒ (x) Let $f: X \to Y$ be an immersion of schemes such that the map f is bijective. According to Proposition (4.4.18), f is the composite of a closed immersion g followed by an open immersion h. Then the maps g and h are bijective. It follows that g and h are isomorphisms and thus f is an isomorphism.

(x) ⇒ (i) Let A be an object. The regular quotient $q_{\text{rad}(A)}: A \to A/\text{rad}(A)$ is such that $\text{Spec}(q_{\text{rad}(A)})$ is an immersion of schemes (Proposition (4.4.15)) whose image is $V(\text{rad}(A)) = \text{Spec}(A)$. Therefore $\text{Spec}(q_{\text{rad}(A)})$ is bijective. Thus $q_{\text{rad}(A)}$ is an isomorphism and $A \simeq A/\text{rad}(A)$ is reduced. ∎

Examples. The categories **RedCRng**, **RedCAlg**(k), **RedCAlg**, **RedRlRng**, **RedRlOrdRng**, and **RedGradRng** are reduced Zariski categories. For any Zariski category **A**, the category **RedA** of reduced objects of **A** is a reduced Zariski category (cf. Theorem (12.3.2) below).

Let us be in a reduced rational Zariski category **A** whose initial object is K. Then any scheme or ultrascheme on **A** is reduced. We shall study the category $\mathbf{UltSchA}_0$ of locally finitely presentable ultraschemes on **A**. Let us denote by $U: \mathbf{UltSchA}_0 \to \mathbf{Set}$ the functor which assigns the set X (the map f) to the object X (the morphism f), and let us call it the *forgetful functor.*

Proposition (6.4.3). *The forgetful functor U: $\mathit{UltSchA}_0 \to$ Set is faithful and has a full and faithful left adjoint.*

Proof.

(i) Let $(f, f^{\#}), (g, g^{\#}): (X, O_X) \rightrightarrows (Y, O_Y)$ be a pair of parallel morphisms in $\mathbf{UltSchA}_0$ such that $f = g$. Using the notation on p. 159, we have $(f, f^{\#})(1_X, \varphi_X) = (1_Y, \varphi_Y)(f, \Phi_f) = (1_Y, \varphi_Y)(g, \Phi_g) = (g, g^{\#})(1_X, \varphi_X)$. Since φ_X is monomorphic (Corollary (6.3.10)), $f^{\#} = g^{\#}$ and so $(f, f^{\#}) = (g, g^{\#})$. As a result, U is faithful.

(ii) For any set E, the discrete modelled space (E, Φ_E) is an object of $\mathbf{UltSchA}_0$. Therefore we can define a functor $F: \mathbf{Set} \to \mathbf{UltSchA}_0$ by

$F(E) = (E, \Phi_E)$ and $F(f) = (f, \Phi_f)$. It satisfies the relation $UF = 1_{\mathbf{Set}}$ and thus it is full and faithful. We define a natural transformation $\varepsilon : FU \to 1_{\mathbf{UltSchA_0}}$ by $\varepsilon_{(X, O_X)} = (1_X, \Phi_{1_X} \varphi_X)$ where $1_X : U(X) \to X$ is the identity continuous map. It provides a co-unit for an adjunction (F, U). ■

Corollary (6.4.4). *UltSchA$_0$ is a concrete category relative to its forgetful functor, and therefore its morphisms are actual maps.*

Proof. See [31], p. 26. ■

The sheaf of regular functions on $X \in$ UltSchA$_0$

The sheaf of regular functions on $X \in \mathbf{UltSchA_0}$ is the subsheaf of the sheaf of K-valued functions on X, i.e. the subobject of Φ_X in the category **Sh[X, A]**, which is represented by the monomorphism $\varphi_X : O_X \to \Phi_X$ (Corollary (6.3.10)). It will be denoted by O_X, following the usual abuse of language for subobjects, and, for any open set U of X, the subobject $O_X(U)$ of $\Phi_X(U) = K^U$ will be called the *object of regular functions on U.*

Proposition (6.4.5). *If X and Y are objects in UltSchA$_0$, a map $f : X \to Y$ is a morphism in UltSchA$_0$ if and only if it is continuous and for any open set V in Y the image of the subobject $O_Y(V)$ of regular functions on V by the morphism $\Phi_{f,V} : K^V \to K^{f^{-1}(V)}$ is included in the subobject $O_X(f^{-1}(V))$ of regular functions on $f^{-1}(V)$.*

Proof. Let us assume that f is a morphism. The morphism $f^\# : O_Y \to f_*(O_X)$ satisfies the relation $f_*(\varphi_X) f^\# = \Phi_f \varphi_Y$. Thus, for any $V \in \Omega(Y)$, we have $\varphi_{X, f^{-1}(V)} f_V^\# = \Phi_{f,V} \varphi_{Y,V}$ so that the image of the subobject represented by $\varphi_{Y,V}$ along the morphism $\varphi_{f,V}$ is included in the subobject represented by $\varphi_{X, f^{-1}(V)}$. Conversely, let us assume that the map f is such that this last property holds. Then for any $V \in \Omega(Y)$ there exists a unique morphism $f_V^\# : O_{Y,V} \to O_{X, f^{-1}(V)}$ such that $\varphi_{X, f^{-1}(V)} f_V^\# = \Phi_{f,V} \varphi_{Y,V}$. These morphisms provide a natural transformation $f^\# : O_Y \to f_*(O_X)$ and a morphism $(f, f^\#) : (X, O_X) \to (Y, O_Y)$ in **UltSchA$_0$**. ■

Proposition (6.4.6). *The category UltSchA$_0$ has concrete finite limits and gluing colimits.*

Proof. The existence of finite limits and gluing colimits is proved in Corollary (6.3.6). The concreteness of finite limits follows from the fact that the forgetful functor has a left-adjoint (Proposition (6.4.3)). The concreteness of gluing colimits follows from their construction via the equivalence of categories **UltSchA$_0$** $\sim$ **SchA$_0$** (Propositions (4.5.2) and (6.3.5)). ■

Corollary (6.4.7). *A morphism in UltSchA_0 is a monomorphism if and only if it is injective.*

Proposition (6.4.8). *An object in UltSchA_0 has a locally noetherian topology.*

Proof. Since any object in $\mathbf{UltSchA}_0$ has an open covering by locally finitely presentable affine ultraschemes, it is sufficient to prove that the ultrascheme $\mathrm{Spec}_{\max}(A)$, where A is a finitely presentable object, has a noetherian topology. Let $\mathrm{ClSpec}(A)$ and $\mathrm{ClSpec}_{\max}(A)$ be the lattices of the closed sets of $\mathrm{Spec}(A)$ and $\mathrm{Spec}_{\max}(A)$ respectively. Since A is a noetherian object, $\mathrm{ClSpec}(A)^{\mathrm{op}} \simeq \mathrm{Cong}(A) \simeq \mathbf{QUOTReg}(A)^{\mathrm{op}}$ is noetherian. Since A is a Jacobson object, $\mathrm{ClSpec}_{\max}(A) \simeq \mathrm{ClSpec}(A)$. Consequently, $\mathrm{ClSpec}_{\max}(A)^{\mathrm{op}}$ is noetherian, i.e. $\mathrm{Spec}_{\max}(A)$ has a noetherian topology. ■

Subultraschemes

The notions of ultrascheme induced, subultrascheme, open or closed subultrascheme, embedding, immersion, and open or closed immersion of ultraschemes are defined in the same way as they were for schemes in § 4.4. Because of the equivalence of categories $\mathbf{UltSchA}_0 \sim \mathbf{SchA}_0$ the properties proved in § 4.4 for schemes are also valid for locally finitely presentable ultraschemes.

Proposition (6.4.9). *In the category UltSchA_0 there are initial structures along injective maps whose images are locally closed (open, closed) subsets and they are precisely the immersions (open immersions, closed immersions).*

Proof. Let Y be an object in $\mathbf{UltSchA}_0$, let X be a set, and let $f: X \to Y$ be an injective map. Let us equip X with the initial topology defined by f and the topology on Y, and let $(f, f^{\#}) : (X, O_X) \to (Y, O_Y)$ be the inverse image of the modelled space (Y, O_Y) along the continuous map $f: X \to Y$ (cf. Proposition (4.1.2) and the notation on p. 113). If f is a bijection, then it is a homeomorphism. Therefore $(f, f^{\#})$ is an isomorphism of modelled spaces, and thus it is an isomorphism in $\mathbf{UltSchA}_0$ and provides an initial structure along f. If f is the inclusion map of an open subset X of Y, then $(X, O_X) = (X, O_Y/X)$ is an object in $\mathbf{UltSchA}_0$ and $(f, f^{\#})$ provides an initial structure along f. Let us assume that Y is an affine ultrascheme of the form $Y = \mathrm{Spec}_{\max}(A)$ and that f is the inclusion map of a closed subset X of Y. Because $\mathrm{Spec}_{\max}(A)$ is very dense in $\mathrm{Spec}(A)$, X is uniquely of the form $V \cap \mathrm{Spec}_{\max}(A)$ where V is a closed subset of $\mathrm{Spec}(A)$. According to Proposition (3.3.11) and (4.4.15), V is the image of a closed immersion of affine schemes $\mathrm{Spec}(q) : \mathrm{Spec}(B) \to \mathrm{Spec}(A)$ where $q : A \to B$ is a regular epimorphism defined uniquely up to isomorphisms. Then X is the image of the closed immersion of affine ultraschemes $\mathrm{Spec}_{\max}(q)$: $\mathrm{Spec}_{\max}(B) \to \mathrm{Spec}_{\max}(A)$. This last morphism provides an initial structure along its underlying map, and we obtain an initial structure along f by composing the previous structure with the initial structure along the bijection $X \simeq \mathrm{Spec}_{\max}(B)$. Then we obtain precisely the initial structure $(f, f^{\#})$ along f. Let us assume that f is the inclusion map of a closed subset X of an arbitrary Y. Let $(Y_i)_{i \in I}$ be an ultra-affine open covering of Y such that each $A_i = O_Y(Y_i)$ is a finitely presentable object (Definition (5.3.5)). Let $X_i = Y_i \cap X$ for each $i \in I$. Then X_i is a closed subset of Y_i. Thus according to the preceding discussion, there is an initial structure along the inclusion map $X_i \to Y_i$ such that X_i is an affine ultrascheme. Then $(X_i)_{i \in I}$ is an open covering of X which makes the modelled space (X, O_X) an ultrascheme and provides the initial structure $(f, f^{\#})$ along f. Finally, let us assume that the image of f is a locally closed subset of Y. Then it is the composite of an open inclusion map, a closed inclusion map, and a bijection. By the previous results and composition of the initial structures, $(f, f^{\#})$ provides an initial structure along f. It is easy to see that initial structures along injective maps whose images are locally closed (open, closed) are precisely immersions (open immersions, closed immersions) of locally finitely presentable ultraschemes. ∎

Proposition (6.4.10). *In the category* ***UltSchA***$_0$, *the class of immersion (open immersions, closed immersions) is universal and closed under composition and finite limits.*

Proof.

(i) Let $f: X \to Y$ be an immersion, let $g : F \to Y$ be any morphism, and let $(h : E \to X,\ k : E \to F)$ be the pullback of (f, g). Then f is initial along $U(f)$ (Proposition (6.4.9)). Since initial structures are universal, k is initial along $U(k)$. On the other hand, $U(f)$ is an injective map with a locally closed image. Since $(U(h), U(k))$ is the pullback of $(U(f), U(g))$, $U(k)$ is an injective map with a locally closed image. According to Proposition (6.4.9), k is an immersion. As a result, the class of immersions is universal.

(ii) The closeness of the class of immersions under compositions follows from that of the class of initial structures and of injective continuous maps with locally closed images, and from Proposition (6.4.9).

(iii) Let $f: X \to Y, g : E \to F$ be a pair of immersions. Then $f \times 1_F$ is an immersion because it is the pullback of f along the projection $Y \times F \to Y$. In the same way $1_X \times g$ is an immersion. Therefore $f \times g = (f \times 1_F)(1_X \times g)$ is an immersion. It follows that the class of immersions is closed under finite products and, as it is universal, it is closed under finite limits.

(iv) It is straightforward to check that the previous results hold for open or closed immersions. ■

6.5 ALGEBRAIC SPACES AND ALGEBRAIC VARIETIES

Let us be in a reduced rational Zariski category **A** e.g. **RedCAlg**(k) where k is an algebraically closed field, or **RedRlAlg**(R) or **RedRlOrdAlg**(R) where R is a real closed field.

Definition (6.5.1). *An* algebraic space *on A is a separated finitely presentable ultrascheme on A.*

The *category* **EspAlgA** *of algebraic spaces on* **A** is the full subcategory of **UltSchA**$_0$ whose objects are the algebraic spaces. It is a concrete category whose forgetful functor U is induced by that on **UltSchA**$_0$.

Proposition (6.5.2). *An algebraic space has a noetherian topology and a base of ultra-affine open sets closed under non-empty finite intersections.*

Proof. Let X be an algebraic space. According to Proposition (6.4.8) it has a locally noetherian topology, and according to Definition (5.3.13) it is compact. Thus it is noetherian. Let U, V be ultra-affine open subsets of X. Then U, V, $U \cap V$ are open subultraschemes of X. Let us denote the associated open immersions by $i: U \to X, j: V \to X$, and $k: U \cap V \to X$, the diagonal of X by $\Delta_X: X \to X \times X$, and the morphism induced by Δ_X by $f: U \cap V \to U \times V$. Then (f, k) is the pullback of $(i \times j, \Delta_X)$ in **UltSchA**$_0$. Since X is separated, Δ_X is a closed immersion and thus f is also a closed immersion (Proposition (6.4.10)). Since U and V are ultra-affine, $U \times V \simeq \mathrm{Spec}_{\max}(O_X(U)) \times \mathrm{Spec}_{\max}(O_Y(V)) \simeq \mathrm{Spec}_{\max}(O_X(U) \amalg O_X(V))$ is ultra-affine. It follows that $U \cap V$ is also ultra-affine. ■

Proposition (6.5.3). *The category EspAlgA is closed in UltSchA$_0$ under immersions and finite limits.*

Proof.

(i) Let $f: X \to Y$ be an immersion in **UltSchA**$_0$ whose domain is an algebraic space. Let Δ_X and Δ_Y be the diagonals of X and Y respectively in **UltSchA**$_0$. Since f is monomorphic, (Δ_X, f) is the pullback of $(\Delta_Y, f \times f)$ and thus $(U(\Delta_X), U(f))$ is the pullback of $(U(\Delta_Y), U(f \times f))$. Since Y is separated, Δ_Y has a closed image and thus Δ_X has a closed image, so that X is separated. On the other hand, X is homeomorphic to a subspace of Y. Since the topology of Y is noetherian (Proposition (6.5.2)), X also has a noetherian topology and thus is compact. It follows that X is an algebraic space. Therefore **EspAlgA** is closed under immersions and in particular under equalizers.

(ii) Let X, Y be a pair of objects in **EspAlgA**. Let $X \times Y$ be their product in $\mathbf{UltSchA_0}$. Since X and Y are compact, they are covered by finitely many ultra-affine open subsets. According to the construction of $X \times Y$ via the equivalence of categories $\mathbf{UltSchA_0} \sim \mathbf{SchA_0}$ (Propositions (4.6.3) and (6.3.5)), $X \times Y$ is covered by finitely many ultra-affine open subsets and thus is compact. On the other hand, $X \times Y$ is separated since its diagonal $\Delta_{X \times Y} \simeq \Delta_X \times \Delta_Y$ is a closed immersion (Proposition (6.4.10)). It follows that $X \times Y$ is an algebraic space and that **EspAlgA** is closed under finite products. ■

Corollary (6.5.4). *The category EspAlgA has concrete finite limits.*

Proof. The proof follows from Proposition (6.4.6). ■

Proposition (6.5.5). *The category EspAlgA has concrete disjunctors, i.e. any parallel pair of morphisms has a disjunctor preserved by the forgetful functor.*

Proof. The empty algebraic space $\varnothing = \mathrm{Spec}_{\max}(1)$ is a strict initial object in **EspAlgA** which is preserved and reflected by the forgetful functor U. Since U also preserves equalizers, U preserves and reflects disjointed pairs of morphisms. Then a pair of morphisms $(f, g) \colon X \rightrightarrows Y$ is disjointed in **EspAlgA** if and only if the pair of maps (f, g) is disjointed, i.e. $\forall x \in X$, $f(x) \neq g(x)$. Let $(f, g) \colon X \rightrightarrows Y$ be an arbitrary pair of morphisms in **EspAlgA**. Let $d \colon D \to U(X)$ be the disjunctor of $(U(f), U(g))$, i.e. d is the inclusion map of the subset $D = \{x \in X : f(x) \neq g(x)\}$. The set D is open in the topological space X because its complement underlies the equalizer of (f, g) which is a closed immersion since Y is separated. If we equip the subspace D of X with the sheaf induced by O_X, we obtain a locally finitely presentable subultrascheme D of X which is an algebraic space according to Proposition (6.5.3). It is easy to check that $d \colon D \to X$ is a disjunctor of (f, g) in **EspAlgA**. ■

Proposition (6.5.6). *On a locally closed subset of an algebraic space there exists an algebraic space initial structure.*

Proof. The proof follows from Propositions (6.4.9) and (6.5.3). ■

Definition (6.5.7). *An* algebraic subspace *of an algebraic space X is an initial algebraic space on a locally closed subset Y of X. It is* open (closed) *if the subset Y is open (closed) in X.*

Proposition (6.5.8). *Locally finitely presentable affine ultraschemes on A are algebraic spaces on A.*

Proof. Affine ultraschemes are separated (Propositions (4.5.7) and (6.3.8) and compact (Theorem 3.1.9). ■

Notation. It follows from Proposition (6.5.8) that locally finitely presentable affine ultraschemes on **A** are precisely the affine algebraic spaces on **A**. Therefore the category $\mathbf{AffUltA}_0$ can be denoted by **AffEspAlgA**.

Proposition (6.5.9). *The category AffEspAlgA of affine algebraic spaces on A is equivalent to the dual of the category A_0 of finitely presentable objects in A.*

Proof. The proof follows from Theorem (5.3.2) and Corollary (5.3.7). ■

Definition (6.5.10). *An* algebraic variety *on A is an irreducible algebraic space.*

The *category* **VarAlgA** *of algebraic varieties on* **A** is the full subcategory of **EspAlgA** whose objects are the algebraic varieties. It contains the *category* **AffVarAlgA** *of affine algebraic varieties* on **A** as a full subcategory. They are concrete categories whose forgetful functors are the restrictions of that on **EspAlgA**. An algebraic variety which is an algebraic subspace of an algebraic space X is called an *algebraic subvariety* of X.

Proposition (6.5.11). *Algebraic varieties on A are precisely finitely presentable integral ultraschemes of A.*

Proof. The proof follows from Propositions (4.3.8) and (5.3.9). ■

Proposition (6.5.12). *Any algebraic space on A is a finite union of its algebraic subvarieties.*

Proof. Let X be an algebraic space on **A**. According to Proposition (6.5.2), X is a noetherian topological space and thus it has finitely many irreducible components $(X_i)_{i\in[1,n]}$. Since each X_i is a closed subset of X, it underlies a unique algebraic subspace of X (Proposition (6.5.6) and Definition (6.5.7)) which is indeed an algebraic subvariety of X. Let Y be a subalgebraic space of X containing all the subvarieties X_i. Then the underlying set of Y is identical to that of X, and, according to Definition (6.5.7), Y is identical to X. It follows that X is the join of its family of algebraic subvarieties. ■

Proposition (6.5.13). *The category AffVarAlgA of affine algebraic varieties on A is equivalent to the dual of the full subcategory of A whose objects are the finitely presentable integral objects.*

Proof. The proof follows from Propositions (2.7.5), (3.3.4), and (6.5.9). ∎

Notation. The simple object $K(X)$ of an algebraic variety X (Definition (5.3.12)) is called the *rational function object* of X.

Proposition (6.5.14). *For an algebraic variety X and any $x \in U \in \Omega(X)$, $O_X(U)$ is a subobject of $O_{X,x}$ and $O_{X,x}$ is a subobject of $K(X)$.*

Proof. The proof follows from Proposition (4.3.14). ∎

The rational function functor $K_{\mathbf{A}} : (\mathbf{VarAlg}_{\mathrm{dom}}\mathbf{A})^{\mathrm{op}} \to \mathbf{SimA}$

Let us denote by $\mathbf{VarAlg}_{\mathrm{dom}}\mathbf{A}$ the subcategory of **VarAlgA** with the same objects and whose morphisms are the dominant ones. Let $f: X \to Y$ be such a morphism. For any non-empty $U \in \Omega(X)$ (or $V \in \Omega(Y)$) let ι_U: $O_X(U) \to K(X)$ (or $\iota_V : O_Y(V) \to K(Y)$) be the canonical induction. Let $K(f) : K(Y) \to K(X)$ be the morphism defined by $K(f)\iota_V = \iota_{f^{-1}(V)} f^{\#}_V$. In this way we obtain a functor $K_{\mathbf{A}} : (\mathbf{VarAlg}_{\mathrm{dom}}\mathbf{A})^{\mathrm{op}} \to \mathbf{SimA}$ called the *rational function functor.*

Rational morphisms of algebraic varieties

The class $\mathfrak{M}$ of morphisms in $\mathbf{VarAlg}_{\mathrm{dom}}\mathbf{A}$ which are inclusions of non-empty open algebraic subvarieties admits a calculus of right-fractions (cf. 4.4.22.). The category $\mathbf{VarAlg}_{\mathrm{dom}}\mathbf{A}[\mathfrak{M}^{-1}]$ is denoted by $\mathbf{VarAlg}_{\mathrm{rat}}\mathbf{A}$. Its morphisms are called *rational morphisms*, its isomorphisms are called *birational equivalences*, and its isomorphic objects are said to be *birationally equivalent*. Since the rational function functor $K_{\mathbf{A}} : (\mathbf{VarAlg}_{\mathrm{dom}}\mathbf{A})^{\mathrm{op}} \to \mathbf{SimA}$ sends morphisms of $\mathfrak{M}$ onto isomorphisms, it extends canonically to a functor $(\mathbf{VarAlg}_{\mathrm{rat}}\mathbf{A})^{\mathrm{op}} \to \mathbf{SimA}$ which is also called the *rational function functor* and denoted by $K_{\mathbf{A}}$.

Theorem (6.5.15). *The rational function functor $K_{\mathbf{A}} : (VarAlg_{\mathrm{rat}}A)^{\mathrm{op}} \to SimA$ is full and faithful.*

Proof. The proof follows from Corollary (4.6.5) and Propositions (5.3.9), (5.3.11), and (6.5.11). ∎

Corollary (6.5.16). *Two algebraic varieties X and Y are birationally equivalent if and only if their rational function objects $K(X)$ and $K(Y)$ are isomorphic.*

7
ZARISKI TOPOSES

The category of schemes on a Zariski category can be fully and left-exactly embedded in a large topos—its large Zariski topos—while the category of locally finitely presentable schemes can be fully and left-exactly embedded in a topos—its Zariski topos. Modelled toposes and locally modelled toposes can be defined relatively to a Zariski category. They efficiently play the role of ringed toposes and locally ringed toposes. The Zariski topos is the classifying topos for locally modelled toposes, and to any modelled topos is associated a universal locally modelled topos—its prime spectrum. In order to describe the relations existing between the different kinds of schemes, modelled spaces, modelled toposes, and Zariski toposes associated to different Zariski categories, it is time to introduce the notion of morphisms of Zariski categories. There are, in fact three notions: morphisms, geometrical morphisms, and cogeometrical morphisms.

7.1 THE ZARISKI TOPOS OF A ZARISKI CATEGORY

Let **A** be an arbitrary Zariski category.

Proposition (7.1.1) *AffA* (*or* $AffA_0$) *is a dense subcategory of SchA* (*or* $SchA_0$).

Proof. Let X be an object in **SchA**. Let **S** denote the comma category $(\mathbf{AffA}, X)$, let $P: \mathbf{S} \to \mathbf{A}$ be the projection functor defined by $P(S, s) = S$ and $P(g) = g$, and let $(s: S \to X)_{(S,s)\in\mathbf{S}}$ be the inductive cone based on P canonically associated to X. Let **G** be the full subcategory of **S** whose objects are pairs (S, s) where $s: S \to X$ is the immersion of an open affine subscheme S of X. We know that the inductive cone $(s: S \to X)_{(S,s)\in\mathbf{G}}$ based on $P/\mathbf{G}$ is a colimit in **SchA**. Let Y be an object in **SchA** and let $(g_{(S,s)}: S \to Y)_{(S,s)\in\mathbf{S}}$ be an inductive cone with vertex Y based on P. Then $(g_{(S,s)}: S \to Y)_{(S,s)\in\mathbf{G}}$ is an inductive cone based on $P/\mathbf{G}$. Therefore there exists a morphism $g: X \to Y$ in **SchA** such that $gs = g_{(S,s)}$ for any $(S, s) \in \mathbf{G}$. Let us prove that this relation is also true for any $(S, s) \in \mathbf{S}$. Let (S, s) be an arbitrary object of **S**. Let $t \in S$. Then $s(t)$ belongs to an open affine subscheme $u: U \to X$ of X. Then $s^{-1}(U)$ is an open subscheme of S containing t. There exists an affine open subscheme $v: V \to S$ of S containing t and included in $s^{-1}(U)$. Let $\sigma: V \to U$ be the morphism induced by s. Then (U, u) is an object of **G**, $(V, u\sigma)$ is an object of **S**, and $v: (V, u\sigma) \to (S, s)$

and $\sigma : (V, u\sigma) \to (U, u)$ are morphisms in **S**. The relations $g_{(S,s)}v = g_{(V,u\sigma)} = g_{(U,u)}\sigma$ follow from the fact that $(g_{(S,s)} : S \to Y)_{(S,s) \in \mathbf{S}}$ is an inductive cone. Thus $g_{(S,s)}(t) = g_{(S,s)}(v(t)) = g_{(U,u)}(\sigma(t)) = g(u(\sigma(t))) = g(s(t))$. Consequently, $g_{(S,s)} = gs$ for any $(S, s) \in \mathbf{S}$. As a result $(s : S \to X)_{(S,s) \in \mathbf{S}}$ is a colimit in **SchA** and thus **AffA** is a dense subcategory of **SchA**. We can prove similarly that $\mathbf{AffA}_0$ is a dense subcategory of $\mathbf{SchA}_0$. ■

Corollary (7.1.2). *The functor* $\Sigma'_{\mathbf{A}} : SchA \to [A, Set]$ *defined by* $\Sigma'_{\mathbf{A}}(—) = \mathrm{Hom}_{\mathbf{SchA}}(\Sigma_{\mathbf{A}}(.), —)$ *is a full embedding.*

Proof: Let us consider the functor $\Sigma_{\mathbf{A}} : \mathbf{A}^{\mathrm{op}} \to \mathbf{SchA}$ (cf. Theorem (4.1.6)). According to Proposition (7.1.1), it is dense. According to [31], p. 242, the functor $\Sigma'_{\mathbf{A}} : \mathbf{SchA} \to [\mathbf{A}, \mathbf{Set}]$ associated to $\Sigma_{\mathbf{A}}$ is full and faithful. Because it is injective on objects, it is a full embedding. ■

Corollary (7.1.3) *The functor* $\Sigma'_{\mathbf{A}_0} : SchA_0 \to [A_0, Set]$ *defined by* $\Sigma'_{\mathbf{A}_0}(—) = \mathrm{Hom}_{\mathbf{SchA}_0}(\Sigma_{\mathbf{A}_0}(.), —)$ *is a full embedding.*

Proof. The proof is similar to the proof of Corollary (7.1.2). ■

We shall now prove that the category **SchA** (or $\mathbf{SchA}_0$) is fully embedded not only in the category of presheaves on $\mathbf{A}^{\mathrm{op}}$ (or $\mathbf{A}_0^{\mathrm{op}}$), but also in a category of sheaves on $\mathbf{A}^{\mathrm{op}}$ (or $\mathbf{A}_0^{\mathrm{op}}$).

Proposition (7.1.4) *The finite families of singular epimorphisms* $(d_i : A \to D_i)_{i \in [1,n]}$ *such that* $(\mathrm{Spec}(d_i) : \mathrm{Spec}(D_i) \to \mathrm{Spec}(A))_{i \in [1,n]}$ *are surjective families of maps define a subcanonical Grothendieck pretopology on the category* A^{op}.

Proof. For any object A in **A** let $\mathrm{Cov}(A)$ be the set of such families $(d_i : A \to D_i)_{i \in [1,n]}$ with domain A.

(i) The identity 1_A is a singular epimorphism which is a codisjunctor of the pair of projections $(p_1, p_2) : A \times A \rightrightarrows A$, and $\mathrm{Spec}(1_A)$ is bijective. Thus $1_A \in \mathrm{Cov}(A)$.

(ii) If $(d_i : A \to D_i)_{i \in [1,n]} \in \mathrm{Cov}(A)$ and, for any $i \in [1, n]$, $(d_{ij} : D_i \to D_{ij})_{j \in [1, mi]} \in \mathrm{Cov}(D_i)$, then each morphism $d_{ij}d_i$ is a singular epimorphism (Proposition (1.8.4)) and

$$(d_{ij}d_i : A \to D_{ij})_{(i,j) \in \amalg_{i=1}^{n} m_i} \in \mathrm{Cov}(A).$$

(iii) Let $(d_i : A \to D_i)_{i \in [1,n]} \in \mathrm{Cov}(A)$ and $f : A \to B$, and let $(\delta_i : B \to \Delta_i)_{i \in [1,n]}$ be the family of morphisms obtained from $(d_i)_{i \in [1,n]}$ by pushing out along f in **A**. According to Proposition (1.8.4), any δ_i is a singular epimorphism. According to the construction of the map $\mathrm{Spec}(f)$, the image of $\mathrm{Spec}(\delta_i)$ is the inverse image by $\mathrm{Spec}(f)$ of the image of $\mathrm{Spec}(d_i)$. It follows that $(\delta_i : B \to \Delta_i)_{i \in [1,n]} \in \mathrm{Cov}(B)$.

(iv) According to Theorem (3.4.1), the functor $\Delta : \mathbf{CongRadCodis}(A)^{\mathrm{op}} \to \mathbf{A}$ is a sheaf with values in $\mathbf{A}$. This fact implies precisely that any $(d_i : A \to D_i)_{i \in [1,n]} \in \mathrm{Cov}(A)$ is a co-universal effective monomorphic family of morphisms in $\mathbf{A}$. Thus the pretolopogy on $\mathbf{A}^{\mathrm{op}}$ determined by the sets $\mathrm{Cov}(A)$ is subcanonical. ■

Definition (7.1.5).

(i) *The Grothendieck topology on A^{op} generated by the pretopology described in Proposition (7.1.4) is called the* Zariski topology on A^{op}.

(ii) *The large topos of sheaves on A^{op} for the Zariski topology is called the* large Zariski topos *of A and is denoted by $ZARA$.*

Proposition (7.1.6). *The functor $\Sigma'_A : SchA \to [A, Set]$ fully embeds the category $SchA$ into the large Zariski topos $ZARA$.*

Proof. The functor $\Sigma_{\mathbf{A}} : \mathbf{A}^{\mathrm{op}} \to \mathbf{SchA}$ sends families of morphisms belonging to $\mathrm{Cov}(A)$ onto effective epimorphic families of morphisms in $\mathbf{SchA}$. Then the image of the full embedding $\Sigma'_{\mathbf{A}} : \mathbf{SchA} \to [\mathbf{A}, \mathbf{Set}]$ is included in $\mathbf{ZARA}$. ■

Proposition (7.1.7). *The embedding $\Sigma'_A : SchA \to ZARA$ is left-exact and preserves gluing colimits.*

Proof. The functor $\Sigma'_{\mathbf{A}} : \mathbf{SchA} \to [\mathbf{A}, \mathbf{Set}]$ preserves obviously finite limits. Since $\mathbf{ZARA}$ is closed in $[\mathbf{A}, \mathbf{Set}]$ under finite limits, the induced functor $\Sigma'_{\mathbf{A}} : \mathbf{SchA} \to \mathbf{ZARA}$ preserves them. Let us equip the category $\mathbf{SchA}$ with the subcanonical Grothendieck pretopology determined by the families of open immersions $(\varphi_i : X_i \to X)_{i \in I}$ such that $(\phi_i(X_i))_{i \in I}$ is a cover of X. Then the Zariski Grothendieck topology on $\mathbf{A}^{\mathrm{op}}$ is precisely the topology induced on $\mathbf{A}^{\mathrm{op}}$ by the topology on $\mathbf{SchA}$ along the full and faithful functor $\Sigma_A : \mathbf{A}^{\mathrm{op}} \to \mathbf{SchA}$. Moreover, any object in $\mathbf{SchA}$ is the codomain of a covering family of morphisms with domains of the form $\Sigma_A(A)$. According to the comparison lemma ([21], Chapter III, § 4.1) used in the right-universe, the functor $(\Sigma_A)^* : \mathbf{Top(SchA)} \to \mathbf{ZARA}$, where $\mathbf{Top(SchA)}$ is the large topos of sheaves on *SchA*, is an equivalence of the categories. The functor $\Sigma'_{\mathbf{A}} : \mathbf{SchA} \to \mathbf{ZARA}$ is thus equivalent to the Yoneda embedding $\mathbf{SchA} \to \mathbf{Top(SchA)}$. Because this last functor preserves the covering families of morphisms, the functor $\Sigma'_{\mathbf{A}} : \mathbf{SchA} \to \mathbf{ZARA}$ also preserve them. But, for a gluing diagram $(X_{ij})_{(i,j) \in \mathbf{G}(I)}$ in $\mathbf{SchA}$, the family of inductions $(f_i : X_i \to X)_{i \in I}$ of its colimits is a covering family in $\mathbf{SchA}$. Thus $(\Sigma'_{\mathbf{A}}(f_i) : \Sigma'_{\mathbf{A}}(X_i) \to \Sigma'_{\mathbf{A}}(X))_{i \in I}$ is a covering family of morphisms in $\mathbf{ZARA}$, and this fact implies that $(\Sigma'_{\mathbf{A}}(f_i))_{i \in I}$ is the colimit of the diagram $(\Sigma'_{\mathbf{A}}(X_{ij}))_{(i,j) \in \mathbf{G}(I)}$

in **ZARA**. As a result, the functor $\Sigma'_A : \mathbf{SchA} \to \mathbf{ZARA}$ preserves gluing colimits. ■

Definition (7.1.8).
(1) *The* Zariski site *of A is the category A_0^{op} equipped with the Zariski topology, i.e. the Grothendieck topology on A_0^{op} induced by the Zariski topology on A^{op}.*
(ii) *The* Zariski topos *of A is the topos $ZarA$ of sheaves on the Zariski site of A.*

Proposition (7.1.9). *The functor $\Sigma'_{A_0} : SchA_0 \to [A_0, Set]$ fully embeds the category $SchA_0$ into the Zariski topos $ZarA$. This embedding is left-exact and preserves gluing colimits.*

Proof. The proof is similar to the proofs of Propositions (7.16) and (7.17). ■

Proposition (7.1.10). *If A is a rational Zariski category, $AffUltA_0$ is a dense subcategory of $UltSchA_0$ and the functor $\Sigma'_{\max}(-) = \mathrm{Hom}_{UltSchA_0}(\Sigma_{\max}(.), -) : UltSchA_0 \to [A_0, Set]$ fully embeds the category $UltSchA_0$ into the Zariski topos $ZarA$ and preserves finite limits and gluing colimits.*

Proof. The proof follows from Propositions (6.3.5), (7.1.1), (7.1.3), and (7.1.9). ■

7.2 MODELLED TOPOSES ON A ZARISKI CATEGORY

The toposes used here are Grothendieck toposes ([21], § IV.1), their morphisms are geometrical morphisms ([21], § IV.3; [26], § 1.16), and their 2-morphisms are natural transformations between the inverse image functors ([23], § 1.1; [26], § 1.16). They constitute the 2-category of toposes denoted by **Topos**. Let **A** be a Zariski category.

Definition (7.2.1). *A* model *of A in a topos E is a left-exact functor $M : A_0^{op} \to E$.*

The *category* **Mod**[**A**, **E**] *of models of* **A** *in* **E** is the full subcategory **Lex**[$\mathbf{A}_0^{op}$, **E**] of the functor category [$\mathbf{A}_0^{op}$, **E**] whose objects are the models of **A** in **E**.

Proposition (7.2.2). *The category $Mod[A, E]$ of models of A in a topos E is equivalent to the category $Sh[E, A]$ of sheaves on E with values in A.*

Proof. The category $\mathbf{Lex}[\mathbf{A}_0^{\mathrm{op}}, \mathbf{E}]$ is isomorphic to the dual of the category $\mathbf{Rex}[\mathbf{A}_0, \mathbf{E}^{\mathrm{op}})$ of right-exact functors $\mathbf{A}_0 \to \mathbf{E}^{\mathrm{op}}$. According to [16], the category $\mathbf{Rex}[\mathbf{A}_0, \mathbf{E}^{\mathrm{op}}]$ is equivalent to the category $\mathbf{Ladj}[\mathbf{A}, \mathbf{E}^{\mathrm{op}}]$ of left-adjoint functors $\mathbf{A} \to \mathbf{E}^{\mathrm{op}}$ via the left Kan extension along the inclusion functor $\mathbf{A}_0 \to \mathbf{A}$ and also to the dual of the category $\mathbf{Radj}[\mathbf{E}^{\mathrm{op}}\, \mathbf{A}]$ of right-adjoint functors $\mathbf{E}^{\mathrm{op}} \to \mathbf{A}$. According to [21], Chapter IV, Proposition 1.4 and Corollary 1.5, the category $\mathbf{Radj}[\mathbf{E}^{\mathrm{op}}, \mathbf{A}]$ is identical to the category $\mathbf{Sh}[\mathbf{E}, \mathbf{A}]$ of sheaves on the site $\mathbf{E}$ with values in $\mathbf{A}$. Consequently, the categories $\mathbf{Mod}[\mathbf{A}, \mathbf{E}]$ and $\mathbf{Sh}[\mathbf{E}, \mathbf{A}]$ are equivalent. ∎

Definition (7.2.3). *The* inverse image of a model $M: A_0^{\mathrm{op}} \to F$ (*a* morphism of models $\alpha: M \to N$) of A in F *by a morphism of toposes* $f: E \to F$ *is the model* $f^*M: A_0^{\mathrm{op}} \to E$ *(the morphism of models* $f^*\alpha: f^*M \to f^*N$) of A in E.

Inverse images of models and morphisms of models, by a morphism of toposes $f: \mathbf{E} \to \mathbf{F}$, determine the *inverse image functor* $f_{\mathbf{A}}^*: \mathbf{Mod}[\mathbf{A}, \mathbf{F}] \to \mathbf{Mod}[\mathbf{A}, \mathbf{E}]$.

Proposition (7.2.4). *The topos* $[A_0, \mathbf{Set}]$ *is the classifying topos for models of* A*, and the Yoneda embedding* $Y: A_0^{\mathrm{op}} \to [A_0, \mathbf{Set}]$ *is the generic model of* A*. More precisely, for any topos* $\mathbf{E}$ *the functor* $\mathbf{Hom}_{\mathbf{Topos}}[E, [A_0, \mathbf{Set}]] \to \mathbf{Mod}[A, E]$ *which assigns to a morphism of toposes* $f: E \to [A_0, \mathbf{Set}]$ *the inverse image of the model* Y *by* f*, and to a 2-morphism of toposes* $\alpha: f \to g$ *the morphism of models* $\alpha Y: f^*Y \to g^*Y$*, is an equivalence of categories.*

Proof. The proof follows from [21], Chapter IV, Proposition 4.9.4. ∎

Definition (7.2.5). *A* model $M: A_0^{\mathrm{op}} \to E$ *of* A *in* E *is* local *if it is a continuous functor for the Zariski topology on* A_0^{op} *and the canonical topology on* E ([21], Chapter III, § 1.1).

For any topos $\mathbf{E}$, let us denote by $\mathbf{ModL}[\mathbf{A}, \mathbf{E}]$ the full subcategory of $\mathbf{Mod}[\mathbf{A}, \mathbf{E}]$ whose objects are the local models.

Proposition (7.2.6). *The inverse image of a local model of* A *in* F *by a morphism of toposes* $f: E \to F$ *is a local model of* A *in* E.

Proof. The proof follows from the fact that the inverse image functor f^* is continuous for the canonical topologies. ∎

Proposition (7.2.7). *The Zariski topos* $\mathbf{Zar}A$ *is the classifying topos for local models of* A *and the Yoneda embedding* $Y: A_0^{\mathrm{op}} \to \mathbf{Zar}A$ *is the generic local* *model. More precisely, for any topos* **E** *the functor which assigns to a morphism of toposes* $f: E \to ZarA$ *the inverse image of the local model* Y by f, *and to a 2-morphism of toposes* $\alpha: f \to g$ *with* $(f, g): E \rightrightarrows ZarA$ *the morphism of models* $\alpha Y: f^* Y \to g^* Y$, *provides an equivalence of categories*

$$\mathbf{Hom_{Topos}}[\mathbf{E}, \mathbf{ZarA}] \sim \mathbf{ModL}[\mathbf{A}, \mathbf{E}].$$

Proof. The proof follows from [21], Chapter IV, Proposition 4.9.4. ■

Definition (7.2.8). *A* morphism $\alpha: M \to N$ of models of A in E *is* local *if the models* M, N *are local and, for any singular epimorphism* $f: A \to B$ *in* A_0, $(\alpha_B, M(f))$ *is the pullback of* $(N(f), \alpha_A)$.

Local models of **A** in **E** and local morphisms form the *category* **ModLoc**[**A**, **E**] *of local models of* **A** *in* **E**.

Proposition (7.2.9). *The inverse image of a local morphism of models of A in F by a morphism of toposes* $f: E \to F$ *is a local morphism of models of A in E.*

Proof. The proof follows from Proposition (7.2.6) and from the fact that f^* is a left-exact functor. ■

Definition (7.2.10).

(i) *A* modelled topos on *A is a topos* **E** *equipped with a model* O_E *of A in E.*

(ii) *A* morphism of modelled toposes on *A from a modelled topos* (E, O_E) *to a modelled topos* (F, O_F) *is a morphism of toposes* $f: E \to F$ *equipped with a natural transformation* $f^{\#}: f^* O_F \to O_E$.

(iii) *A* 2-morphism of modelled toposes on *A from a morphism of modelled toposes* $f: (E, O_E) \to (F, O_F)$ *to a morphism of modelled toposes* $g: (E, O_E) \to (F, O_F)$ *is a natural transformation* $\varphi: f^* \to g^*$ *such that* $g^{\#}(\varphi O_F) = f^{\#}$.

The modelled toposes on **A**, their morphisms, and their 2-morphisms constitute the 2-category **ToposMod A** *of modelled toposes on* **A**.

The *forgetful 2-functor* **ToposModA** → **Topos** assigns the topos **E** to a modelled topos $(\mathbf{E}, O_{\mathbf{E}})$, the morphism of toposes $f: \mathbf{E} \to \mathbf{F}$ to a morphism of modelled toposes $(f, f^{\#}): (\mathbf{E}, O_{\mathbf{E}}) \to (\mathbf{F}, O_{\mathbf{F}})$, and the 2-morphism of toposes $\varphi: f \to g$ to a 2-morphism of modelled toposes $\varphi: f \to g$.

Proposition (7.2.11) *The forgetful 2-functor ToposModA → Topos is a 2-fibration and a 2-cofibration* [23].

Proof.

(i) Let $\mathbf{E}$ be a topos, let $\mathbf{F}$ be a modelled topos on $\mathbf{A}$, and let $f: \mathbf{E} \to \mathbf{F}$ be a morphism of toposes. Let us equip $\mathbf{E}$ with the model $O_{\mathbf{E}} = f^* O_{\mathbf{F}}: \mathbf{A}^{\mathrm{op}} \to \mathbf{E}$. The natural transformation $f^{\bullet} = 1_{O_{\mathbf{E}}}$ provides a morphism of modelled toposes $(f, f^{\bullet}): (\mathbf{E}, O_{\mathbf{E}}) \to (\mathbf{F}, O_{\mathbf{F}})$. It is straightforward to check that this morphism of modelled toposes $f: \mathbf{E} \to \mathbf{F}$ is a 2-inverse image of the modelled topos $\mathbf{F}$ along the morphism of toposes f.

(ii) Let $\mathbf{E}$ be a modelled topos, let $\mathbf{F}$ be a topos, and let $f: \mathbf{E} \to \mathbf{F}$ be a morphism of toposes. Let us equip $\mathbf{F}$ with the model $O_{\mathbf{F}} = f_* O_{\mathbf{E}}: \mathbf{A}_0^{\mathrm{op}} \to \mathbf{F}$. If we denote the co-unit of the adjunction (f^*, f_*) by $\varepsilon_*: f^* f_* \to 1_{\mathbf{E}}$ and if $f^{\bullet} = \varepsilon_* O_{\mathbf{E}}$, we obtain a morphism of modelled toposes $f: \mathbf{E} \to \mathbf{F}$. It is straightforward to check that this morphism of modelled toposes is a 2-direct image of the modelled topos $\mathbf{E}$ along the morphism of toposes f. ■

The *singleton functor* $J: \mathbf{A}^{\mathrm{op}} \to \mathbf{ToposModA}$ assigns to an object A the modelled topos $(\mathbf{Set}, \mathrm{Hom}_{\mathbf{A}}(J_{\mathbf{A}_0}(—), A))$, and to a morphism $f: A \to B$ the morphism $(1_{\mathbf{Set}}, \mathrm{Hom}_{\mathbf{A}}(J_{\mathbf{A}_0}(—), f))$.

The *global section functor* $\Gamma: \mathbf{ToposModA} \to \mathbf{A}^{\mathrm{op}}$ assigns to a modelled topos $(\mathbf{E}, O_{\mathbf{E}})$ the object $\Gamma(\mathbf{E}, O_{\mathbf{E}})$ of $\mathbf{A}$ which is represented by the left-exact functor $\mathrm{Hom}_{\mathbf{E}}(1, O_{\mathbf{E}}(—)): \mathbf{A}_0^{\mathrm{op}} \to \mathbf{Set}$, and to a morphism of modelled toposes $(f, f^{\bullet}): (\mathbf{E}, O_{\mathbf{E}}) \to (\mathbf{F}, O_{\mathbf{F}})$ the morphism $\Gamma(f, f^{\bullet}): \Gamma(\mathbf{F}, O_{\mathbf{F}}) \to \Gamma(\mathbf{E}, O_{\mathbf{E}})$ which is represented by the natural transformation $\mathrm{Hom}_{\mathbf{F}}(1, O_{\mathbf{F}}) \to \mathrm{Hom}_{\mathbf{E}}(1, O_{\mathbf{E}})$ which is the composite of the three following natural transformations: the natural transformation $\mathrm{Hom}_{\mathbf{F}}(1, \eta_* O_{\mathbf{F}}): \mathrm{Hom}_{\mathbf{F}}(1, O_{\mathbf{F}}) \to \mathrm{Hom}_{\mathbf{F}}(1, f^* f_* O_{\mathbf{F}})$, where $\eta_*: 1_{\mathbf{F}} \to f_* f^*$ denotes the unit of the adjunction (f^*, f_*), the natural isomorphism $\mathrm{Hom}_{\mathbf{F}}(1, f_* f^* O_{\mathbf{F}}) \simeq \mathrm{Hom}_{\mathbf{E}}(1, f^* O_{\mathbf{F}})$, and the natural transformation $\mathrm{Hom}_{\mathbf{E}}(1, f^{\bullet}): \mathrm{Hom}_{\mathbf{E}}(1, f^* O_{\mathbf{F}}) \to \mathrm{Hom}_{\mathbf{F}}(1, O_{\mathbf{E}})$.

Proposition (7.2.12). *The singleton functor $J: A^{\mathrm{op}} \to$ ToposModA is full and faithfull and right-adjoint to the global section functor Γ : ToposModA $\to A^{\mathrm{op}}$.*

Proof. The proof is straightforward. ■

Definition (7.2.13).

(i) *A* locally modelled topos *on A is a modelled topos (E, O_E) on A such that O_E is a local model.*

(ii) *A* local morphism of locally modelled toposes on *A from a locally modelled topos (E, O_E) to a locally modelled topos (F, O_F) is a morphism of modelled toposes $(f, f^{\bullet}): (E, O_E) \to (F, O_F)$ such that $f^{\bullet}: f^* O_F \to O_E$ is a local morphism of models.*

(iii) *A* 2-morphism of locally modelled toposes *on A is a 2-morphism of the underlying modelled toposes on A.*

The locally modelled toposes on A, their local morphisms, and their 2-morphisms constitute the 2-category **ToposModLocA** of locally modelled toposes on **A**.

Theorem (7.2.14). *The inclusion 2-functor ToposModLocA → ToposModA has a right 2-adjoint.*

Proof. We shall use the results of Coste [5,6]. We obtain a localization triple in the sense of [5], Chapter VI, by taking the category **A**, the set V of singular epimorphisms in $\mathbf{A}_0$ (Proposition (1.8.4)), and the Zariski topology $\mathfrak{Z}$ on $\mathbf{A}_0^{op}$. Then the 2-category denoted by $\mathbf{TopT_0Mod}$ (*or* **AdmTopTMod**) in [5] is precisely the 2-dual of our category **ToposModA** (or **ToposModLocA**). According to Theorem 5.7 of [5], Chapter VII, or Corollary 3.6.5 of [6], the inclusion 2-functor **ToposModLocA** → **ToposModA** has a right 2-adjoint. ■

Definition (7.2.15). *The* prime spectrum 2-functor for modelled toposes on *A is the 2-functor* Σ_{ModA} : *ToposModA → ToposModLocA right 2-adjoint to the inclusion 2-functor ToposModLocA → ToposModA.*

7.3 MORPHISMS OF ZARISKI CATEGORIES

Notation [11]

Let us consider an arbitrary functor $U: \mathbf{A} \to \mathbf{B}$ and an arbitrary pair of parallel morphisms $(g, h) : C \rightrightarrows A$ in **A**.

The *functor U preserves codisjointed pairs* if: (g, h) is codisjointed in $\mathbf{A} \Rightarrow (Ug, Uh)$ is codisjointed in **B**.

The *functor U reflects codisjointed pairs if*: (Ug, Uh) is codisjoined in $\mathbf{B} \Rightarrow (g, h)$ is codisjointed in **A**.

The *functor U preserves codisjunctors* if: f is the codisjunctor of (g, h) in $\mathbf{A} \Rightarrow Uf$ is the codisjunctor of (Ug, Uh) in **B**. Let us assume that U has a left-adjoint $F: \mathbf{B} \to \mathbf{A}$ and defines the adjunction (isomorphism) φ : $\mathrm{Hom}_{\mathbf{A}}(F(.),—) \to \mathrm{Hom}_{\mathbf{B}}(.,U(—))$. Let A be an arbitrary object in **A**, B be an arbitrary object in **B**, and let $(g, h) : FB \rightrightarrows A$ be an arbitrary pair of parallel morphisms in **A**.

The *adjunction φ preserves codisjointed pairs* if: $(g, h) : FB \rightrightarrows A$ is codisjointed in $\mathbf{A} \Rightarrow (\varphi_{B,A}(g), \varphi_{B,A}(h)) : B \rightrightarrows UA$ is codisjointed in **B**.

The *adjunction φ reflects codisjointed pairs* if: $(\varphi_{B,A}(g), \varphi_{B,A}(h))$ is codisjointed in $\mathbf{B} \Rightarrow (g, h)$ is codisjointed in **A**.

Definition (7.3.1). *A* morphism of Zariski categories *is a functor* $U: A \to \mathbf{B}$

(i) *whose domain A and codomain **B** are Zariski categories,*

(ii) *which preserves filtered colimits,*

(iii) *which has a left adjoint F, and*
(iv) *which defines an adjunction* $\varphi : \mathrm{Hom}_{\mathbf{A}}(F(.),\text{—}) \to \mathrm{Hom}_{\mathbf{B}}(.,U(\text{—}))$ *which preserves and reflects codisjointed pairs.*

If **A** is a Zariski category, a *Zariski subcategory* of **A** is a subcategory of **A** such that the inclusion functor is a morphism of Zariski categories.

Examples The forgetful functors **CAlg** → **Mod** and **RlOrdRng** → **RlRng** are morphisms of Zariski categories. The functors **CAlg** → **CRng** and **Mod** → **CRng** which take the coefficient ring part are morphisms of Zariski categories. The functors **CRng** → **CAlg** and **CRng** → **Mod** whose value are the zero algebras or modules are morphisms of Zariski categories. The functors **GradCRng** → **CRng** and **GradMod** → **Mod** which take the zero degree part are morphisms of Zariski categories. The category **RlRng** is a Zariski subcategory of **CRng**.

Proposition (7.3.2). *A morphism of Zariski categories preserves simple, integral, reduced, regular, and local objects, and prelocal and local morphisms, and reflects the terminal object and codisjointed pairs of morphisms. Its left-adjoint preserves finite products and codisjunctors.*

Proof. Let $U : \mathbf{A} \to \mathbf{B}$ be a morphism of Zariski categories, let $F : \mathbf{B} \to \mathbf{A}$ be its left-adjoint, let $\varphi : \mathrm{Hom}_{\mathbf{A}}(F(.),\text{—}) \to \mathrm{Hom}_{\mathbf{B}}(.,U(\text{—}))$ be the associated adjunction, let $\eta : 1_{\mathbf{B}} \to UF$ be its unit, and let $\varepsilon : FU \to 1_{\mathbf{A}}$ be its co-unit.

Let $(g, h) : C \rightrightarrows A$ be a pair of morphisms in **A** such that (Ug, Uh) is codisjointed in **B**. Because φ reflects codisjointed pairs, the pair $(\varphi^{-1}_{UC,A}(Ug),\ \varphi^{-1}_{UC,A}(Uh)) : FUC \rightrightarrows A$ is codisjointed. But this pair is equal to the pair $(g\,\varepsilon_C, h\,\varepsilon_C)$. Thence the pair (g, h) is codisjointed. It follows that U reflects codisjointed pairs. As a consequence, U reflects the terminal object because if A is an object of **A** such that UA is terminal, then the pair $(U(1_A), U(1_A)) = (1_{UA}, 1_{UA})$ is codisjointed. Thus the pair $(1_A, 1_A)$ is codisjointed and hence A is terminal. The existence of the unit morphism $\eta_1 : 1 \to UF1$ implies that $UF1$ is terminal and hence $F1$ is terminal.

Let K be a simple object in **A**. Then K is not terminal, and thus UK is not terminal. Let $(g, h) : B \rightrightarrows UK$ be a pair of distinct morphisms in **A**. Then $(\varphi^{-1}_{B,K}(g), \varphi^{-1}_{B,K}(h)) : FB \rightrightarrows K$ is a pair of distinct morphisms. According to proposition (2.1.2), this pair is codisjointed. Thence the pair (g, h) is codisjointed. According to Proposition (2.1.2), UK is simple. As a result, U preserves simple objects. It follows that U preserves integral, reduced, and regular objects because it preserves monomorphisms, regular monomorphisms, and products.

Let A be a local object in **A**. As A is not terminal, UA is not terminal.

Let $(r = (r_1, r_2) : R \rightrightarrows UA,\ s = (s_1, s_2) : S \rightrightarrows UA)$ be a pair of congruences on UA such that $r \overset{c}{\vee} s = 1_{UA \times UA}$. Then $q_r \wedge q_s = O_{UA}$ in QuotReg(UA). The pairs $(Fr_1, Fr_2) : FR \rightrightarrows FUA$ and $(Fs_1, Fs_2) : FS \rightrightarrows FUA$ in $\mathbf{A}$ have coequalizers $F(q_r)$ and $F(q_s)$ respectively which satisfy $F(q_r) \wedge F(q_s) = O_{FUA}$ in QuotReg(FUA). Let p_1 and p_2 be the pushouts of $F(q_r)$ and $F(q_s)$ respectively along the co-unit morphism $\varepsilon_A : FUA \to A$. Then p_1, p_2 are regular quotients of A satisfying $p_1 \wedge p_2 = O_A$. As A is local, $p_1 = O_A$ or $p_2 = O_A$. Let us assume, for example, that $p_1 = O_A$. Because p_1 is the coequalizer of $(\varepsilon_A(Fr_1), \varepsilon_A(Fr_2))$, it follows that this last pair of morphisms is codisjointed. Thence the pair $(\varphi_{R,A}(\varepsilon_A(Fr_1)), \varphi_{R,A}(\varepsilon_1(Fr_2)))$ is codisjointed. But this is the pair (r_1, r_2). It follows that r is codisjointed, i.e. $r = 1_{A \times A}$. As a result, the object UA is local (Proposition (2.12.2)).

Let $(g, h) : B \rightrightarrows C$ be a codisjointed pair in $\mathbf{B}$. Let $q : FC \to Q$ be the coequalizer of the pair (Fg, Fh). Then $\varphi_{C,Q}(q)g = \varphi_{C,Q}(q)h$. Thus $UQ \simeq 1$ and hence $Q \simeq 1$. It follows that F preserves codisjointed pairs. Let $(g, h) : B \rightrightarrows C$ be a pair of morphisms in $\mathbf{B}$ with a codisjunctor $d : C \to D$. Then d codisjoints (g, h). Thus $Fd : FC \to FD$ codisjoints (Fg, Fh). Because F preserves epimorphisms, the morphism Fd is epimorphic. Let $f : FC \to A$ be a morphism which codisjoints (Fg, Fh). Then the pair $(f(Fg), f(Fh))$ is codisjointed. Thus the pair $(\varphi_{B,A}(f(Fg)), \varphi_{B,A}(f(Fh)))$ is codisjointed. But the latter is the pair $((\varphi_{C,A}(f)g, \varphi_{C,A}(f)h))$. Thus $\varphi_{C,A}(f)$ codisjoints (g, h). Therefore there exists a morphism $k : D \to UA$ such that $kd = \varphi_{C,A}(f)$. Then the morphism $\varphi^{-1}_{D,A}(k) : FD \to A$ satisfies $(\varphi^{-1}_{D,A}(k))(Fd) = f$. Thence Fd is a codisjunctor of (Fg, Fh). As a result, the functor F preserves codisjunctors.

Let $f : A \to B$ be a prelocal morphism in $\mathbf{A}$ and let $d : UA \to D$ be a singular epimorphism in $\mathbf{B}$ through which Uf factors in the form $Uf = gd$. Then $\varphi^{-1}_{D,B}(g)\ (Fd) = f\varepsilon_A$. Since F preserves codisjunctors, Fd is a singular epimorphism. Since f is prelocal, there is a uniqu morphism $h : FD \to A$ such that $h(Fd) = \varepsilon_A$ and $fh = \varphi^{-1}_{D,A}(g)$. Then the morphism $\varphi_{D,A}(h) : D \to UA$ is such that $\varphi_{D,A}(h)d = 1_{UA}$. Therefore d is an isomorphism. Consequently, Uf is prelocal. It follows that U preserves prelocal and local morphisms.

Let $(p : B \to P, q : B \to Q)$ be a product in $\mathbf{B}$. According to Theorem (3.9.1), the morphisms p and q are both regular and singular epimorphisms. Moreover, if $r = (r_1, r_2)$ and $s = (s_1, s_2)$ are the kernel pairs of p and q respectively, then p is the codisjunctor of s and q is the codisjunctor of r. Because F preserves coequalizers, codisjunctors, and pushouts, Fp is the coequalizer of (Fr_1, Fr_2) and the codisjunctor of (Fs_1, Fs_2), Fq is the coequalizer (Fs_1, Fs_2) and the codisjunctor of (Fr_1, Fr_2), and we have $Fp \wedge Fq = O_{FB}$. According to the construction of the structure sheaf $\widetilde{FB}$, the pair (Fp, Fq) is a product. As a result, F preserves finite products.

■

Corollary (7.3.3). *A morphism of Zariski categories is a morphism of locally indecomposable categories. Thus it preserves indecomposable objects.*

Proof. The proof follows from Theorem (3.10.1) and Proposition (7.3.2). ■

Proposition (7.3.4). *A faithful morphism of Zariski categories reflects simple objects.*

Proof. Let $U : \mathbf{A} \to \mathbf{B}$ be such a morphism. Then U reflects the terminal object (Proposition (7.3.2)) and monomorphisms. Let A be an object in $\mathbf{A}$ such that UA is simple. If $q : A \to Q \neq 1$ is a regular quotient of A, then $Uq : UA \to UQ$ is not terminal and thus it is monomorphic. Hence q is monomorphic and thus is an isomorphism. Consequently, A is simple. ■

Proposition (7.3.5). *A faithful morphism of Zariski categories preserving regular epimorphisms reflects noetherian objects.*

Proof. Let $U : \mathbf{A} \to \mathbf{B}$ be such a morphism and let A be an object in $\mathbf{A}$ such that UA is noetherian. Since U reflects monomorphisms, the order-preserving map $\mathrm{QuotReg}(A) \to \mathrm{QuotReg}(UA)$ induced by U is injective. Since $\mathrm{QuotReg}(UA)$ satisfies the descending-chain condition, $\mathrm{QuotReg}(A)$ does also, i.e. A is noetherian. ■

Proposition (7.3.6). *A full and faithful morphism of Zariski categories reflects algebraically closed simple objects.*

Proof. Let $U : \mathbf{A} \to \mathbf{B}$ be such a morphism with left-adjoint F. We can assume that $FU = 1_{\mathbf{A}}$. Let K be an object in $\mathbf{A}$ such that UK is algebraically closed simple in $\mathbf{B}$. Then K is simple (Proposition (7.3.4)). Let $m : K \to M$ be a finitely presentable monomorphism. In the category $(UK)/\mathbf{B}$ the object (UM, Um) is a filtered colimit of finitely presentable objects, say $((UM, Um), (f_i)_{i\in\mathbf{I}}) = \xrightarrow{lim}_{i\in\mathbf{I}}(B_i, g_i)$. Since F preserves colimits, we have $((M, m), (Ff_i)_{i\in\mathbf{I}}) = \xrightarrow{lim}_{i\in\mathbf{I}}(FB_i, Fg_i)$ in $K/\mathbf{A}$. Since (M, m) is finitely presentable, there exist some $i \in \mathbf{I}$ and some morphism $h_i : M \to FB_i$ such that $(Ff_i)h_i = 1_M$ and $h_i m = Fg_i$. According to Proposition (6.1.2), the monomorphism g_i has a retraction $r_i : B_i \to UK$. Then $(Fr_i)h_i m = (Fr_i)(Fg_i) = F(r_i g_i) = F(1_{UK}) = 1_K$. Consequently, m has a retraction. According to Proposition (6.1.2), K is algebraically closed. ■

Proposition (7.3.7). *A full and faithful morphism of Zariski categories whose left-adjoint reflects the terminal object preserves algebraically closed simple objects.*

Proof. Let $U: \mathbf{A} \to \mathbf{B}$ be such a morphism with left-adjoint F and adjunction isomorphism φ. We assume that $FU = 1_{\mathbf{A}}$. Let K be an algebraically closed simple object in $\mathbf{A}$. The object UK is simple (Proposition (7.3.2)). Let $f: UK \to B$ be a finitely presentable monomorphism. Then B is not terminal and neither is FB. Thus $Ff: K \to FB$ is a finitely presentable monomorphism. According to Proposition (6.1.2), Ff has a retraction g. Then the morphism $\varphi_{B,K}(g)$ is a retraction of f. According to Proposition (6.1.2), UK is algebraically closed. ■

Corollary (7.3.8). *For any Zariski category A, the categories of simple (algebraically closed simple) objects of A, **RedA**, and **RegA** are identical.*

Proof. The inclusion functor **RedA** $\to$ **A** is a full and faithful morphism of Zariski categories whose reflector reflects the terminal object, since a non-terminal object A has a simple quotient $s: A \to S$ which factors through the universal reduced object associated to A. Similarly, the inclusion functor **RegA** $\to$ **A** is a full and faithful morphism of Zariski categories whose reflector reflects the terminal object. Then the result follows from Propositions (7.3.2), (7.3.4), (7.3.6), and (7.3.7). ■

Examples. (cf. examples on p. 177). The functors **RlRng** $\to$ **CRng**, **RlOrdRng** $\to$ **RlRng**, **CAlg** $\to$ **Mod**, **CRng** $\to$ **CAlg**, and **CRng** $\to$ **Mod** preserve and reflect simple objects. The functor **RlRng** $\to$ **CRng** reflects algebraically closed simple objects. The functors **CRng** $\to$ **Mod** and **CRng** $\to$ **CAlg** preserve and reflect algebraically closed simple objects.

Proposition (7.3.9). *If A is a simply noetherian Zariski category, **RedA** is also.*

Proof. Let K be a simple object in **A** and let $f: K \to A$ be a finitely presentable morphism in **RedA**. In the category $K/\mathbf{A}$ the object (A,f) is a filtered colimit of finitely presentable objects, say $(A,f) = \xrightarrow{lim}_{i \in \mathbf{I}} (A_i, f_i)$, with inductions $g_i: (A_i, f_i) \to (A,f)$. For each $i \in \mathbf{I}$, let $g_i = m_i q_i$ be the regular factorization of g_i with $q_i: A_i \to B_i$ and $m_i: B_i \to A$. The object B_i is reduced as a subobject of A, and it is noetherian in **A** and in **RedA** as a regular quotient of the noetherian object A_i. Then we obtain a filtered monomorphic colimit $(m_i: (B_i, q_i f_i) \to (A,f))_{i \in \mathbf{I}}$ in **RedA**. Since f is finitely presentable, there exists some $i \in \mathbf{I}$ such that m_i is an isomorphism. It follows that A is noetherian in **RedA**. As a result, K is transnoetherian in **RedA**, and **RedA** is simply noetherian. ■

Definition (7.3.10). *The* direct image functor U_{sp}: ***SpModA*** $\to$ ***SpModB*** *determined by a morphism of Zariski categories $U: A \to B$ is defined by $U_{sp}(X, O_X) = (X, UO_X)$ and $U_{sp}(f, f^{\#}) = (f, Uf^{\#})$. As the functor U*

preserves local objects and morphisms (Proposition (7.3.2)), the functor U_{sp} *induces a functor* U_{spl} : ***SpModLocA*** $\to$ ***SpModLocB*** *which is also called direct image functor determined by U.*

Proposition (7.3.11). *The direct image functor* U_{sp} : ***SpModA*** $\to$ ***SpModB*** *satisfies the relations* $U^{op}\Gamma_A = \Gamma_B U_{sp}$ *and* $\Gamma_B U_{sp}\Sigma_A = U^{op}$, *and determines a natural transformation* $\Sigma_U : U_{sp}\Sigma_A \to \Sigma_B U^{op}$.

Proof. For any object (X, O_X) of **SpModA**, we have $U^{op}\Gamma_{\mathbf{A}}(X, O_X) = U(O_X(X)) = (UO_X)(X) = \Gamma_{\mathbf{B}}(X, UO_X) = \Gamma_{\mathbf{B}} U_{sp}(X, O_X)$. For any morphism $(f, f^{\#}) : (X, O_X) \to (Y, O_Y)$ of **SpModA** we have $U^{op}\Gamma_{\mathbf{A}}(f, f^{\#}) = U(f^{\#}_X) = (Uf^{\#})_X = \Gamma_{\mathbf{B}}(f, Uf^{\#}) = \Gamma_{\mathbf{B}}(U_{sp}(f, f^{\#}))$. Thus the relation $U^{op}\Gamma_{\mathbf{A}} = \Gamma_{\mathbf{B}} U_{sp}$ holds. The equality $\Gamma_{\mathbf{B}} U_{sp}\Sigma_{\mathbf{A}} = U^{op}\Gamma_{\mathbf{A}}\Sigma_{\mathbf{A}} = U^{op}$ follows from the equality $1_{\mathbf{A}^{op}} = \Gamma_{\mathbf{A}}\Sigma_{\mathbf{A}}$. Let us denote by $\Sigma^{l}_{\mathbf{A}} : \mathbf{A}^{op} \to$ **SpModLocA**, $\Sigma^{l}_{\mathbf{B}} : \mathbf{B}^{op} \to$ **SpModLocB**, $\Gamma^{l}_{\mathbf{A}}$: **SpModLocA** $\to \mathbf{A}^{op}$, and $\Gamma^{l}_{\mathbf{B}}$: **SpModLocB** $\to \mathbf{B}^{op}$ the functors induced by $\Sigma_{\mathbf{A}}$, $\Sigma_{\mathbf{B}}$, $\Gamma_{\mathbf{A}}$, and $\Gamma_{\mathbf{B}}$ respectively. Because the functor U_{spl} : **SpModLocA** $\to$ **SpModLocB** is induced by U_{sp}, the relations $U^{op}\Gamma^{l}_{\mathbf{A}} = \Gamma^{l}_{\mathbf{B}} U_{spl}$ and $\Gamma^{l}_{\mathbf{B}} U_{spl}\Sigma^{l}_{\mathbf{A}} = U^{op}$ hold. Let us denote the inclusion functor by $J_{\mathbf{B}}$: **SpModLocB** $\to$ **SpModB** and the unit of the adjunction defined by $\Sigma^{l}_{\mathbf{B}}$ by $\eta_{\mathbf{B}} : 1_{\mathbf{SpModLocB}} \to \Sigma^{l}_{\mathbf{B}}\Gamma^{l}_{\mathbf{B}}$. Then the natural transformation $\Sigma_U : U_{sp}\Sigma_{\mathbf{A}} \to \Sigma_{\mathbf{B}} U^{op}$ is defined by $\Sigma_U = J_{\mathbf{B}}\eta_{\mathbf{B}} U_{spl}\Sigma^{l}_{\mathbf{A}} : J_{\mathbf{B}} U_{spl}\Sigma^{l}_{\mathbf{A}} \to J_{\mathbf{B}}\Sigma^{l}_{\mathbf{B}}\Gamma^{l}_{\mathbf{B}} U_{spl}\Sigma^{l}_{\mathbf{A}}$. The value of Σ_U at A in $\mathbf{A}$ is the morphism $((\Sigma_U)_A, (\Sigma_U)^{\#}_A) : (\mathrm{Spec}(A), U\tilde{A}) \to (\mathrm{Spec}(U(A)), \widetilde{UA})$, where $(\Sigma_U)_A : \mathrm{Spec}(A) \to \mathrm{Spec}(U(A))$ is defined by $(\Sigma_U)_A(p) = U(p)$, and $(\Sigma_U)^{\#}_A : \widetilde{UA} \to ((\Sigma_U)_A)_*(U\tilde{A})$ is defined on an affine open set $D(r)$ of $\mathrm{Spec}(U(A))$ as follows. Let F be the left-adjoint to U, let φ be its adjunction isomorphism, and let η and ε be the unit and co-unit of the adjunction. Let $\delta : UA \to \Delta$ be the codisjunctor of $r = (r_1, r_2) : R \rightrightarrows UA$. Then $F\delta : FUA \to F\Delta$ is the codisjunctor of $(F(r_1), F(r_2))$. Let $(\delta_1 : A \to \Delta_1, \sigma : F(\Delta) \to \Delta_1)$ be the pushout of $(\varepsilon_A, F\delta)$. *Then* $((\Sigma_U)^{\#}_A)_{D(r)} = \varphi_{\Delta,\Delta_1}(\sigma) : \Delta \to U(\Delta_1)$. ∎

Definition (7.3.12). *A morphism of Zariski categories* $U : \mathbf{A} \to \mathbf{B}$ *is called* geometrical *if the associated natural transformation* Σ_U *is an isomorphism.*

A morphism of Zariski categories is said to *lift prime congruences* if, for any object A in **A** and any prime congruence p on UA, there exists a prime congruence r on A such that $Ur = p$.

Proposition (7.3.13). *For a morphism of Zariski categories* $U : \mathbf{A} \to \mathbf{B}$, *the following assertions are equivalent.*

(i) *U is geometrical.*

(ii) *There exists an isomorphism* $U_{sp}\Sigma_A \simeq \Sigma_B U^{op}$.

(iii) U_{sp} *preserves affine schemes.*

(iv) *U preserves codisjunctors of congruences and lifts prime congruences.*

Then the functor U_{sp} preserves schemes and induces a functor U_{sch}: $\mathbf{Sch A} \to \mathbf{Sch B}$ and a functor $U_{\mathrm{aff}} : \mathbf{Aff A} \to \mathbf{Aff B}$.

Proof.

(i) ⇒ (ii) is obvious.

(ii) ⇒ (iii) Let (X, O_X) be an affine scheme on $\mathbf{A}$. There is an object A in $\mathbf{A}$ such that $(X, O_X) \simeq (\mathrm{Spec}(A), \tilde{A})$. Then $U_{\mathrm{sp}}(X, O_X) \simeq U_{\mathrm{sp}}(\mathrm{Spec}(A), \tilde{A}) = U_{\mathrm{sp}}\Sigma_{\mathbf{A}}(A) \simeq \Sigma_{\mathbf{B}}(UA) = (\mathrm{Spec}(UA), \widetilde{UA})$ is an affine scheme.

(iii) ⇒ (i) According to Proposition (7.3.11), the value of the natural transformation Σ_U at A is $(\Sigma_U)_A = (\eta_{\mathbf{B}})_{(\mathrm{Spec}(A), U\tilde{A})}$ where $\eta_{\mathbf{B}} : 1_{\mathbf{SpModLocB}} \to \Sigma_{\mathbf{B}}\Gamma_{\mathbf{B}}$ is the unit of the adjunction associated to the adjoint pair $(\Gamma_{\mathbf{B}}, \Sigma_{\mathbf{B}})$. Thus U is geometrical ⇔ for every object A in $\mathbf{A}$, $(\Sigma_U)_A$ is an isomorphism

⇔ for every object A in $\mathbf{A}$, $(\eta_{\mathbf{B}})_{(\mathrm{Spec}(A), U\tilde{A})}$ is an isomorphism

⇔ for every A in $\mathbf{A}$, $(\mathrm{Spec}(A), U\tilde{A})$ is affine

⇔ U_{sp} preserves affine schemes.

(iii) ⇒ U_{sp} preserves schemes. Let (X, O_X) be a scheme on $\mathbf{A}$ and let $(X_i)_{i \in I}$ be an open affine covering of X. For any $i \in I$, $U_{\mathrm{sp}}(X_i, O_X/X_i) = (X_i, UO_X/X_i) = U_{\mathrm{sp}}(X, O_X)/X_i$ is an affine scheme. Thus $U_{\mathrm{sp}}(X, O_X)$ is a scheme on $\mathbf{B}$.

(i) ⇒ (iv) Let A be an object in $\mathbf{A}$. The map $(\Sigma_U)_A : \mathrm{Spec}(A) \to \mathrm{Spec}(UA)$ is an order-preserving bijection. Thus the assignment $r \mapsto Ur$ is an order-preserving bijection $\mathrm{CongRad}(A) \to \mathrm{CongRad}(UA)$. According to Proposition (3.1.2) and Definition (3.1.4), the assignment $D(r) \mapsto D(Ur)$ is an order-preserving bijection $\Omega(\mathrm{Spec}(A)) \to \Omega(\mathrm{Spec}(UA))$. Let $r = (r_1, r_2) : R \rightrightarrows A$ be a codisjunctable congruence on an object A in $\mathbf{A}$ and let $\delta : A \to \Delta$ be its codisjunctor. According to Proposition (4.4.11), $\Sigma_{\mathbf{A}}(\delta) : \Sigma_{\mathbf{A}}(\Delta) \to \Sigma_{\mathbf{A}}(A)$ is an open immersion of schemes on $\mathbf{A}$. Then $U_{\mathrm{sp}}\Sigma_{\mathbf{A}}(\delta) : U_{\mathrm{sp}}\Sigma_{\mathbf{A}}(\Delta) \to U_{\mathrm{sp}}\Sigma_{\mathbf{A}}(A)$ is an open immersion of schemes on $\mathbf{B}$. As Σ_U is an isomorphism, $\Sigma_{\mathbf{B}}U(\delta) : \Sigma_{\mathbf{B}}U(\Delta) \to \Sigma_{\mathbf{B}}U(A)$ is an open immersion of schemes on $\mathbf{B}$. According to Proposition (4.4.11), the morphism $U\delta$ is a singular epimorphism. Moreover, the image of the map $\mathrm{Spec}(U\delta)$: $\mathrm{Spec}(U\Delta) \to \mathrm{Spec}(UA)$ is $\mathrm{Spec}(U\delta)(((\Sigma_U)_\Delta(\mathrm{Spec}(\Delta))) = (\Sigma_U)_A(D(r)) = D(Ur)$. According to Proposition (3.1.11) the congruence Ur is codisjunctable, and according to Proposition (3.3.10) $U\delta$ is its codisjunctor. As a result, the functor U preserves codisjunctors of congruences. On the other hand, the fact that the map $(\Sigma_U)_A : \mathrm{Spec}(A) \to \mathrm{Spec}(UA)$ is surjective implies that the functor U lifts prime congruences.

(iv) ⇒ (i) Let us prove first that for any object A in $\mathbf{A}$ and any congruence r on A, $(\Sigma_U)_A^{-1}(D(U(r))) = D(r)$. Let $p \in \mathrm{Spec}(A)$ and let q_p, j_p, l_p, and k_p be the morphisms canonically associated to p. If $p \in D(r)$, then $j_p q_p = k_p l_p$ codisjoints r (Proposition (3.1.1)). Thus $U(j_p)U(q_p)$ codisjoints Ur, so that neither $U(j_p)U(q_p)$ nor $U(q_p)$ coequalizes Ur. Therefore $Up \in D(Ur)$. If $p \notin D(Ur)$, then $r \leqslant p$. Thus $Ur \leqslant Up$, so that $Up \notin D(Ur)$. As a result, $(\Sigma_U)_A^{-1}(D(Ur)) = D(r)$. Let r be a codisjunctable congruence on A with codisjunctor $\delta : A \to \Delta$. Then Ur is a codisjunctable congruence on UA with codisjunctor $U\delta$. According to the definition of $\Sigma_U^{\#}$, the morphism $((\Sigma_U^{\#})_A)_{D(Ur)} : \widetilde{UA}(D(Ur)) \to U\tilde{A}(D(r))$ is such that $((\Sigma_U^{\#})_A)_{D(Ur)}(U\delta) = U\delta$. Thus it is an isomorphism. It follows that $\Sigma_U^{\#}$ is an isomorphism. According to the hypothesis, the map $(\Sigma_U)_A : \mathrm{Spec}(A) \to \mathrm{Spec}(UA)$ is surjective. Consequently, $(\Sigma_U)_A(D(r)) = (\Sigma_U)_A(\Sigma_U)_A^{-1}(D(Ur)) = D(Ur)$. It follows that $(\Sigma_U)_A$ is a continuous open map. Let us prove that it is injective. Let $p_1, p_2 \in \mathrm{Spec}(A)$ be such that $(\Sigma_U)_A(p_1) = (\Sigma_U)_A(p_2)$. For any affine open set $D(r)$ of $\mathrm{Spec}(A)$ containing p_1, the open set $(\Sigma_U)_A(D(r)) = D(Ur)$ contains $Up_1 = Up_2$. Thus $p_2 \in (\Sigma_U)^{-1}D(Ur) = D(r)$. It follows that p_1 and p_2 have the same open neighbourhoods. Because the topological space $\mathrm{Spec}(A)$ is T_0 (Theorem (3.1.9)), $p_1 = p_2$. As a result, the map $(\Sigma_U)_A$ is a homeomorphism and $(\Sigma_U, \Sigma_U^{\#})$ is an isomorphism so that U is a geometrical morphism. ■

Proposition (7.3.14). *A geometrical morphism of Zariski categories U: $\mathbf{A} \to \mathbf{B}$ determines a geometric morphism of toposes $(U_*, U^*) : \mathbf{ZARA} \to \mathbf{ZARB}$.*

Proof. We use the notations in [21], Chapters III and IV, for the large sites $\mathbf{A}^{\mathrm{op}}$ and $\mathbf{B}^{\mathrm{op}}$ equipped with their Zariski topologies. Let $F : \mathbf{B} \to \mathbf{A}$ be the left-adjoint to U and let φ be the associated adjunction. Let $(d_i : B \to D_i)_{i \in I} \in \mathrm{Cov}(B)$. According to Proposition (7.3.2), the morphisms $Fd_i : FB \to FD_i$ are singular epimorphisms. Let us prove that $(Fd_i)_{i \in I} \in \mathrm{Cov}(FB)$. Let $p \in \mathrm{Spec}(FB)$. The morphism $\varphi_{B,(FB)_p}(l_p) : B \to U((FB)_p)$ has a local codomain (Proposition (7.3.2)), and thus it factors in the form gl_q with $q \in \mathrm{Spec}(B)$ (Theorem (2.13.3)). Because $(\mathrm{Spec}(d_i) : \mathrm{Spec}(D_i) \to \mathrm{Spec}(B))_{i \in I}$ is a surjective family of open immersions, there exists $i \in I$ and $q_i \in \mathrm{Spec}(D_i)$ such that $l_{q_i} d_i = l_q$. Then the morphism $f = \varphi^{-1}_{D_i,(FB)_p}(gl_{q_i}) : FD_i \to (FB)_p$ is such that $f(Fd_i) = l_p$. It follows that $p \in \mathrm{ImSpec}(Fd_i)$. As a result, the family $(\mathrm{Spec}(Fd_i))_{i \in I}$ is surjective, and so $(Fd_i)_{i \in I} \in \mathrm{Cov}(FB)$. According to [21], Chapter III, Proposition 1.6, the functor F^{op} is continuous. According to [21], Chapter III, Definition 1.1, the functor F^* : $[\mathbf{A}, \mathbf{Set}] \to [\mathbf{B}, \mathbf{Set}]$ induces a functor $F^* : \mathbf{ZARA} \to \mathbf{ZARB}$. According to [21], Chapter I, Proposition 5.5, the functor $U_* : [\mathbf{A}, \mathbf{Set}] \to [\mathbf{B}, \mathbf{Set}]$ is identical to the functor F^* and so it induces a functor $U_* : \mathbf{ZARA} \to \mathbf{ZARB}$. On the other hand, it follows from Proposition (7.3.13) that the functor U preserves singular epimorphisms and their pushouts and that U^{op} preserves covering families. It follows that U^{op} is continuous. Consequently, the functor $U^* : [\mathbf{B}, \mathbf{Set}] \to [\mathbf{A}, \mathbf{Set}]$ induces a functor U^* :

ZARB $\rightarrow$ **ZARA** which preserves limits and is left-adjoint to U_*. As a result (U_*, U^*) : **ZARA** $\rightarrow$ **ZARB** is a geometric morphism of toposes. ■

Definition (7.3.15). *The* inverse image functor U^{sp} : ***SpModB*** $\rightarrow$ ***SpModA*** *determined by a morphism of Zariski categories* $U : A \rightarrow B$ *with left-adjoint* $F : B \rightarrow A$ *is defined by* $U^{sp}(Y, O_Y) = (Y, a(FO_Y))$, *where* $a(FO_Y)$ *denotes the sheaf on Y universally associated to the presheaf* FO_Y, *and by* $U^{\mathrm{sp}}(g, g^{\#}) = (g, a(Fg^{\#}))$, *where* $a(Fg^{\#})$ *denotes the morphism of sheaves universally associated to the morphism of presheaf* $Fg^{\#}$.

Proposition (7.3.16). *The inverse image functor* U^{sp} : ***SpModB*** $\rightarrow$ ***SpModA*** *is right-adjoint to the direct image functor* U_{sp} : ***SpModA*** $\rightarrow$ ***SpModB***.

Proof. Let ε be the co-unit of the adjunction (F, U). Let us define the natural transformation $\theta : \mathrm{Hom}_{\mathbf{SpModB}}(U_{\mathrm{sp}}(-), .) \rightarrow \mathrm{Hom}_{\mathbf{SpModA}}(-, U^{\mathrm{sp}}(.))$ by $\theta_{X,Y}(g, g^{\#}) = (g, a(\varepsilon(Fg^{\#})))$. For given (X, O_X) in **SpModA**, (Y, O_Y) in **SpModB**, and $g : X \rightarrow Y$ continuous, the correspondence $g^{\#} \mapsto a(\varepsilon(Fg^{\#}))$ is the composite of the bijections $\mathrm{Hom}_{\mathbf{Sh}[Y,\mathbf{B}]}(O_Y, U(g_*(O_X))) \simeq \mathrm{Hom}_{\mathbf{PrSh}[Y,\mathbf{A}]}(FO_Y, g_*(O_X)) \simeq \mathrm{Hom}_{\mathbf{Sh}[Y,\mathbf{A}]}(a(FO_Y), g_*(O_X))$. Therefore $\theta_{X,Y}$ is a bijection and thus θ is an isomorphism. Hence U_{sp} is left adjoint to U^{sp}. ■

Definition (7.3.17). *A morphism of Zariski categories* $U : A \rightarrow B$ *is called* cogeometrical *if there exists an isomorphism* $U^{\mathrm{sp}}\Sigma_B \simeq \Sigma_A F^{\mathrm{op}}$.

Proposition (7.3.18). *For a morphism of Zariski categories* $U : A \rightarrow B$ *with left-adjoint F the following assertions are equivalent.*

(i) *U is cogeometrical.*

(ii) *F preserves local objects and local morphisms.*

Then the functor U^{sp} *preserves affine schemes and schemes and induces the functors* U^{aff} : ***AffB*** $\rightarrow$ ***AffA***, U^{sch} : ***SchB*** $\rightarrow$ ***SchA***, and U^{spl} : ***SpModLocA*** $\rightarrow$ ***SpModLocB***.

Proof.

(i) $\Rightarrow$ (ii) Let B be an object in **B**. Then $U^{\mathrm{sp}}(\mathrm{Spec}(B), \tilde{B}) = U^{\mathrm{sp}}\Sigma_{\mathbf{B}}(B) \simeq \Sigma_{\mathbf{A}}F^{\mathrm{op}}(B) = (\mathrm{Spec}(FB), \widetilde{FB})$. It follows that U^{sp} preserves affine schemes and therefore preserves schemes. Moreover, the topological space $\mathrm{Spec}(B)$ is homeomorohic to $\mathrm{Spec}(FB)$. Consequently, according to Proposition (3.3.2), B is local $\Leftrightarrow$ $\mathrm{Spec}(B)$ has a unique closed point $\Leftrightarrow$ $\mathrm{Spec}(FB)$ has a unique closed point $\Leftrightarrow$ FB is local. Similarly, if $f : B \rightarrow C$ is any morphism in **B**, the continuous map $\mathrm{Spec}(f)$ is isomorphic in **TopSpec** to the continuous map $\mathrm{Spec}(Ff)$, and f local $\Leftrightarrow$ $\mathrm{Spec}(f)$ preserves the closed point $\Leftrightarrow$ Spec(Ff) preserves the closed point $\Leftrightarrow$ Ff is local.

(ii) ⇒ (i) The functor U^{sp} induces a functor $U^{\mathrm{spl}}: \mathbf{SpModLocB} \to \mathbf{SpModLocA}$. According to Proposition (7.3.16), U^{spl} is right-adjoint to U_{spl}. According to Proposition (7.3.11), the functor U_{spl} satisfies the relation $U^{\mathrm{op}}\Gamma^{\mathrm{l}}_{\mathbf{A}} = \Gamma^{\mathrm{l}}_{\mathbf{B}} U_{\mathrm{spl}}$. Thus the right-adjoints F^{op}, $\Sigma^{\mathrm{l}}_{\mathbf{A}}$, $\Sigma^{\mathrm{l}}_{\mathbf{B}}$, and U^{spl} to the functors U^{op}, $\Gamma^{\mathrm{l}}_{\mathbf{A}}$, $\Gamma^{\mathrm{l}}_{\mathbf{B}}$, and U_{spl} respectively satisfy the relation $\Sigma^{\mathrm{l}}_{\mathbf{A}} F^{\mathrm{op}} \simeq U^{spl} \Sigma^{\mathrm{l}}_{\mathbf{B}}$. Consequently, the relation $U^{\mathrm{sp}} \Sigma_{\mathbf{B}} \simeq \Sigma_{\mathbf{A}} F^{\mathrm{op}}$ holds. As a result, the morphism U is cogeometrical. ■

8
NEAT OBJECTS AND MORPHISMS

Neat objects are introduced in order to describe for arbitrary objects in a Zariski category **A** the analogue of the separability property for field extensions. Therefore, for a finite field extension $k \to K$, the object K in $\mathbf{CAlg}(k)$ will be neat if and only if K is a separable extension of k. The classical definition of neat algebras by means of nilpotent ideals cannot be used in an arbitrary Zariski category because of lack of such ideals, for example, in the category **RedCRng** of reduced rings. We shall use another feature of the separability property, namely that two roots of a separable irreducible polynomial $P \in k[X]$ in a local k-algebra L are either equal or relatively prime. If K is the splitting field of P, this property is equivalent to the property that any pair of morphisms $(f, g) : K \rightrightarrows L$ in $\mathbf{CAlg}(k)$ is either equal or codisjointed. Thus we have the definition of a neat object: a finitely presentable object A such that any pair of morphisms $(f, g) : A \rightrightarrows L$ into a local object L is either equal or codisjointed. Indeed, there are many other equivalent definitions of neat objects, including the following: a neat object is a finitely presentable codecidable object, i.e. an object whose codiagonal is a direct factor. Naturally, in any category of commutative algebras $\mathbf{CAlg}(A)$ over some ring A, neat objects are precisely classical neat algebras.

Arbitrary colimits of neat objects are called neatish objects. For example, infinite algebraic separable field extensions of k are neatish objects in $\mathbf{CAlg}(k)$. They inherit many properties of neat objects except the finite presentability, and to any object is associated a co-universal neatish object. They are the objects of a full subcategory **NtshA** of **A** which plays a central role here. **NtshA** is a locally finitely presentable category whose finitely presentable objects are exactly the neat objects of **A**. Furthermore, it is a locally simple category, i.e. a Zariski category in which all objects are von Neumann regular, and it is indeed the universal locally simple category associated to the Zariski category **A**.

Purely unneat objects correspond to purely inseparable field extensions. They are the objects whose associated co-universal neatish object is the initial object of the category. This notion and the previous ones extend naturally to morphisms in the following way.

A morphism $f: A \to B$ in a Zariski category **A** is required to satisfy a property P if the object (B, f) in the Zariski category $A/\mathbf{A}$ satisfies this property P. Thence we have the notions of neat, neatish, and purely unneat morphisms, and many connections between these morphisms and the objects defined previously. These classes of morphisms are closed under composition, and any morphism can be factorized in an essentially unique way as the composite of a neatish morphism followed by a purely unneat one.

Neatly closed objects correspond to separably closed field extensions. Several equivalent definitions are given : an object A such that any neat morphism $f: A \to B$ is a diagonal $A \to A^n$, a simple object which is injective with respect to interminable neat morphisms, or a simple object whose simple extensions are purely unneat. Under reasonable assumptions, any simple object has a neat closure which is essentially unique and equipped with a Galois group of automorphisms. Indeed, it is the essentially universal neatly closed object associated to it or, equivalently, its neatish hull.

8.1 PRENEAT OBJECTS

Definition (8.1.1). *An object A is* preneat *if, for any local object L, any pair of morphisms $(f, g): A \rightrightarrows L$ is equal or codisjointed.*

Proposition (8.1.2). *For an object A, the following assertions are equivalent.*

(i) *A is preneat.*

(ii) *For any local object L with residue simple quotient $q_m: L \to L/m$, the map* $\mathrm{Hom}_{\mathbf{A}}(A, q_m)$ *is injective.*

(iii) *For any object B with Jacobson quotient $q_{\mathrm{Jac}(B)}: B \to B/Jac(B)$, the map* $\mathrm{Hom}_{\mathbf{A}}(A, q_{\mathrm{Jac}(\mathbf{B})})$ *is injective.*

(iv) *For any local morphism f, the map* $\mathrm{Hom}_{\mathbf{A}}(A, f)$ *is injective.*

(v) *The codiagonal of A is a local isomorphism.*

(vi) *The codiagonal of A is a semisingular epimorphism.*

(vii) *The coequalizer of any pair of morphisms $(f, g): A \rightrightarrows B$ is a semisingular epimorphism.*

Proof.

(i) ⇒ (ii) Any pair of morphisms $(f, g): A \rightrightarrows L$ which satisfies $q_m f = q_m g$ is not codisjointed and thus is equal.

(ii) ⇒ (iii) Let $q = q_{\mathrm{Jac}(B)}: B \to B/\mathrm{Jac}(B)$ and $(f, g): A \rightrightarrows B$ such that $qf = qg$. Let $p \in \mathrm{Spec}(B)$ and $m \in \mathrm{Spec}_{\max}(B)$ such that $p \leqslant m$. Let $u: B/\mathrm{Jac}(\mathbf{B}) \to B/m$ and $v: B_m \to B_p$ be the unique morphisms such that $uq = q_m: B \to B/m$ and $l_p = v l_m$. Then $k_m l_m f = j_m q_m f = j_m uqf = j_m uqg = j_m q_m g = k_m l_m g$. Thus $l_m f = l_m g$ and hence $l_p f = v l_m f = v l_m g = l_p g$. Since the family of morphisms $(l_p: B \to B_p)_{p \in \mathrm{Spec}(B)}$ is monomorphic, we obtain $f = g$. As a result $\mathrm{Hom}_{\mathbf{A}}(A, q)$ is injective.

(iii) ⇒ (iv) Let $f: L \to M$ be a local morphism and let $(g, h): A \rightrightarrows L$ be such that $fg = fh$. Let $q_n: L \to L/n$ and $q_m: M \to M/m$ be the residue simple objects of L and M respectively, and let $\bar{f}: L/n \to M/m$ be the residue morphism of f. Then $\bar{f}q_n g = q_m fg = q_m fh = \bar{f}q_n h$. Thus $q_n g = q_n h$. Since $q_n = q_{\mathrm{Jac}(L)}$, we obtain $g = h$.

(iv) ⇒ (v) Let $(\alpha : A \to A \amalg A, \beta : A \to A \amalg A)$ be the coproduct of A by A, let $\nabla : A \amalg A \to A$ be the codiagonal of A, which is the coequalizer of (α,β), $p \in \mathrm{Spec}(A)$, $q = \mathrm{Spec}(\nabla)(p)$, and let ∇_p be the local morphism of ∇ at p. The relations $\nabla_p l_q \alpha = l_p \nabla \alpha = l_p \nabla \beta = \nabla_p l_q \beta$ and the fact that ∇_p is local entail $l_q \alpha = l_q \beta$. Hence there is a morphism $f : A \to (A \amalg A)_q$ such that $f\nabla = l_q$. Since l_q is a presingular epimorphism and ∇ is epimorphic, f is a presingular epimorphism. The relations $\nabla_p f \nabla = \nabla_p l_q = l_p \nabla$ imply $\nabla_p f = l_p$. Since l_p is a presingular epimorphism, the local morphism ∇_p must be an isomorphism. As a result ∇ is a local isomorphism.

(v) ⇒ (vi) According to Proposition (3.7.3), the epimorphic local isomorphism ∇ is a semisingular epimorphism.

(vi) ⇒ (vii) The coequalizer $q : B \to Q$ of (f, g) is the pushout along the morphism $\langle f, g\rangle : A \amalg A \to B$ of the codiagonal $\nabla : A \amalg A \to A$ of A. Since semisingular epimorphisms are co-universal (Proposition (3.7.4)), q is a semisingular epimorphism.

(vii) ⇒ (i) Let L be a local object, let $(f, g) : A \rightrightarrows L$, and let $q : L \to Q$ be the coequalizer of (f, g). If (f, g) is not codisjointed, Q is not terminal. Thus Q and q are local. But q is a local isomorphism (Proposition (3.7.3)) and, hence q is an isomorphism. Thus $f = g$. ■

Examples.

(i) **CRng**. Any finite power $\mathbf{Z}^n$ of $\mathbf{Z}$ and any quotient ring $\mathbf{Z}/n\mathbf{Z}$ of $\mathbf{Z}$ is preneat. Any ring of fractions of $\mathbf{Z}$ and in particular $\mathbf{Q}$ is preneat. Any number field and in particular the algebraic closure of $\mathbf{Q}$ is preneat. Any ring of the form $(\mathbf{Z}[X]/(P))[P'^{-1}]$ with $P \in \mathbf{Z}[X]$ is preneat. We shall now prove this last result directly. Let us denote this ring by A. Let L be a local ring with maximal ideal m and residue field $q : L \to L/m$, and let $(f, g) : A \rightrightarrows L$ be such that $qf = qg$. Let $\bar{X} \in A$ be the class of the polynomial X, let $\alpha = f(\bar{X})$, let $\beta = g(\bar{X})$, and let $\bar{\alpha} = q(\alpha) = q(\beta)$. Let $\Phi(X, Y) \in \mathbf{Z}[X, Y]$ be the polynominal such that $P(X) - P(Y) = (X - Y)\Phi(X, Y)$. Then $\Phi(X, X) = P'(X)$. Hence $q(\Phi(\alpha, \beta)) = \Phi(q(\alpha), q(\beta)) = \Phi(\bar{\alpha}, \bar{\alpha}) = P(\bar{\alpha}) = P'(q(f(\bar{X}))) = q(f(P'(\bar{X})))$ is invertible in L/m. Consequently, $\Phi(\alpha, \beta)$ is invertible in L. Then the relations $(\alpha - \beta)\Phi(\alpha, \beta) = P(\alpha) - P(\beta) = P(f(\bar{X})) - P(g(\bar{X})) = f(P(\bar{X})) - g(P(\bar{X})) = 0$ entail $\alpha = \beta$. It follows that $f = g$. As a result A is preneat.

(ii) **CAlg**(k) with k a field. Any algebra of the form $(k[X]/(P))[P'^{-1}]$ with $P \in k[X]$ is preneat. Any separable algebraic field extension K of k is a preneat object. The former statement is proved directly as previously. Let us now prove the latter. Let L be a local k-algebra and let $(f, g) : K \rightrightarrows L$ be a pair of distinct morphisms. Let $\alpha \in K$ such that $f(\alpha) \neq g(\alpha)$. Let $k[\alpha]$ be the subalgebra of K generated by α and let $j : k[\alpha] \to \mathbf{K}$ be the inclusion morphism. If P is the minimal polynomial of α, then

$k[\alpha] \simeq k[X]/(P)$. Since α is separable, P and P' are relatively prime. Therefore $P'(\bar{\alpha})$ is invertible in $k[\alpha]$ so that $k[\alpha] \simeq (k[X]/P)[P'^{-1}]$. According to the former result, the pair (fj, gj) is codisjointed. Thus (f, g) is codisjointed. As a result, K is a preneat object in **CAlg**(k).

(iii) **RedCAlg**(k) with k a field. Any reduced algebraic k-algebra is preneat. Let us prove this result. By using finite products and colimits (Proposition (8.1.3) below), it is sufficient to prove that a field extension of the form $K = k[X]/(P)$ with $P \in k[X]$ is a preneat object. If P is a separable polynomial, then according to (ii) K is a preneat object in the category **CAlg**(k), and thus it is a preneat object in **RedCAlg**(k). Let us assume that K is inseparable. Then k has characteristic $p \neq 0$ and there is an integer $\mu > 0$ and a separable polynomial $Q \in k[X]$ such that $P(X) = Q(X^{p^\mu})$ [3]. *Let* L be a reduced local k-algebra with residue field $q: L \to L/m$ and let $(f, g): K \rightrightarrows L$ be a pair of morphisms such that $qf = qg$. If $\alpha = f(\bar{X})$ and $\beta = g(\bar{X})$, then $q(\alpha) = q(\beta)$ and hence $q(\alpha^{p^\mu}) = q(\beta^{p^\mu})$. Since α^{p^μ} and β^{p^μ} are roots of Q in L and Q is separable, $\alpha^{p^\mu} = \beta^{p^\mu}$. But in any field extension of k, and thus in any reduced k-algebra, p^μ roots are unique. Thus $\alpha = \beta$, and so $f = g$. As a result K is a preneat object in **RedCAlg**(k).

(iv) **RlRng**. Any real number field is preneat. This follows from (ii) and the facts that **Q** is of characteristic zero and local objects are local rings.

(v) **RlAlg**(R) with R a real field. Any real algebraic extension field of R is preneat.

(vi) **RegCRng**. Any object is preneat since any local object is simple.

Proposition (8.1.3). *Colimits, quotients, and finite products of preneat objects are preneat.*

Proof: Let $(\alpha_i : Ai \to A)_{i \in I}$ be a colimit of preneat objects, let L be a local object, and let $(f, g): A \rightrightarrows L$ be a pair of morphisms. If there exists some $i \in I$ such that $(f\alpha_i, g\alpha_i)$ is codisjointed, then (f, g) is codisjointed. Otherwise each pair $(f\alpha_i, g\alpha_i)$ is equal, so that (f, g) is equal. As a result A is preneat. Let A be a preneat object, let $f: A \to B$ be an epimorphism, let L be a local object, and let $(g, h): B \rightrightarrows L$. If $g \neq h$ then $gf \neq hf$. Thus (gf, hf) is codisjointed and hence (g, h) is codisjointed. As a result, B is preneat. The terminal object is obviously preneat. Let A, B be a pair of preneat objects, let $(p: A \times B \to A, q: A \times B \to B)$ be their product, let L be a local object, and let $(f, g): A \times B \rightrightarrows L$. Since the object L is indecomposable, the morphisms f and g factor through p or q. If f factors through p and g factors through q, then (f, g) is codisjointed. Let us assume that f and g both factor through p or through q, say p, in the form $f = \bar{f}p, g = \bar{g}p$. Then $(\bar{f}, \bar{g})$ is equal or codisjointed and hence (f, g) is equal or codisjointed. As a result, $A \times B$ is preneat. ∎

Proposition (8.1.4). *An object is preneat if and only if its local objects are preneat.*

Proof. If an object is preneat, its local objects are preneat as they are quotient objects. Let A be an object whose local objects are preneat, let L be a local object with residue simple object $q_m : L \to L/m$, and let $(f, g) : A \rightrightarrows L$ be a pair of morphisms such that $q_m f = q_m g$. Then $\mathrm{Spec}(f)(m) = \mathrm{Spec}(g)(m)$ is a prime congruence p on A and $(f, g) = (f_m l_p, g_m l_p)$. Then $q_m f_m l_p = q_m g_m l_p$, and hence $q_m f_m = q_m g_m, f_m = g_m$, and finally $f = g$. As result, A is preneat. ■

Definition (8.1.5). *The category* **PNtA** *(or* **PNtLocA**, **PNtSimA***) of preneat (or preneat local, preneat simple) objects of* **A** *is the full subcategory of* **A** *(or* **LocA, SimA***) whose objects are the preneat ones.*

Theorem (8.1.6). *To any object of* **A** *is associated a co-universal preneat object, i.e.* **PNtA** *is a coreflective subcategory of* **A**.

Proof. Let A be an object in **A**. Let $u_A : P(A) \to A$ be the union of the set of preneat subobjects of A. The object $P(A)$ is preneat as it is a quotient of a colimit of preneat objects. Let $f: B \to A$ be a morphism with a preneat domain B. Let $f = me$ where $e: B \to E$ is a regular epimorphism and $m : E \to A$ is a monomorphism. Then E is preneat, so that m is a preneat subobject of A. Thence m factors through u_A and f factors also. As a result, $u_A : P(A) \to A$ is a co-universal preneat object associated to A. ■

Proposition (8.1.7). **PNtSimA** *is a full reflective subcategory of* **PNtLocA** *with a faithful reflector.*

Proof. If L is a preneat local object, the residue simple quotient $q_m : L \to L/m$ is obviously the universal preneat simple object associated to it. If $(f, g) : M \rightrightarrows L$ is a pair of morphisms in **PNtLocA** whose residue morphisms are equal, then $q_m f = q_m g$ and thus $f = g$ (Proposition (8.1.2)). Therefore the reflector faithful. ■

Proposition (8.1.8). *The left-adjoint to a morphism of Zariski categories preserves preneat objects.*

Proof: Let $U: \mathbf{A} \to \mathbf{B}$ be a morphism of Zariski categories with left-adjoint F and adjunction isomorphism $\varphi : \mathrm{Hom}_{\mathbf{A}}(F(\cdot), -) \to \mathrm{Hom}_{\mathbf{B}}(\cdot, U(-))$. Let B be a preneat object in **B**, let L be a local object in **A**, and let $(f, g) : F(B) \rightrightarrows L$. According to Proposition (7.3.2), the object $U(L)$ is local in **B**. Then the pair of morphisms $(\varphi_{B,L}(f), \varphi_{B,L}(g)) : B \rightrightarrows U(L)$ is equal or codisjointed. According to Definition (7.3.1), (f, g) is equal or codisjointed. As a result, $F(B)$ is preneat. ■

For any Zariski category **A**, let us denote by $J_{\mathbf{A}} : \mathbf{PNtA} \to \mathbf{A}$ the inclusion functor and by $P_{\mathbf{A}} : \mathbf{A} \to \mathbf{PNtA}$ its coreflector.

Proposition (8.1.9). *Let $U : A \to \mathbf{B}$ be a morphism of Zariski categories with left-adjoint F. The functor $U_p = P_B U J_A : PNtA \to \mathbf{PNtB}$ has a left-adjoint induced by F and defines an adjunction which preserves and reflects codisjointed pairs of morphisms (cf. notation on p. 176).*

Proof. According to Proposition (8.1.8), the functor F induces a functor $F_p : \mathbf{PNtB} \to \mathbf{PNtA}$. Then $P_{\mathbf{B}} U$ is right-adjoint to $F J_{\mathbf{B}} = J_{\mathbf{A}} F_p$. Since $J_{\mathbf{A}}$ is full and faithful, $P_{\mathbf{B}} U J_{\mathbf{A}}$ is right-adjoint to F_p. Then F_p is left-adjoint to U_p. Let $A \in \mathbf{PNtA}$, let $u : U_P(A) \to U(A)$ be the co-unit monomorphism at $U(A)$ of the adjunction defined by $P_{\mathbf{B}}$, let $B \in \mathbf{PNtB}$, and let $(f, g) : F_P(B) \rightrightarrows A$. Let $(f', g') : B \rightrightarrows U_P(A)$ be the pair of morphisms associated to (f, g) by the adjunction defined by U_P and let $(f'', g'') : B \rightrightarrows U(A)$ be the pair associated to (f, g) by the adjunction defined by U. Then $(uf', ug') = (f'', g'')$. If (f', g') is codisjointed, then (f'', g'') is codisjointed and thus (f, g) is codisjointed. Conversely, let us assume that (f, g) is codisjointed. Then (f'', g'') is codisjointed. Let $q : U_P(A) \to Q$ be the coequalizer of (f', g'). Then q is a semisingular epimorphism (Proposition (8.1.2)) and thus is a flat morphism (Proposition (3.7.4)). Let $(\bar{u} : Q \to \bar{Q}, \bar{q} : U(A) \to \bar{Q})$ be the pushout of (q, u). Since u is monomorphic, $\bar{u}$ is monomorphic. But $\bar{q}$ is the coequalizer of (f'', g'') so that $\bar{Q}$ must be terminal. Therefore, Q is terminal and (f', g') is codisjointed. ■

8.2 NEAT OBJECTS

Definition (8.2.1). *An object is* neat *if it is preneat and finitely presentable.*

Notation. An object is said to be *codecidable* if its codiagonal is a direct factor.

Proposition (8.2.2). *For a finitely presentable object A, the following assertions are equivalent.*

(i) *A is neat.*

(ii) *Any local object of A is preneat.*

(iii) *A is codecidable.*

(iv) *The coequalizer of any pair of morphisms $(f, g) : A \rightrightarrows B$ is a direct factor.*

(v) *For any indecomposable object B, any pair of morphisms $(f,g): A \rightrightarrows B$ is equal or codisjointed.*

(vi) *For any prelocal morphism f, the map* $\mathrm{Hom}_{\mathbf{A}}(A,f)$ *in injective.*

Proof.

(i) $\Leftrightarrow$ (ii) follows readily from Proposition (8.1.4).

(i) $\Rightarrow$ (iii) Since the objects A and $A \amalg A$ are finitely presentable, the codiagonal $\nabla_A : A \amalg A \to A$ is a finitely presentable morphism. According to Corollary (3.8.14) and Proposition (8.1.2), ∇_A is a singular epimorphism. According to Theorem (3.9.1), ∇_A is a direct factor. Then A is codecidable.

(iii) $\Rightarrow$ (iv) The coequalizer $q: B \to Q$ of (f,g) is the pushout of the codiagonal ∇_A of A along the morphism $\langle f,g\rangle : A \amalg A \to B$. Since direct factors are co-universal, q is a direct factor.

(iv) $\Rightarrow$ (v) The coequalizer q of (f,g) is a direct factor, and thus is an isomorphism or is terminal. Therefore (f,g) is equal or codisjointed.

(v) $\Rightarrow$ (i) follows from Proposition (8.1.2), for a local object is indecomposable.

(iv) $\Rightarrow$ (vi) Let $f: B \to C$ be a prelocal morphism and let $(g,h): A \rightrightarrows B$ be such that $fg = fh$. The morphism f factors through the coequalizer q of (g,h). Since q is a singular epimorphism (Theorem (3.9.1)), q is an isomorphism. Thence $g = h$.

(vi) $\Rightarrow$ (i) follows from Proposition (8.1.2), for local morphisms are prelocal. ■

Examples.

(i) **CRng**. Any finite power $\mathbf{Z}^n$ and any quotient ring $\mathbf{Z}/n\mathbf{Z}$ of $\mathbf{Z}$ is neat. For any $P \in \mathbf{Z}[X]$, the ring $(\mathbf{Z}[X]/(P))\,[P'^{-1}]$ is neat.

(ii) **CAlg**(k). Any algebra of the form $(k[X]/(P))\,[P'^{-1}]$ is neat and any finite separable field extension of k is neat.

(iii) **RedCAlg**(k). Any finitely presentable reduced algebraic k-algebra is a neat object.

(iv) **CAlg**(A). The neat objects are precisely the neat A-algebras ([34], Chapter III, Proposition 9).

(v) **RlRng**. Any real finite number field is a neat object.

(vi) **RlAlg**(R). The simple neat objects are precisely the real finite algebraic field extension of R.

Proposition (8.2.3). *Finite colimits, finitely presentable quotients, finitely generated regular quotients, singular quotients, and finite products of neat objects are neat.*

Proof: The proof follows from Proposition (8.1.3) and the fact that the class of finitely presentable objects is closed under these computations. ■

Proposition (8.2.4). *If A is a neat object, any morphism $f: A \to Z$ from A to the initial object Z is a direct factor.*

Proof: Let $u: Z \to A$ be the unique morphism. The morphism f is a split epimorphism which is the coequalizer of the pair of morphisms $(1_A, uf)$: $A \rightrightarrows A$. According to Proposition (8.2.2), f is a direct factor. ■

Definition (8.2.5). *The* category *NtA* of neat objects of A *is the full subcategory of A whose objects are the neat ones.*

Proposition (8.2.6). *The left-adjoint to a morphism of Zariski categories preserves neat objects.*

Proof: The proof follows from Proposition (8.1.8) and the fact that the left-adjoint preserves finitely presentable objects. ■

8.3 NEATISH OBJECTS

Defintion (8.3.1). *An object is* neatish *if it is a colimit of neat objects.*

Proposition (8.3.2). *Neatish objects are preneat.*

Proof: The proof follows from Proposition (8.1.3). ■

Examples.

(i) **CAlg**(k). An object is neatish if and only if it is a separable algebraic k-algebra, i.e. its elements are algebraic over k and have separable minimal polynomials.

(ii) **RedCAlg**(k) Any reduced algebraic k-algebra is a neatish object.

(iii) **CAlg**(A) Any local object A_p of A, its henselianzation $\tilde{A}_p$ [34], its strict henselianzation $\tilde{\tilde{A}}_p$[34], and its residue field $K(p)$ are neatish objects.

(iv) **RlAlg**(R). Any real algebraic field extension of R is neatish.

(v) **RegCRng**. Any object is neatish.

Definition (8.3.3). *The* category *NtshA* of neatish objects of A *is the full subcategory of A whose objects are the neatish ones.*

Theorem (8.3.4) *To any object of A is associated a co-universal neatish object i.e. NtshA is a coreflective subcategory of A.*

Proof. Let A be an object in **A**. Let $(N(A), \alpha)$ be the colimit of the neat objects above A, i.e. $(N(A), \alpha) = \xrightarrow{lim}_{(B,f) \in (\mathbf{NtA}, A)} B$, and let n_A : $N(A) \to A$ be the morphism defined by $n_A \alpha_f = f$ for any $(B,f) \in (\mathbf{NtA}, A)$. Then $N(A)$ is neatish. Let C be a neatish object, which is the colimit of neat objects $(C, \gamma) = \xrightarrow{lim}_{i \in \mathbf{I}} C_i$, and let $g: C \to A$. For any $i \in \mathbf{I}$, let $g_i = g\gamma i$ and $h_i = \alpha_{gi}$. For any morphism $u: i \to i'$ in $\mathbf{I}$, the relation $g_i C_u = g_{i'}$ entails $\alpha_{gi} C_u = \alpha_{gi'}$, i.e. $h_i C_u = h_{i'}$. Then we obtain an inductive cone $(h_i: C_i \to N(A))_{i \in I}$ based on $(C_i)_{i \in I}$ and thus a morphism $h: C \to N(A)$ such that $h\gamma i = h_i$ for any $i \in I$. Moreover, the relations $n_A h \gamma i = n_A h_i = n_A \alpha_{gi} = g\gamma i$ entail $n_A h = g$. Let $h': C \to N(A)$ be another morphism such that $n_A h' = g$. Let $i \in \mathbf{I}$ and $h_i' = h'\gamma i$. Since C_i is finitely presentable and $(N(A), \alpha)$ is a filtered colimit, there exist $(B, f) \in (\mathbf{NtA}, A)$ and $v: C_i \to B$ such that $\alpha_f v = h_i'$. Then $fv = n_A \alpha_f v = n_A h_i' = n_A h' \gamma_i = g\gamma_i = g_i$ and hence $\alpha_f v = \alpha_{gi}$, i.e. $h_i' = h_i$. It follows that $h' = h$. As a result, $n_A: N(A) \to A$ is a co-universal neatish object associated to A.

■

Corollary (8.3.5.) *NtshA is closed in A under colimits.*

Theorem (8.3.6). *NtshA is locally simple category [10] whose category of finitely presentable objects is NtA.*

Proof. Since **NtshA** is a coreflective subcategory of the cocomplete category **A**, it is cocomplete, and the colimits are computed in **NtshA** in the same way as they were in **A**. Since the objects of **NtA** are finitely presentable in **A**, they are finitely presentable in **NtshA**. Since any object in **NtshA** is a colimit of objects of **NtA**, the set of objects of **NtA** is a strong generating set in **NtshA**. Consequently **NtshA** is a locally finitely presentable category. Since **NtA** has finite colimits (Proposition (8.2.3)), **NtA** is precisely the category of finitely presentable objects of **NtshA**. Since **NtA** is closed in **A** under finite products (Proposition (8.2.3)) and finite products commute with filtered colimits, **NtshA** is closed under finite products in **A**. Since the square $Z \times Z$ of the initial object is in **NtA**, the category **NtshA** is locally indecomposable ([10], Definition 1.1.1). According to Proposition (8.2.2)), any object of **NtA** is codecidable in **A**, and thus is also codecidable in **NtshA**. Therefore the category **NtshA** is locally simple ([10], Proposition 2.1.2). ■

Proposition (8.3.7). *Finitely presentable quotients, singular quotients, local objects, regular quotients, and finite products of neatish objects are neatish.*

Proof. Let A be a neatish object. According to Theorem (8.3.6), A is a filtered colimit of neat objects, say $(A, \alpha) = \xrightarrow{lim}_{i \in \mathbf{I}} A_i$. Let $f: A \to B$ be a finitely presentable epimorphism. There exist an epimorphism $f_0: A_0 \to B_0$ with a finitely presentable domain and codomain and a morphism $u: A_0 \to A$ such that f is the pushout of f_0 along u ([5], Chapter V,7.4). Then there exist some $i \in I$ and some morphism $v: A_0 \to A_i$ such that $u = \alpha_i v$. Let $f_i: A_i \to B_i$ be the pushout of f_0 along v. Then f_i is a finitely presentable epimorphism. According to Proposition (8.2.3), B_i is neat. It follows that B

is a pushout of the neatish objects A, A_i, B_i. Accordingly to Corollary (8.3.5), B is neatish. As a result, finitely presentable quotients of A are neatish. Since singular quotients of A are finitely presentable (Proposition (1.8.5)), they are also neatish. Since any local object of A is a colimit of singular quotients of A, it is also neatish by Corollary (8.3.5). On the other hand, regular quotients of A are neatish since they are colimits of finitely generated quotients which are finitely presentable morphisms. The finite products property is proved in Theorem (8.3.6). ■

Corollary (8.3.8). *The co-unit morphisms $N(A) \to A$ are prelocal monomorphisms.*

Proposition (8.3.9). *The simple objects in NtshA are precisely the indecomposable neatish objects.*

Proof. According to Proposition (8.3.7), a finite product of objects in **A** is neatish if and only if each factor is neatish. Therefore a neatish object is indecomposable in **NtshA** if and only if it is indecomposable in **A**. But according to [10], Proposition 2.1.1, indecomposable objects in **NtshA** are identical to simple objects in **NtshA**. ■

Proposition (8.3.10). *The left-adjoint to a morphism of Zariski categories preserves neatish objects.*

Proof. It preserves neat objects (Proposition (8.2.6)) and colimits. ■

Proposition (8.3.11). *The coreflector $N: A \to NtshA$ is a morphism of Zariski categories.*

Proof. Let I: **NtshA** → **A** be the inclusion functor and let $n: IN \to I_{\mathbf{A}}$ be the co-unit of the adjunction.

(i) Let us prove that N preserves filtered colimits. Let $(\alpha_i : A_i \to A)_{i \in \mathbf{I}}$ be a filtered colimit in **A**, let $(\beta_i : N(A_i) \to B)_{i \in \mathbf{I}}$ be the colimit in **A** of the diagram $(N(A_i))_{i \in \mathbf{I}}$, and let $f: B \to N(A)$ be the morphism defined by $f\beta_i = N(\alpha_i)$ for any $i \in I$. The morphism $n_A f: B \to A$ is the filtered colimit of the diagram of monomorphisms $(n_{Ai})_{i \in \mathbf{I}}$, and therefore it is a monomorphism. Consequently, f is monomorphic. Let C be a neat object and let $g: C \to N(A)$ be a morphism. Since C is finitely presentable, there exist an index $i \in \mathbf{I}$ and a morphism $h: C \to A_i$ such that $n_A g = \alpha_i h$. Then there exists a morphism $g_i: C \to N(A_i)$ such that $n_{Ai} g_i = h$. Then $n_A N(\alpha_i) g_i = \alpha_i n_{Ai} g_i = \alpha_i h = n_A g$ and hence $f\beta_i g_i = N(\alpha_i) g_i = g$. Since neat objects form a strong generating set in **NtshA**, it follows that f is an isomorphism and hence $(N(\alpha_i) : N(A_i) \to N(A))_{i \in I}$ is a colimit in **NtshA**.

(ii) Let us prove that the adjunction isomorphism associated to N preserves and reflects codisjointed pairs of morphisms. Let A be an object in **A**, let B be an object in **NtshA**, and let $(f, g) : B \rightrightarrows N(A)$ be a pair of morphisms in **NtshA**. The adjunction isomorphism associates to (f, g) the pair of morphisms $(n_A f, n_A g)) : B \rightrightarrows A$. Let $q : N(A) \to Q$ be the coequalizer of (f, g) in **A**. It is also the coequalizer of (f, g) in **NtshA** (Corollary (8.3.5)). Let $(\bar{q} : A \to \bar{Q}, \bar{n} : Q \to \bar{Q})$ be the pushout of (n_A, q). Then $\bar{q}$ is the coequalizer of $(n_A f, n_A g)$. Since B is preneat (Proposition (8.3.2)), q is a semisingular epimorphism (Proposition (8.1.2)) and thus a flat morphism (Proposition (3.7.4)). Consequently, $\bar{n}$ is monomorphic. It follows that Q is terminal if and only if $\bar{Q}$ is terminal, i.e. (f, g) is codisjointed in **NtshA** if and only if $(n_A f, n_A g)$ is codisjointed in **A**. ∎

Theorem (8.3.12). *The category NtshA is the universal locally simple category associated to the Zariski category A.*

Proof. Let I: **NtshA** → **A** be the inclusion functor and let N: **A** → **NtshA** be the coreflector. Let **B** be a locally simple category and let U: **A** → **B** be a morphism of locally Zariski categories with left-adjoint F. Let V: **NtshA** → **B** be the restriction of U and let G : **B** → **NtshA** be the functor induced by F (Proposition (8.3.10)). Then V preserves filtered colimits, G preserves finite products, and V is right-adjoint to G. Therefore V is a morphism of locally simple categories [10], Definition (2.8.0). This morphism satisfies the relation $VN \simeq U$ since $IG = F$. Let V' : **NtshA** → **B** be another morphism of locally simple categories such that $V'N \simeq U$. Then the left-adjoint G' of V' satisfies the relation $IG' \simeq F \simeq IG$. Since I is full and faithful, $G' \simeq G$ and hence $V' \simeq V$. As a result, N: **A** → **NtshA** is the universal locally simple category associated to **A**. ∎

Example. The universal locally simple category associated to the Zariski category **CAlg**(k) is the category **SepAlgCAlg**(k) of separable algebraic k-algebras, and the co-universal neatish object associated to an algebra A is its subalgebra of separable algebraic elements.

Definition (8.3.13). *The* category *NtshLocA* (*NtshSimA*) of neatish local (simple) objects of *A is the full subcategory of* **LocA** (**SimA**) *whose objects are the neatish ones.*

Proposition (8.3.14).

(i) *NtshLocA is a coreflective subcategory of* **LocA**.

(ii) *NtshSimA is a coreflective subcategory of* **SimA**.

(iii) *NtshSimA is a reflective subcategory of NtshLocA with a faithful reflector.*

Proof. The proof follows from Propositions (8.1.7) and (8.3.7), and Corollary (8.3.8). ■

8.4 PURELY UNNEAT OBJECTS

Definition (8.4.1). *An object is* purely unneat *if the co-universal neatish object associated to it is the initial object.*

Examples. The initial object is purely unneat and it is the only one which is both purely unneat and neatish. In **CAlg**(k) the purely unneat objects are precisely the purely inseparable k-algebras. In **RlAlg**(R), purely unneat objects are purely unalgebraic real R-algebras, but the converse is in question.

Proposition (8.4.2). *Subobjects of purely unneat objects are purely unneat.*

Proof. Let B be a purely unneat object, let $f: A \to B$ be a monomorphism, and let $n_A : N(A) \to A$ be the co-universal neatish object associated to A. Then fn_A is monomorphic (Corollary (8.3.8)) and so is the morphism $N(f): N(A) \to N(B) = Z$. It follows that $N(f)$ is an isomorphism and that A is purely unneat. ■

Definition (8.4.3). *The* category ***PuntA*** of purely unneat objects of *A is the full subcategory of* **A** *whose objects are the purely unneat ones.*

8.5 NEATNESS OF MORPHISMS

Definition (8.5.1). *A morphism $f: A \to B$ is* preneat (neat, neatish, purely unneat) *if the object (B, f) in the category $A/\mathbf{A}$ is preneat (neat, neatish, purely unneat).*

Proposition (8.5.2). *Preneat, neat, and neatish morphisms are co-universal.*

Proof. For any morphism $f: A \to B$, the pushout functor $A/\mathbf{A} \to B/\mathbf{A}$ along f is left-adjoint to a morphism of Zariski categories $f_* : B/\mathbf{A} \to A/\mathbf{A}$ (cf. § 12.1 below). According to Propositions (8.1.8), (8.2.6), and (8.3.10), this functor preserves preneat, neat, and neatish objects. ■

Proposition (8.5.3). *For a morphism $f: A \to B$, the following assertions are equivalent.*

(i) *f is preneat.*
(ii) *For any $p \in \mathrm{Spec}(A)$, the pushout of f along l_p is preneat.*
(iii) *For any $q \in \mathrm{Spec}(B)$, the local morphism of f at q is preneat.*

Proof.

(i) $\Rightarrow$ (ii) follows from Proposition (8.5.2).

(ii) $\Rightarrow$ (iii) Let $p = \mathrm{Spec}(f)\,(q)$, let $f_q : A_p \to B_q$ be the local morphism of f at q, let $(\bar{f} : A_p \to \bar{B}, \bar{l}_p : B \to \bar{B})$ be the pushout of (l_p, f), and let $g : \bar{B} \to B_q$ be the morphism such that $g\bar{l}_p = l_q$ and $g\bar{f} = f_q$. Then $\bar{f}$ is preneat and g is preneat as it is an epimorphism. Therefore f_q is preneat.

(iii) $\Rightarrow$ (i) For any $q \in \mathrm{Spec}(B)$, the morphism $l_q f = f_q l_p$ is preneat, i.e. $(B_q, l_q f)$ is a preneat object in $A/\mathbf{A}$. According to proposition (8.1.4) applied to the object (B, f) in $A/\mathbf{A}$, (B, f) is preneat, i.e. f is a preneat morphism. ■

Proposition (8.5.4). *Any neat morphism is the pushout, along some morphism, of a neat morphism with a finitely presentable domain and codomain.*

Proof.

(i) Let us consider a neat morphism of the form $f : A \to A \amalg B$ where A is an arbitrary object, B is a finitely presentable object, and f is the canonical induction. If $\nabla : B \amalg B \to B$ is the codiagonal of B, then $1_A \amalg \nabla : A \amalg B \amalg B \to A \amalg B$ is the codiagonal of f in $A/\mathbf{A}$. Since f is neat, $1_A \amalg \nabla$ is a direct factor, which is the coequalizer of a direct factor congruence r on $A \amalg B \amalg B$ (Definition (3.9.2)). The object A is a filtered colimit of finitely presentable objects, say $(A, (\alpha_i)_{i \in \mathbf{I}}) = \xrightarrow{lim}_{i \in \mathbf{I}} A_i$. Then $(A \amalg B \amalg B, (\alpha_i \amalg 1_B \amalg 1_B)_{i \in I}) = \xrightarrow{lim}_{i \in \mathbf{I}} A_i \amalg B \amalg B$. For any $i \in \mathbf{I}$, let $f_i : A_i \to A_i \amalg B$ be the canonical induction. The codiagonal of f_i is $1_{A_i \amalg \nabla} \langle \mathrm{A}_\iota \amalg B \amalg B \to A_i \amalg B$, which is the coequalizer of some congruence r_i on $A_i \amalg B \amalg B$. Then r is the direct image of r_i along $\alpha_i \amalg 1_B \amalg 1_B$. On the other hand, according to Proposition (3.9.7), there exist some $i \in \mathbf{I}$ and a direct factor congruence r_i' on $A_i \amalg B \amalg B$ whose direct image along $\alpha_i \amalg 1_B \amalg 1_B$ is r. According to Proposition (1.5.5), there exists some morphism $u : i \to j$ in $\mathbf{I}$ such that the direct images of r_i and r_i' along the morphism $A_u \amalg 1_B \amalg 1_B : A_i \amalg B \amalg B \to A_j \amalg B \amalg B$ are identical. Then r_j is a direct factor congruence, and hence $1_{A_j} \amalg B$ is a direct factor. Consequently, f_j is neat. As a result, f is the pushout along α_j of the neat morphism f_j which has a finitely presentable domain and codomain.

(ii) Now let us consider an arbitrary neat morphism $f : A \to B$. We know that there exist a morphism $f_0 : A_0 \to B_0$ with a finitely presentable domain and codomain and morphisms $g : A_0 \to B$, $h : B_0 \to B$ such that (f, h) is the pushout of (g, f_0). In the category $A_0/\mathbf{A}$, the morphism $f : (A, g) \to (B, fg)$ is the neat canonical induction $(A, g) \to (A, g) \amalg (B_0, f_0)$. By applying the result obtained in (i) in the category $A_0/\mathbf{A}$, there exist a finitely presentable object (A_1, g_1) in $A_0/\mathbf{A}$ such that the canonical induction $f_1 : (A_1, g_1) \to (A_1, g_1) \amalg (B_0, f_0)$ is neat and a morphism $\alpha : (A_1, g_1) \to (A, g)$ such that the pushout of f_1 along α is $f : (A, g) \to (B, fg)$. Then

f_1 is a neat morphism in $\mathbf{A}$ with a finitely presentable domain and codomain, and its pushout along $\alpha : A_1 \to A$ is f. ■

Proposition (8.5.5). *Any retraction of a neat monomorphism is a direct factor morphism.*

Proof. The proof follows readily from proposition (8.2.4). ■

Proposition (8.5.6). *If B is a preneat (neatish) object, then any morphism $f : A \to B$ is preneat (neatish).*

Proof. Let $(\alpha : A \to A \amalg B, \beta : B \to A \amalg B)$ be the coproduct of A with B. Then α is the pushout along the morphism $Z \to A$ of the morphism $Z \to B$. According to Proposition (8.5.2), α is preneat (neatish), i.e. $(A \amalg B, \alpha)$ is a preneat (neatish) object in $A/\mathbf{A}$. If $g = \langle f, 1_B \rangle : A \amalg B \to B$, then $g\beta = 1_B$, and hence g is a regular epimorphism in $\mathbf{A}$ and $g : (A \amalg B, \alpha) \to (B, f)$ is a regular epimorphism in $A/\mathbf{A}$. According to Propositions (8.1.3) and (8.3.7), (B, f) is preneat (neatish) in $A/\mathbf{A}$, i.e. f is a preneat (neatish) morphism. ■

Proposition (8.5.7) *Let $f : A \to B$ be any morphism. If A and f are preneat (neat, neatish, purely unneat), then B is also.*

Proof.

(i) Let A and f be preneat. Let L be a local object and let $(g, h) : B \rightrightarrows L$. Then $(gf, hf) : A \rightrightarrows L$ is equal or codisjointed. If (gf, hf) is codisjointed, (g, h) is codisjointed. If $gf = hf$, then the pair of morphisms $(g, h) : (B, f) \rightrightarrows (L, gf)$ is equal or codisjointed in $A/\mathbf{A}$. Thus $(g, h) : B \rightrightarrows A$ is equal or codisjointed in $\mathbf{A}$. It follows that B is preneat.

(ii) If A and f are finitely presentable, then B is finitely presentable. Consequently if A and f are neat, B is neat.

(iii) Let A be neatish and f neat. The object A is filtered colimit of neat objects, say $(A, (\alpha_i)_{i \in \mathbf{I}}) \xrightarrow{lim}_{i \in I} A_i$. According to Proposition (8.5.4), there exist some neat morphism $f_0 : A_0 \to B_0$ with a finitely presentable domain and codomain and some morphism $g : A_0 \to A$ such that f is the pushout of f_0 along g. There exist some $i_0 \in \mathbf{I}$ and some morphism $g_0 : A_0 \to A_{i_0}$ such that $\alpha_{i_0} g_0 = g$. We can assume that i_0 is the initial object in $\mathbf{I}$ by substituting the category $i_0/\mathbf{I}$ to $\mathbf{I}$. For any $i \in \mathbf{I}$, let $u(i) : i_0 \to i$ be the unique morphism in $\mathbf{I}$, let $A_{u(i)} : A_{i_0} \to A_i$ be the morphism in the diagram $(A_i)_{i \in \mathbf{I}}$, and let $f_i : A_i \to B_i$ be the pushout of f_0 along $A_u(i) g_0$. According to Proposition (8.5.2), each morphism f_i is neat, and according to (ii) each object B_i is neat. Then $B = \xrightarrow{lim}_{i \in \mathbf{I}} B_i$ is neatish.

(iv) Let A and f be neatish. Then (B, f) is a colimit of neat objects in $A/\mathbf{A}$, say $(B, f) = \xrightarrow{lim}_{i \in \mathbf{I}} (B_i, f_i)$. Then each f_i is neat, and according to (iii), each B_i is neatish. Consequently $B = \xrightarrow{lim}_{i \in I} B_i$ is neatish (Corollary (8.3.5)).

(v) Let A and f be purely unneat. Let $u : Z \to A$ be the unique morphism, let $n : N \to B$ be the co-universal neatish object associated to B, and let $v : Z \to N$ be the unique morphism. Let $(\alpha : A \to A \amalg N, \beta : N \to A \amalg N)$ be the coproduct of A and N, and let $g = \langle f, n \rangle : A \amalg N \to B$. According to Proposition (8.5.2), $(A \amalg N, \alpha)$ is a neatish object in $A/\mathbf{A}$. Then there exists a morphism $h : (A \amalg N, \alpha) \to (A, 1_A)$ such that $fh = g$. Since N is neatish, there exists a morphism $w : N \to Z$ such that $uw = h\beta$. Then $nvw = fuw = fh\beta = g\beta = n$, and hence $vw = 1_N$. But $wv = 1_Z$; thus v is an isomorphism and B is purely unneat. ∎

8.6 THE NEATISH FACTORIZATION

Proposition (8.6.1). *The classes of preneat, neat, neatish, and purely unneat morphisms are closed under composition.*

Proof. Let $f : A \to B$ and $g : B \to C$ be two preneat morphisms. Then (B, f) is a preneat object in $A/\mathbf{A}$ and (C, g) is a preneat object in $B/\mathbf{A}$. But the category $B/\mathbf{A}$ is isomorphic to the category $(B, f)/(A/\mathbf{A})$ by the isomorphism which assigns to (C, g) the object $((C, gf), g)$. Then $((C, gf), g)$ is a preneat object in the latter category, i.e. $g : (B, f) \to (C, gf)$ is a preneat morphism in $A/\mathbf{A}$. According to Proposition (8.5.7), (C, gf) is a preneat object in $A/\mathbf{A}$, i.e. $gf : A \to C$ is a preneat morphism in $\mathbf{A}$. The same proof applies to neat, neatish, or purely unneat morphisms. ∎

Proposition (8.6.2). *Let $f : A \to B$ and $g : B \to C$ be two morphisms.*

(i) *If gf is preneat (neatish), then g is also.*

(ii) *If f is epimorphic and gf is neat, g is neat.*

(iii) *If g is monomorphic and gf is purely unneat, then f is purely unneat.*

Proof.

(i) follows from Proposition (8.5.6) applied in the category $A/\mathbf{A}$.

(ii) follows from (i) and the fact that g is finitely presentable as a pushout of the finitely presentable morphism gf.

(iii) follows from Proposition (8.4.2) applied in the category $A/\mathbf{A}$. ∎

Proposition (8.6.3). *Purely unneat morphisms are prelocal monomorphisms.*

Proof. Let $f : A \to B$ be a purely unneat morphism. Let $e : A \to E$ be a regular (singular) epimorphism through which f factors. According to Proposition (8.3.7), (E, e) is a neatish object in $A/\mathbf{A}$. Thus there exists a morphism $g : (E, e) \to (A, 1_A)$. Then $ge = 1_A$ and e is an isomorphism. Consequently, f is a prelocal monomorphism. ■

Proposition (8.6.4). *Let $f : A \to B$ be a purely unneat morphism.*
(i) *If B is local, A is local.*
(ii) *If B is simple, A is simple.*

Proof. The proof follows from Propositions (2.1.4), (2.12.4), and (8.6.3). ■

Proposition (8.6.5). *Purely unneat morphisms are precisely the morphisms which are right-orthogonal to neat, or neatish, morphisms.*

Proof. Let us note first that, since neatish morphisms are colimits of neat morphisms, a morphism is right-orthogonal to neatish morphisms if and only if it is right-orthogonal to neat morphisms.

(i) Let $f : A \to B$ be a neat morphism, let $g : C \to D$ be a purely unneat morphism, and let $u : A \to C$, $v : B \to D$ be such that $vf = gu$. Let $(\bar{f} : C \to \bar{B}, \bar{u} : B \to \bar{B})$ be the pushout of (u, f) and let $h : \bar{B} \to D$ be the morphism such that $h\bar{u} = v$ and $h\bar{f} = g$. According to Proposition (8.5.2), $\bar{f}$ is neat, i.e. $(\bar{B}, \bar{f})$ is a neat object in $C/\mathbf{A}$. Therefore there exists a morphism $w : (\bar{B}, \bar{f}) \to (C, 1_C)$ such that $h = gw$. Then the morphism $t = w\bar{u} : B \to C$ is such that $tf = w\bar{u}f = w\bar{f}u = u$ and $gt = gw\bar{u} = h\bar{u} = v$. Let $t' : B \to C$ be another morphism such that $t'f = u$ and $gt' = v$. Then we obtain in the category $A/\mathbf{A}$ a neat object (B, f), a prelocal morphism $g : (C, u) \to (D, fv)$, and a pair of morphisms $(t, t') : (B, f) \rightrightarrows (C, u)$ such that $gt = gt'$. According to Proposition (8.2.2), we have $t = t'$. As a result, g is right-orthogonal to f.

(ii) Let $g : C \to D$ be a morphism that is right-orthogonal to neatish morphisms. Let $u : (B, f) \to (D, g)$ be a morphism in $C/\mathbf{A}$ with a neatish domain. Then f is neatish, and so there exists a unique morphism $v : B \to C$ such that $vf = 1_C$ and $gv = u$. In other words, there exists a unique morphism $v : (B, f) \to (C, 1_C)$ such that $gv = u$, i.e. (D, g) is purely unneat in $C/\mathbf{A}$, i.e. g is purely unneat. ■

Proposition (8.6.6). *The co-unit morphisms $N(A) \to A$ are purely unneat.*

Proof. Let $u_A : N(A) \to A$ be the co-universal neatish object associated to A and let $f : (N, n) \to (A, u_A)$ be the co-universal neatish object associated to the object (A, u_A) in the category $N(A)/\mathbf{A}$. According to Proposition (8.5.7), N is neatish. Thus there exists a morphism $g : N \to N(A)$ such that $u_A g = f$. Then the relation $u_A gn = fn = u_A$ entails $gn = 1_{N(A)}$. We obtain a pair of morphisms $(ng, 1_N) : (N, n) \rightrightarrows (N, n)$ in $N(A)/\mathbf{A}$ such that

$fng = u_A g = f1_N$. Consequently $ng = 1_N$. It follows that n is an isomorphism and that u_A is purely unneat. ■

Theorem (8.6.7). *Any morphism factors in an essentially unique way as the composite of a neatish morphism followed by a purely unneat morphism.*

Proof. Let $f: A \to B$ be a morphism. Let $u: (N, n) \to (B, f)$ be the co-universal neatish object associated to the object (B, f) in the category $A/\mathbf{A}$. Then $n: A \to N$ is neatish and $f = un$. According to Proposition (8.6.6), u is a purely unneat morphism in $A/\mathbf{A}$ and thus is a purely unneat object in $(N, n)/(A, \mathbf{A}) \simeq N/\mathbf{A}$. Hence $u: N \to B$ is purely unneat in $\mathbf{A}$. Let $f = vm$ be another factorization of f where $m: A \to M$ is neatish and $v: M \to B$ is purely unneat. Since u is right-orthogonal to m (Proposition (8.6.5)), there exists some morphism $g: M \to N$ such that $gm = n$ and $ug = v$. According to Proposition (8.6.2), g is both neatish and purely unneat. Hence it is an isomorphism. ■

Corollary (8.6.8). *Neatish morphisms are precisely the morphisms which are left-orthogonal to purely unneat morphisms.*

Proof. According to Proposition (8.6.5), neatish morphisms are left-orthogonal to unneat morphisms. Let $f: A \to B$ be a morphism which is left-orthogonal to unneat morphisms, and let $f = un$ where $n: A \to N$ is neatish and $u: N \to B$ is purely unneat. Then there is some morphism $g: B \to N$ such that $gf = n$ and $ug = 1_B$. Since u is monomorphic, u is an isomorphism. Then f is neatish. ■

Corollary (8.6.9). *An object A is neatish if and only if, for any purely unneat morphism f, the map* $\mathrm{Hom}_A(A, f)$ *is bijective.*

Proof. The proof follows from Corollary (8.6.8) applied to the morphism $Z \to A$. ■

8.7 NEATLY CLOSED OBJECTS

Definition (8.7.1). *An object K is* neatly closed *if it is not terminal and any neat morphism $f: K \to A$ is a diagonal $\Delta: K \to K^n$ for some $n \in \mathbb{N}$.*

Proposition (8.7.2). *A neatly closed object is simple.*

Proof. Let K be a neatly closed object and let $q: K \to Q$ be a finitely generated regular quotient of K. Then q is neat and hence is a diagonal. If Q is not terminal, then q must be a monomorphism and thus an isomorphism. It follows that any regular quotient of K is trivial, i.e. K is simple. ■

Proposition (8.7.3). *For a simple object K, the following assertions are equivalent.*

(i) *K is neatly closed.*

(ii) *Any simple extension of K is a purely unneat morphism.*

(iii) *Any neatish extension of K splits, i.e. has a retraction.*

(iv) *Any neatish simple extension of K is an isomorphism.*

Proof.

(i) ⇒ (ii) Let $g: K \to L$ be a simple extension of K, let $f: A \to B$ be a neat morphism, and let $u: A \to K, v: B \to L$ be such that $vf = gu$. Let ($\bar{f}$: $K \to \bar{B}, \bar{u}: B \to \bar{B}$) be the pushout of (u, f) and let $h: \bar{B} \to L$ be the morphism such that $v = h\bar{u}$ and $h\bar{f} = g$. Then $\bar{f}$ is neat, and hence is a diagonal $\nabla: K \to K^n$ with $n \in \mathbb{N}^*$. Since L is indecomposable, the morphism h: $K^n \to L$ factors through some projection $p: K^n \to K$ in the form $h = gp$. It follows that the morphism $w = p\bar{u}$ satisfies the relations $wf = p\bar{u}f = pu\bar{f}$ $= u$ and $gw = gp\bar{u} = h\bar{u} = v$. According to Proposition (8.6.5), g is purely unneat.

(ii) ⇒ (iii) Let $f: K \to A$ be a neatish monomorphism. The non-terminal object A has a simple quotient $q: A \to L$. Then $qf: k \to L$ is a neatish simple extension of K (Proposition (8.6.1)). Since it must be purely unneat, it is an isomorphism. Then the morphism $(qf)^{-1}q$ is a retraction of f.

(iii) ⇒ (iv) Any neatish simple extension of K has a retraction which must be monomorphic. Thus it is an isomorphism.

(iv) ⇒ (i) Let $f: K \to A$ be a neat morphism. Let $p \in \operatorname{Spec}(A)$. The morphism $k_p l_p f: K \to K(p)$ is a neatish simple extension of K. Thus it is an isomorphism. Therefore we can assume that $K(p) = K$. Then $k_p l_p$ is a retraction of the neat morphism f. According to Proposition (8.5.5), $k_p l_p$ is a direct factor morphism. Since l_p is epimorphic and direct factor morphisms are co-universal, k_p is a direct factor morphism. Since A_p is indecomposable, k_p is an isomorphism. Then we can assume that $A_p = K$. Then l_p is a retraction of the neat morphism f. According to Proposition (8.5.5), l_p is a direct factor morphism and hence is both a singular and a regular epimorphism. According to Propositions (3.3.10) and (3.3.11), the image $\{p\}$ of $\operatorname{Spec}(l_p)$ is a clopen subset in $\operatorname{Spec}(A)$. It follows that $\operatorname{Spec}(A)$ is a discrete space and, since it is compact (Theorem (3.1.9)), it is finite. Then A is of the form K^n with $n \in \mathbb{N}$ and f is the diagonal $K \to K^n$. As a result, A is neatly closed. ■

Proposition (8.7.4). *For a transnoetherian simple object K, the following assertions are equivalent.*

(i) *K is neatly closed.*

(ii) *Any neat simple extension of K is an isomorphism.*
(iii) *Any neat extension of K splits.*

Proof.
(i) ⇒ (ii) follows readily from Proposition (8.7.3).
(ii) ⇒ (iii) Let $f: K \to A$ be a neat monomorphism. The non-terminal object A has a simple quotient $s: A \to S$. Since K is transnoetherian and f is finitely presentable, (A, f) is a finitely presentable noetherian object in $A/\mathbf{A}$. Then its regular quotient (S, sf) is a finitely presentable object in $A/\mathbf{A}$ and hence sf is a neat simple extension of K. Therefore sf is an isomorphism and f is a split monomorphism.
(iii) ⇒ (i) Let $f: K \to A$ be a neat morphism. Since K is transnoetherian and f is finitely presentable, A is noetherian, i.e. the ordered set QuotReg(A) satisfies the descending-chain condition. We shall prove by noetherian induction that any $q: A \to Q \in \text{QuotReg}(A)$ is such that $qf: K \to Q$ is a diagonal $K \to K^n$ for some $n \in \mathbb{N}$. Let $q: A \to Q \in \text{QuotReg}(A)$ be such that any $q' \in \text{QuotReg}(A)$ strictly less than q is such that $q'f$ is a diagonal. If Q is terminal, qf is diagonal. Let us assume that Q is non-terminal. Then $qf: K \to Q$ is an extension of K. Since A is noetherian, q is finitely presentable and hence is neat. Thus qf is a neat extension of K. Then qf has a retraction $g: Q \to K$. According to Proposition (8.5.5), g is a direct factor, i.e. there is a morphism $g': Q \to Q'$ such that $(g: Q \to K, g': Q \to Q')$ is a product. According to the hypothesis, $g'qf: K \to Q'$ is a diagonal $K \to K^n$. Then $qf = (gqf, g'qf): K \to K^{n+1}$ is a diagonal. As a result, K is neatly closed. ■

Corollary (8.7.5). *A transnoetherian algebraically closed simple object is neatly closed.*

Proof. The proof follows from Proposition (6.1.5).

Proposition (8.7.6.). *For a simple object* K *in an amalgamative Zariski category the following assertions are equivalent.*
(i) K is neatly closed.
(ii) K is injective with respect to interminable neatish morphisms.

Proof.
(i) ⇒ (ii) Let $f: A \to B$ be an interminable neatish morphism, let $g: A \to K$ be a morphism, and let $(u: K \to C, v: B \to C)$ be the pushout of (g, f). The non-terminal object C has a simple quotient $s: C \to S$. Then u and s are neatish morphisms, and hence su is a neatish simple extension of K. According to Proposition (8.7.3), su is an isomorphism. Then the morphism $h = (su)^{-1}sv: B \to K$ is such that $hf = g$. Hence K is injective with respect to f.

(ii) ⇒ (i) Since any neatish simple extension $f: K \to L$ is an interminable morphism, there exists some morphism $g: L \to K$ such that $gf = 1_K$. Then g is monomorphic and thus f is an isomophism. According to Proposition (8.7.3), K is neatly closed. ■

Proposition (8.7.7). *For a transnoetherian simple object* K *in an amalgamative Zariski category, the following assertions are equivalent.*

(i) K *is neatly closed.*

(ii) K *is injective with respect to interminable neat morphisms.*

Proof.

(i) ⇒ (ii) follows from Proposition (8.7.6.).

(ii) ⇒ (i) Any neat simple extension $f: K \to L$ of K is interminable. Thus there exists a morphism $g: L \to K$ such that $gf = 1_K$, and since g is monomorphic, f is an isomorphism. According to Proposition (8.7.4), K is neatly closed. ■

Examples. The neatly closed simple objects in **CRng** are the separably closed fields. In **CAlg**(k), they are the separably closed field extensions of k. In **RedCRng**, they are the algebraically closed fields. In **RedCAlg**(k), they are the algebraically closed field extensions of k. In **RlRng** they are the real closed fields. In **RlAlg**(R) they are the real closed field extensions of R.

Proposition (8.7.8). *In an amalgamative Zariski category, purely unneat subobjects of neatly closed objects are neatly closed objects.*

Proof. Let K be a neatly closed object and let $f: A \to K$ be a purely unneat monomorphism. According to Proposition (8.6.4), A is simple. Let $m: M \to N$ be an interminable neatish morphism and $g: M \to A$. Since K is injective with respect to m (Proposition (8.7.6)), the morphism fg factors through m in the form $fg = nm$. According to Proposition (8.6.5), there is some morphism $h: N \to A$ such that $hm = g$ and $fh = n$. It follows that A is injective with respect to m. According to Proposition (8.7.6), A is neatly closed. ■

Definition (8.7.9). *The* category *NtclA* (*NtclshA*) of neatly closed (neatly closed neatish) objects of *A is the full subcategory of A whose objects are the neatly closed (neatly closed neatish) ones.*

Theorem (8.7.10). *The category NtclshA is a groupoid equivalent to the disjoint union of the automorphism groups of a class of simple objects of A.*

Proof. Any morphism in **NtclshA** is both neatish (Proposition (8.6.2)) and purely unneat (Proposition (8.7.3)), and thus is an isomorphism. Therefore

NtclshA is a groupoid. Since neatly closed objects are simple, **NtclshA** is a full subcategory of **SimA**. The latter is equivalent to the disjoint union of its connected components, each of which is equivalent to the automorphism group of a simple object of **A**. ■

Theorem (8.7.11). *In an amalgamative Zariski category A, to any neatly closed object is associated a co-universal neatly closed neatish object, i.e.* ***NtclshA*** *is a coreflective subcategory of* ***NtclA***.

Proof. Let K be a neatly closed object and let $f: N(K) \to K$ be the co-universal neatish object associated to it (theorem (8.3.4)). According to Proposition (8.6.6) f is purely unneat, and according to Proposition (8.7.8), $N(K)$ is neatly closed. It follows that $f: N(K) \to K$ is a co-universal neatly closed neatish object associated to K. ■

Theorem (8.7.12). *In an amalgamative simply noetherian Zariski category A, to any simple object is associated an essentially universal neatly closed object, i.e.* ***NtclA*** *is an essentially reflective subcategory of* ***SimA***.

Proof. Let K be a simple object. According to Theorem (6.1.7). K has an algebraically closed simple extension, say $h: K \to N$. Let $h = gf$ be the neatish factorization of h, where $f: K \to L$ is neatish and $g: L \to N$ is purely unneat (Theorem (8.6.7)). The object N is neatly closed (Corollary (8.7.5)) and, according to Proposition (8.7.8), L is neatly closed. Let $m: K \to M$ be a morphism with a neatly closed codomain. According to Theorem (8.7.6), M is injective with respect to the interminable neatish morphism f. Therefore there exists a morphism $u: L \to M$ such that $uf = m$. Let $v: L \to M$ be another morphism such that $vf = m$. According to proposition (8.7.3), u and v are purely unneat morphisms. According to Theorem (8.6.7), there is an automorphism $w: L \to L$ such that $wf = f$ and $uw = v$. As a result, $f: K \to L$ is an essentially universal neatly closed object associated to K.

■

Notation. The essentially universal neatly closed object $f: K \to \bar{K}$ associated to a simple object K is the *neat closure* of K. It is defined up to isomorphisms. The group of automorphisms $\alpha: \bar{K} \to \bar{K}$ such that $\alpha f = f$ is the *Galois group of the neat closure of* K. It is defined up to isomorphisms.

Theorem (8.7.13). *For a simple extension $f: K \to L$ in an amalgamative simply noetherian Zariski category, the following assertions are equivalent.*

(i) *f is a neat closure of K.*

(ii) *f is neatish and L is neatly closed.*

(iii) *f is a neatish hull of K, i.e. f factors through any neatish monomorphism $n: K \to N$.*

Proof.

(i) ⇒ (ii) See the proof of Theorem (8.7.12).

(ii) ⇒ (iii) Let $n: K \to N$ be a neatish monomorphism. According to the amalgamation property, n is interminable. According to Theorem (8.7.6), L is injective with respect to n. Therefore f factors through n.

(iii) ⇒ (i) Let $n: K \to N$ be a neat closure of K. Then n is neatish, and hence f factors through n in the form $f = gn$. Since N is neatly closed, g is purely unneat. It is an isomorphism according to the essential uniqueness of the neatish factorization. As a result $f: K \to L$ is a neat closure of K. ■

Examples. In **CRng** and **CAlg**(k) the neat closure of a field is its separable closure. In **RedCAlg**(k) the neat closure of a field is its algebraic closure. In **RlRng** or **RlOrdRng** the neat closure of a real field does not necessarily exist because the category is not amalgamative, and indeed a real closure does not satisfy the required universal property.

9

FLATNESS PROPERTIES

Flatness properties are among the basic properties of Zariski categories. Up to now we have used the concept of flat morphism, but the first concept is that of flat object. An object A in **A** is flat if the endofunctor $A \amalg (-)$ of **A** preserves monomorphisms. Then a morphism $f: A \to B$ in **A** is flat if (B, f) is a flat object in $A/\mathbf{A}$, and an object A in **A** is absolutely flat if any morphism $f: A \to B$ is flat in **A**. In an arbitrary Zariski category simple objects need not be absolutely flat, and therefore the two notions of von Neumann regular objects and absolutely flat objects are generally different. Some Zariski categories can have special flatness properties such as the following. Any object is flat, as in the category **CAlg**(k) of commutative algebras over a field k. Any object is absolutely flat, as in the category **RegCRng** of von Neumann regular commutative rings. Any simple object is flat, as in the category **CRng** of commutative rings. Any local flat epimorphism is an isomorphism, as in the category **CAlg**(A) of commutative algebra over a ring A. In an arbitrary Zariski category, a flat object A is faithfully (reflexively, properly) flat if the endofunctor $A \amalg (-)$ reflects monomorphisms (isomorphisms, the terminal object). These three notions are generally different, but in some Zariski categories, such as **CAlg**(A), they coincide.

9.1 FLAT, ABSOLUTELY FLAT, AND SIMPLY FLAT OBJECTS

Definition (9.1.1). *An object A is*

(i) flat *if the endofunctor $A \amalg (-)$ preserves monomorphisms,*
(ii) absolutely flat *if any morphism $f: A \to B$ is flat, and*
(iii) simply flat *if any residue simple object of A is absolutely flat.*

Proposition (9.1.2).

(i) *The class of flat objects is closed under coproducts, filtered colimits, and finite products.*

(ii) *The class of absolutely flat objects is closed under quotients, filtered colimits and finite products.*

Proof. The proof follows from Proposition (1.6.2).

Proposition (9.1.3). *The class of non-flat objects is closed under filtered colimits.*

Proof. Let $(A, (\alpha_i)_{i\in\mathbf{I}}) = \xrightarrow{lim}_{i\in\mathbf{I}} A_i$ be a filtered colimit of non-flat objects. For each $i\in\mathbf{I}$, there is some monomorphism $f_i: B_i \to C_i$ such that $1_{A_i} \amalg f_i: A_i \amalg B_i \to A_i \amalg C_i$ is not monomorphic. Let $B = \Pi_{i\in\mathbf{I}} B_i$, $C = \Pi_{i\in\mathbf{I}} C_i$, and $f = \Pi_{i\in\mathbf{I}} f_i: B \to C$. Then f is monomorphic. Since finite products are co-universal, for each $i\in\mathbf{I}$ we have

$$1_{A_i} \amalg f = 1_{A_i} \amalg (\prod_{j\in I-i} f_j \times f_i) = (1_{A_i} \amalg (\prod_{j\in I-i} f_j)) \times (1_{A_i} \amalg f_i).$$

According to Proposition (1.4.6), $1_{A_i} \amalg f$ is not monomorphic. It follows that the morphism $1_A \amalg f$ is a filtered colimit of the diagram of non-monomorphisms $(1_{A_i} \amalg f_i)_{i\in\mathbf{I}}$. Since this colimit is cocartesian, i.e. any induction diagram is a pushout, it follows that $1_A \amalg f$ is not a monomorphism. Thus A is not flat. ■

Corollary (9.1.4). *Let* $A = \xrightarrow{lim}_{i\in I} A_i$ *be a filtered colimit. If* A *is flat, there is some* $i\in \boldsymbol{I}$ *such that* A_j *is flat for any* $j\in \boldsymbol{I}$ *under* i.

Proof. Let $\mathbf{J}$ be the full subcategory of $\mathbf{I}$ whose objects are those $i\in\mathbf{I}$ such that A_i is not flat. If $\mathbf{J}$ were cofinal in $\mathbf{I}$, then $(A_i)_{i\in\mathbf{I}}$ would be a filtered colimit of non-flat objects, whose colimit A would be non-flat. Consequently, $\mathbf{J}$ is not cofinal in $\mathbf{I}$. ■

Proposition (9.1.5). *An object is flat (absolutely flat, simply flat) if and only if its local objects are.*

Proof. Let A be an object, let $f: Z \to A$ be the unique morphism, let $p\in \mathrm{Spec}(A)$, and let $q = Spec(f)(p)$. Since the localization morphism $l_p: A \to A_p$ is flat, the local object A_p is flat whenever A is flat. Let us suppose that A_p is flat. Then the morphism $f_p l_q: Z \to A_p$ is flat. According to Proposition (1.6.2), f_p is flat. It follows that the local morphisms of f are flat. Therefore f is flat (Proposition (3.6.5)), i.e. A is flat. If A is absolutely flat, its local objects are absolutely flat since they are quotient objects. Let A be an object whose local objects are absolutely flat, and let $g: A \to B$ be a morphism. Let $p\in \mathrm{Spec}(B)$ and $q = \mathrm{Spec}(f)(p)$. Since A_q is absolutely flat, the local morphism $g_p: A_q \to B_p$ is flat. According to Proposition (3.6.5), g is flat. Consequently A is absolutely flat. The result for simply flat objects is obvious. ■

Proposition (9.1.6).

(i) *Absolutely flat objects are simply flat.*
(ii) *Simply flat regular objects are absolutely flat.*
(iii) *Reduced absolutely flat objects are regular.*

Proof.

(i) follows from Proposition (9.1.2), since a residue simple object is a quotient object.

(ii) follows from Proposition (9.1.5) since the local objects of a regular object are identical to its residue simple objects.

(iii) Let A be a reduced absolutely flat object. Let $p \in \mathrm{Spec}(A)$. The object A_p is absolutely flat (Proposition (9.1.5)). Then the morphism $k_p : A_p \to K(p)$ is flat and, since it is local, it is interminable (Proposition (11.3.4) below. Since A_p is reduced, k_p is a monomorphism (Proposition (11.3.2) below) and thus is an isomorphism. Consequently, the local objects of A are simple, and therefore A is regular (Proposition (3.6.1)). ■

Proposition (9.1.7). *The prime spectrum of an absolutely flat object is a boolean space.*

Proof. Let A be an absolutely flat object. The reduced object $A/\mathrm{rad}(A)$ is absolutely flat (Proposition (9.1.2)) and thus is regular (Proposition (9.1.6)). Then $\mathrm{Spec}(A/\mathrm{rad}(A))$ is a boolean space (Proposition (3.2.3)). But $\mathrm{Spec}(A) \simeq \mathrm{Spec}(A/\mathrm{rad}(A))$ (Proposition (3.3.3)). ■

Proposition (9.1.8). *Let $f: A \to B$ be a morphism.*

(i) *If A and f are flat, B is flat.*

(ii) *If A is neatish and B is flat, then f is flat.*

Proof.

(i) follows readily from Proposition (1.6.2).

(ii) Let us assume that A is neat. Let $m : A \to M$ be a morphism, let $n : M \to N$ be a monomorphism, let $(u : M \to M', m' : B \to M')$ be the pushout of (m, f), and let $(v : N \to N', n' : M' \to N')$ be the pushout of (n, u). Let $(\gamma_1 : A \to A \amalg A, \gamma_2 : A \to A \amalg A)$ be the coproduct of A by A, let $\nabla : A \amalg A \to A$ be the codiagonal of A, and let $\nabla_1 : A \amalg A \to A_1$ be the morphism such that (∇, ∇_1) is a product, i.e. $A \amalg A = A \times A_1$. Let $\alpha = \nabla_1 \gamma_1$ and $\beta = \nabla_1 \gamma_2$. Let $(\mu : M \to M_1, m_1 : A_1 \to M_1)$ be the pushout of (m, α), let $\nu : N \to N_1, n_1 : M_1 \to N_1)$ be the pushout of (n, μ), let $(f_1 : A_1 \to B_1, \beta_1 : B \to B_1)$ be the pushout of (β, f), let $(u_1 : M_1 \to M_1',$ $m_1' : B_1 \to M_1')$ be the pushout of (m_1, f_1), and let $(v_1 : N_1 \to N_1', n_1' :$ $M_1' \to N_1')$ be the pushout of (n_1, u_1). Because finite products are co-universal, we obtain the following string of pushouts: $((1_M, \mu) : M \to$ $M \times M_1, m \times m_1 : A \times A_1 \to M \times M_1)$ is the pushout of $(m : A \to M,$ $(1_A, \alpha) : A \to A \times A_1)$; $((1_N, \nu) : N \to N \times N_1, n \times n_1 : M \times M_1 \to N \times N_1)$ is the pushout of $(n : M \to N, (1_M, \mu) : M \to M \times M_1)$; $(u \times u_1, m' \times m_1')$ is the pushout of $(m \times m_1, f \times f_1)$; $(v \times v_1, n' \times n_1')$ is the pushout of $(n \times n_1, u \times u_1)$. But if we note that the morphism $(1_A, \alpha) : A \to A \times A_1$ is identical to the morphism $\gamma_1 : A \to A \amalg A$, and the morphism $(1_A, \beta) :$ $A \to A \times A_1$ is identical to $\gamma_2 : A \to A \amalg A$, then we realize that the morphism $n' \times n_1' : M' \times M_1' \to N' \times N_1'$ is identical to the morphism $B \amalg n : B \amalg M \to B \amalg N$. Since B is flat, the morphism $B \amalg n = n' \times n_1'$ is monomorphic, and according to Proposition (1.4.6) n' is also monomorphic. Consequently, f is flat. If we assume that A is neatish, then A is a filtered colimit of neat objects, say $(A, (\alpha_i)_{i\in I}) = \xrightarrow{lim}_{i\in I} A_i$. Then each morphism $f\alpha_i : A_i \to B$ is flat according to the preceding discussion. Therefore $f = \xrightarrow{lim}_{i\in I} f_i$ is flat (Proposition (1.6.2). ■

Proposition (9.1.9). *Let $f: A \to B$ be a neatish morphism.*

(i) *A morphism $g : B \to C$ is flat if the morphism gf is flat.*

(ii) *If A is absolutely flat, B is absolutely flat.*

(iii) *If A is simply flat, B is simply flat.*

Proof.

(i) follows from Proposition (9.1.8) applied in the category $A/\mathbf{A}$.

(ii) follows from (i).

(iii) Let $p \in \operatorname{Spec}(B)$ and $q = \operatorname{Spec}(f)(p)$, and let $g: K(q) \to K(p)$ be the residue morphism of f at p. The morphism $gk_q l_q = k_p l_p f$ is neatish; hence the morphism g is also neatish (Proposition (8.6.2)). Since $K(q)$ is absolutely flat, $K(p)$ is absolutely flat. As a result, B is simply flat. ■

Proposition (9.1.10). *Let $U: \mathbf{A} \to \mathbf{B}$ be a full and faithful morphism of Zariski categories whose left-adjoint F preserves monomorphisms. Then F preserves flat objects and flat morphisms.*

Proof. There is an isomorphism of functors $FU \simeq Id_{\mathbf{A}}$. Let B be a flat object in $\mathbf{B}$. We have the following isomorphisms of endofunctors of $\mathbf{A}$: $F(B) \amalg (-) = F(B) \amalg Id_{\mathbf{A}}(-) \simeq F(B) \amalg (FU(-)) = F(B \amalg U(-))$ $\simeq F(B \amalg (-))U$. Since the functors $U, F, B \amalg (-)$ preserve monomorphisms, the functor $F(B) \amalg (-)$ does also. Thus $F(B)$ is flat in $\mathbf{B}$. Let $g : B \to C$ be a flat morphism in $\mathbf{B}$. Let us consider the Zariski categories $FB/\mathbf{A}$ and $B/\mathbf{B}$, and the morphism of Zariski categories $V: F(B)/\mathbf{A} \to B/\mathbf{B}$ defined by $V(A, \alpha) = (U(A), \varphi_{B,A}(\alpha))$ and $V(f) = U(f)$, where φ: $\operatorname{Hom}_{\mathbf{A}}(F(.),-) \to \operatorname{Hom}_{\mathbf{B}}(.,U(-))$ is the adjunction isomorphism. The left-adjoint to V is the functor $G: B/\mathbf{B} \to F(B)/\mathbf{A}$ defined by $G(C, \gamma) = (F(C), F(\gamma))$ and $G(g) = F(g)$. The functor V is full and faithful, and the functor G preserves monomorphisms. By the previous result G preserves flat objects. Then $G(C, g) = (F(C), F(g))$ is flat in $FB/\mathbf{A}$, and thus $F(g)$ is flat in $\mathbf{A}$. ■

9.2 FLAT, ABSOLUTELY FLAT, AND SIMPLY FLAT ZARISKI CATEGORIES

Definition (9.2.1). *A Zariski category is* flat (absolutely flat, simply flat) *if any object in it is flat* (*absolutely flat, simply flat*).

Proposition (9.2.2). *For a Zariski category A, the following assertions are equivalent.*

(i) *A is flat.*
(ii) *The initial object of A is absolutely flat.*
(iii) *The neatish objects of A are absolutely flat.*

Proof.
(i) $\Leftrightarrow$ (ii) is obvious.
(ii) $\Rightarrow$ (iii) follows from Proposition (9.1.9).
(iii) $\Rightarrow$ (ii) follows from the fact that the initial object is neatish. ∎

Proposition (9.2.3). *For a Zariski category A, the following assertions are equivalent.*

(i) *A is absolutely flat.*
(ii) *Any morphism is flat.*
(iii) *Monomorphisms are co-universal.*

Proof. The proof is immediate. ∎

Proposition (9.2.4). *For a Zariski category A, the following assertions are equivalent.*

(i) *A is simply flat.*
(ii) *Any simple object is absolutely flat.*
(iii) *Any regular object is absolutely flat.*

Proof. The proof follows from Proposition (9.1.6). ∎

Proposition (9.2.5). *In a simply flat Zariski category, regular objects are precisely reduced absolutely flat objects.*

Proof. The proof follows from Propositions (2.11.3) and (9.1.6). ∎

Proposition (9.2.6). *Simply flat Zariski categories are amalgamative.*

Proof. Let $(f: K \to L, g: K \to M)$ be a pair of simple extensions with pushout $(u: M \to A, v: L \to A)$. Since f is flat, v is monomorphic, and hence A is not terminal. If $q: A \to Q$ is a simple quotient of A, then $qv: L \to Q$ and $qu: M \to Q$ are simple extensions fulfilling the amalgamation property. ∎

Definition (9.2.7). *The* cofibre of a morphism $f: A \to B$ *at* $p \in \mathrm{Spec}(A)$ *is the object* $B(p)$ *arising in the pushout* $(i_p: K(p) \to B(p), \varphi(p): B \to B(p))$ *of the pair of morphisms* $(k_p l_p: A \to K(p), f: A \to B)$. *The morphism* $\varphi(p)$ *is the* cofibre morphism *and the morphism* i_p *is called canonical.*

Proposition (9.2.8). *In a simply noetherian simply flat Zariski category, cofibres of neat (neatish) morphisms are finite products of pseudo-simple objects (are absolutely flat objects).*

Proof. Let $f: A \to B$ be a neatish morphism and $p \in \mathrm{Spec}(A)$. The canonical morphism $i_p: K(p) \to B(p)$ is neatish (Proposition (8.5.2)). Since $K(p)$ is absolutely flat, $B(p)$ is absolutely flat (Proposition (9.1.9)). Let us suppose that f is neat. Since simple extensions satisfy the amalgamation property (Proposition (9.2.6)), the object $K(p)$ has an algebraically closed simple extension $k: K(p) \to L$ (Theorem (6.1.7)). Let $(g: L \to C, h: B(p) \to C)$ be the pushout of (k, i_p). Then g is neat. Since L is neatly closed (Corollary (8.7.5)), g is a diagonal $L \to L^n$ with $n \in \mathbb{N}$. According to the amalgamation property, the morphism k is interminable, and thus the morphism h is also interminable. According to Proposition (11.6.5) below, the map $\mathrm{Spec}(h)$ is surjective. Since $\mathrm{Spec}(L^n)$ is finite, $\mathrm{Spec}(B(p))$ is finite. Since $B(p)$ is absolutely flat, $\mathrm{Spec}(B(p))$ is a boolean space (Proposition (9.1.7)). Consequently, $\mathrm{Spec}(B(p))$ is a finite discrete space. Then $B(p)$ is a finite product of pseudosimple objects (Proposition (2.6.2)). ■

Theorem (9.2.9). *If $f: A \to B$ is a neat (neatish) morphism in a simply noetherian simply flat Zariski category, the continuous map* $\mathrm{Spec}(f)$: $\mathrm{Spec}(B) \to \mathrm{Spec}(A)$ *has finite discrete (boolean) fibres.*

Proof. The proof follows from Propositions (9.1.7) and (9.2.8), and the fact that the cofibre morphism $\varphi(p): B \to B(p)$ of f at p is such that the continuous map $\mathrm{Spec}(\varphi(p)): \mathrm{Spec}(B(p)) \to \mathrm{Spec}(B)$ is a topological embedding whose image is the fibre of the map $\mathrm{Spec}(f)$ at p. ■

Proposition (9.2.10). *If $f: A \to B$ is a neatish local morphism in a simply noetherian simply flat Zariski category, then a morphism $g: B \to C$ is local if and only if the morphism gf is local.*

Proof. If g is local, the composite gf is obviously local. Let us suppose that gf is local, and let m, n, and p be the maximal congruences of A,B, and C respectively. In $\mathrm{Spec}(B)$, the prime congruence $\mathrm{Spec}(g)(p)$ is less than n, and they are both in the fibre of $\mathrm{Spec}(f)$ at m. Since this fibre is a boolean space (Theorem (9.2.9)), $\mathrm{Spec}(g)(p) = n$. Then g is local. ■

Examples. The categories **RegCRng** and **RegCAlg**(k) are absolutely flat. The categories **CAlg**(k) and **RedCAlg**(k) are flat. The categories **CRng**, **RedCRng**, **Mod**, and **CAlg** are simply flat. The categories **RlRng** and **RlOrdRng** are not simply flat because the simple object $\mathbf{Q}[\sqrt{2}]$ in **RlRng**, and in **RlOrdRng** if equipped with the sum of squares order, is not absolutely flat (cf. examples on p. 207).

9.3 FAITHFULLY (REFLEXIVELY, PROPERLY) FLAT MORPHISMS

Definition (9.3.1). *A morphism $f: A \to B$ is* faithfully (reflexively, properly) flat *if it is flat and the pushout functor $A/A \to B/A$ along f reflects monomorphisms (isomorphisms, terminal objects).*

In what follows, we denote the pushout functor along $f: A \to B$ by $F: A/\mathbf{A} \to B/\mathbf{A}$.

Proposition (9.3.2). *For a flat morphism f, the following assertions are equivalent.*

(i) *f is faithfully flat.*

(ii) *The pushout functor along f is faithful.*

(iii) *f is a co-universal monomorphism.*

Then f is properly flat.

Proof.

(i) $\Rightarrow$ (ii) Let (g, h) be a pair of parallel morphisms in $A/\mathbf{A}$ such that $F(g) = F(h)$. Let q be the coequalizer of (g, h). Then $F(q)$ is the coequalizer of $(F(g), F(h))$. Hence $F(q)$ is an isomorphism. Therefore q is a monomorphism and hence is an isomorphism. Consequently $g = h$. Thus F is faithful.

(ii) $\Rightarrow$ (iii) Let $u: A \to C$ be an arbitrary morphism, let $(v: B \to E, h: C \to E)$ be the pushout of (f, u), and let $(w: E \to G, k: E \to G)$ be the pushout of (h, h). Then k is a split monomorphism. The image by F of the morphism $h: (C, u) \to (E, hu)$ is precisely the morphism $k: (E, v) \to (G, kv)$. Since F is faithful, F reflects monomorphisms. Thus h is monomorphic. As a result f is a co-universal monomorphism.

(iii) $\Rightarrow$ (i) Let $g: (C, u) \to (D, gu)$ be a morphism in $A/\mathbf{A}$ such that $F(g)$ is monomorphic. Let $(h: C \to E, v: B \to E)$ be the pushout of (u, f) and let $(w: D \to G, k: E \to G)$ be the pushout of (g, h). Then h is monomorphic and $F(g)$ is the morphism $k: (E, v) \to (G, kv)$. Thus $wg = kh$ is monomorphic and hence g is monomorphic, so that f is faithfully flat.

Let us assume that f is faithfully flat and let (C, u) in $A/\mathbf{A}$ be such that $F(C, u)$ is the terminal object in $B/\mathbf{A}$. Let $(h: C \to E, v: B \to E)$ be the pushout of (u, f). Then h is monomorphic and $E = 1$. According to Axiom (1.2.4) of Zariski categories, h is an isomorphism. Thus $C = 1$. As a result F reflects the terminal object, i.e. f is properly flat. ■

Proposition (9.3.3). *For a flat morphism f, the following assertions are equivalent.*

(i) *f is reflexively flat.*

(ii) *f is a co-universal regular monomorphism.*

Then f is faithfully flat.

Proof.

(i) ⇒ (ii) First let us prove that the functor F reflects regular monomorphisms. Let g be a morphism in $A/\mathbf{A}$ such that $F(g)$ is a regular monomorphism. Let (m, n) be the cokernel pair of g, let e be the equalizer of (m, n), and let h be the morphism such that $eh = g$. Then $(F(m), F(n))$ is the cokernel pair of $F(g)$ which is the equalizer of $(F(m), F(n))$. Since F preserves monomorphisms, $F(e)$ is monomorphic. The relation $F(e)F(h) = F(g)$ implies $F(g) \leqslant F(e)$, and the relation $F(m)F(e) = F(n)F(e)$ implies $F(e) \leqslant F(g)$. Consequently, $F(g) \simeq F(e)$, so that $F(h)$ is an isomorphism. Therefore h is an isomorphism and hence g is a regular monomorphism. Let us consider a morphism $u: A \to C$, the pushout $(h: C \to E, v: B \to E)$ of (u, f), and the pushout $(w: E \to G, k: E \to G)$ of (h, h). Then k is a split monomorphism, and thus is a regular monomorphism. As the morphism $k: (E, v) \to (G, kv)$ is precisely the image by F of the morphism $h: (C, u) \to (E, hu)$, it follows that h is a regular monomorphism. As a result, f is a co-universal regular monomorphism.

(ii) ⇒ (*i*) *Let* $g: (C, u) \to (D, gu)$ be a morphism in $A/\mathbf{A}$ such that $F(g)$ is an isomorphism. Let (h, v) be the pushout of (u, f), and (w, k) that of (g, h). Then h is a regular monomorphism and k is an isomorphism for it is $F(g)$. It follows that wg is a regular monomorphism. Since w is also a regular monomorphism, g is a regular monomorphism. But g is also an epimorphism because $F(g)$ is an isomorphism and F is faithful according to Proposition (9.3.2). Consequently, g is an isomorphism. As a result F reflects isomorphisms and f is reflexively flat. ■

Proposition (9.3.4). *For a flat morphism $f: A \to B$, the following assertions are equivalent.*

(i) *f is properly flat.*

(ii) *For any pair of morphisms $(g, h): C \rightrightarrows A$, (fg, fh) is codisjointed ⇒ (g, h) is codisjointed.*

(iii) *Any maximal congruence on A is the image by* Spec(f) *of a maximal congruence on B.*

(iv) *f_{*c} preserves proper congruences.*

(v) *For any prime congruence p on A, $f^{*c}(f_{*c}(p)) = p$.*

(vi) Spec(f) *is surjective.*

Proof.

(i) ⇒ (ii) Let $q: A \to Q$ be the coequalizer of (g, h). The pushout of q along f is the coequalizer of (fg, fh), and thus is $O_B: B \to 1$. Thence $Q \simeq 1$ and (g, h) is codisjointed.

(ii) ⇒ (iii) Let $m = (m_1, m_2)$ be a maximal congruence on A and let q_m be the quotient of A by m. Let r be the congruence $f_{*c}(m)$ on B and let q_r

be the quotient of B by r. Then q_r is the coequalizer of (fm_1, fm_2). Because m is proper, q_m is not null; thus (m_1, m_2) is not codisjointed. Therefore (fm_1, fm_2) is not codisjointed; thus q_r is not null and r is proper. According to Corollary (2.1.10), there exists a maximal congruence n on B such that $r \leqslant n$. Then $f_{*c}(m) \leqslant n$ so that $m \leqslant f^{*c}(n)$. But $f^{*c}(n)$ is proper because $f_{*c}(f^{*c}(n)) \leqslant n$. Thus $m = f^{*c}(n) = \operatorname{Spec}(f)(n)$.

(iii) ⇒ (iv) Let r be a proper congruence on A. According to Corollary (2.1.10), there exists a maximal congruence m on A such that $r \leqslant m$. By hypothesis, there exists a maximal congruence n on B such that $f^{*c}(n) = m$. Then $r \leqslant f^{*c}(n)$ and thus $f_{*c}(r) \leqslant n$. It follows that $f_{*c}(r)$ is proper.

(iv) ⇒ (i) Let $g : A \to C$ be a morphism such that the pushout of (g, f) is $(O_C : C \to 1, O_B : B \to 1)$. Let r be the kernel pair of g, let $q_r : A \to A/r$ be the quotient of A by r, and let $m : A/r \to C$ be the monomorphism such that $mq_r = g$. The pushout of q_r along f is the quotient $q_s : B \to B/s$ of B by $s = f_{*c}(r)$. The image of m by the pushout functor along f is a monomorphism, and is the morphism $O_{B/s} : B/s \to 1$. Thus $B/s \simeq 1$ and s is improper. Therefore r is improper, $A/r \simeq 1$, and $C \simeq 1$. It follows that f is properly flat.

(iv) ⇔ (v) ⇔ (vi) follows from Propositions (3.3.5) and (3.3.7). ■

Corollary (9.3.5). *Any properly flat morphism is prelocal and any flat local morphism is properly flat.*

Proof. Let $f : A \to B$ be a properly flat morphism and let d be a singular epimorphism through which f factors in the form $f = gd$. Let $(m, n) : C \rightrightarrows A$ be a pair of morphisms whose codisjunctor is d. Then f codisjoints (m, n), i.e. (fm, fn) is codisjointed. Thus (m, n) is codisjointed and hence d is isomorphic. Thus f is prelocal. It follows from Proposition (2.12.4) that any flat local morphism is properly flat. ■

Theorem (9.3.6). *For a Zariski category, the following properties are equivalent.*

(i) *Any properly flat epimorphism is an isomorphism.*
(ii) *Any local flat epimorphism is an isomorphism.*
(iii) *Any flat epimorphism is a local isomorphism.*
(iv) *Any flat epimorphism is semisingular.*

Proof.

(i) ⇒ (ii) According to Corollary (9.3.5) any local flat epimorphism is a properly flat epimorphism and thus is an isomorphism.

(ii) ⇒ (iii) Any local morphism of a flat epimorphism is a local flat epimorphism (Proposition (3.6.5)) and thus is an isomorphism.

(iii) ⇔ (iv) follows from Proposition (3.7.3).

(iii) ⇒ (i) Let $f: A \to B$ be a properly flat epimorphism. According to Proposition (3.3.8) the map Spec(f) is a topological embedding, and according to Proposition (9.3.4) it is surjective. Thus Spec(f) is bijective. Because the morphisms f_p ($p \in \mathrm{Spec}(B)$) are isomorphisms, the morphism

$$\varphi = \prod_{p \in \mathrm{Spec}(B)} f_p : \prod_{q \in \mathrm{Spec}(A)} A_q \to \prod_{p \in \mathrm{Spec}(B)} B_p$$

is an isomorphism. According to Proposition (3.5.1) the morphisms $l = (l_q) : A \to \Pi_{q \in \mathrm{Spec}(A)} A_q$ and $m = (l_p) : B \to \Pi_{p \in Spec(B)} f_p$ are regular monomorphisms. Then the relation $\varphi l = mf$ implies that f is a regular monomorphism and that f is an isomorphism. ■

Example. The category **CAlg**(A), where A is a commutative ring, satisfies property (i) of Theorem (9.3.6) [30] and thus it satisfies the other properties [12].

10

ETALE OBJECTS AND MORPHISMS

An etale morphism in a Zariski category **A** is a morphism $f: A \to B$ whose spectrum $\mathrm{Spec}(f) : \mathrm{Spec}(B) \to \mathrm{Spec}(A)$ makes $\mathrm{Spec}(B)$ a kind of etale space over $\mathrm{Spec}(A)$. Because $\mathrm{Spec}(A)$ and $\mathrm{Spec}(B)$ are not mere topological spaces but modelled spaces, the usual definition of an etale continuous map cannot be used as such. Nevertheless, a close look at the universal properties of etale continuous maps in categories of topological spaces will help us to understand etale morphisms in **A**. In the meantime, we have to bring in to question the fact that $\mathrm{Spec}(f)$ must be a local homeomorphism. It is easy to go from etale objects to etale morphisms, and vice versa, by identifying a morphism $f: A \to B$ in **A** with an object in the Zariski category $A/\mathbf{A}$.

A continuous map $\varphi : E \to F$ is etale if and only if it is locally injective and open. The local injectivity means that any element in E has a neighbourhood on which φ is injective. This property is equivalent to the fact that the diagonal morphism $\Delta\varphi : E \to E \times {}_F E$ of φ, computed in the category **Esp**, has an open image or is an open embedding. This last property can easily be carried into the category **AffA** of affine schemes on **A**. It states that the diagonal of $\mathrm{Spec}(f)$, computed in the category **AffA**, is an open immersion. For a morphism $f: A \to B$ in **A** we get the following property: the codiagonal of f is a singular epimorphism. For a finitely presentable morphism f, this property is exactly the neatness of f. On the other hand, the openness property of the map $\mathrm{Spec}(f)$ appears to be related in some way to the fact that f is finitely presentable. Therefore the first property of an etale morphism f in **A** will be that it is neat.

The second property of an etale morphism $f: A \to B$ in **A** is that it is flat. This property is not so easy to justify a priori. Let us just say that in the category of Hausdorff spaces etale continuous maps are coflat morphisms and that coflat morphisms are necessarily generizing maps, while in the category of topological spaces any morphism is coflat. Therefore a morphism f in **A** is said to be etale if it is neat and flat. Then the map $\mathrm{Spec}(f)$ is a generizing map.

From a categorical point of view, the central role is played by the etalish objects, arbitrary colimits of etale objects, and objects in the category **EtshA**. This category is a full coreflective subcategory of **A**. It is a locally simple category whose finitely presentable objects are precisely the etale objects of **A**. Etalish morphisms provide an etalish factorization system: any morphism in **A** factors in an essentially unique way as the composite of an etalish morphism followed by a purely unetale morphism.

The strictly local objects are the 'etale closed' objects. They play the same role for etale morphisms that neatly closed objects play for neat morphisms, or the same role that local objects play for singular epimorphisms. In a Zariski category in which any simple object is absolutely flat, any local object has a strict localization, which is essentially unique and equipped with a Galois group of automorphisms. It is the essentially universal strictly local object associated to it, which is also called its strict henselianzation. On the other hand, henselian local objects are introduced and the henselianzation process is built up.

10.1 ETALE OBJECTS

Definition (10.1.1). *An object is* pre-etale *if it is preneat and flat.*

Examples. In **CRng**, any finite power $\mathbf{Z}^n$ of $\mathbf{Z}$, any ring of fractions of $\mathbf{Z}$, any number field, and any ring of the form $\mathbf{Z}[X]/(P)\,[P'^{-1}]$ with $P \in \mathbf{Z}[X]$ are pre-etale objects. In $\mathbf{CAlg}(k)$, any preneat object is pre-etale for k is absolutely flat. In $\mathbf{RlAlg}(R)$, for a non-null element $a \in R$ sum of squares, the object $R[\sqrt{a}] = R[X]/(X^2 - a)$ is pre-etale.

Proposition (10.1.2). *Colimits, finite products, and flat quotients of pre-etale objects are pre-etale.*

Proof. Let $(f, g) : A \rightrightarrows B$ be a pair of morphisms between pre-etale objects and let $q : B \to C$ be its coequalizer. According to Proposition (8.1.2), q is a semisingular epimorphism and hence a flat morphism. Then C is flat (Proposition (9.1.8)). The proposition then follows from Propositions (8.1.3) and (9.1.2). ∎

Proposition (10.1.3). *An object is pre-etale if and only if its local objects are pre-etale.*

Proof. The proof follows from Propositions (8.1.4) and (9.1.5). ∎

Definition (10.1.4). *An object is* etale *if it is pre-etale and finitely presentable or, equivalently, neat and flat.*

Examples. In **CRng** etale objects are precisely etale rings, i.e. $\mathbf{Z}$-algebras ([34], Corollary 2, Theorem 2, Chapter V). In **RedCRng** etale reduced rings are etale objects, but the converse is in question. In $\mathbf{CAlg}(k)$ and $\mathbf{RedCAlg}(k)$ etale objects are identical to neat objects. In **RlRng** the object $Q[\sqrt{2}]$ is etale, but we do not know if, for example, the object $Q[\sqrt[3]{2}]$ is etale. In **RlOrdRng** the object $\mathbf{Q}[\sqrt{2}]$, equipped with the sum of squares order, is etale.

Proposition (10.1.5). *Finite colimits, finite products, and singular quotients of etale objects are etale.*

Proof. The proof follows from Proposition (10.1.2). ■

Proposition (10.1.6). *An object is etale if and only if it is finitely presentable and its local objects are pre-etale.*

Proof. The proof follows from Proposition (10.1.3). ■

Definition (10.1.7). *The* category **EtA** of etale objects of **A** *is the full subcategory of* **A** *whose objects are the etale ones.*

10.2 ETALISH OBJECTS

Definition (10.2.1). *An object is* etalish *if it is a colimit of etale objects.*

Proposition (10.2.2). *Etalish objects are pre-etale.*

Proof. The proof follows from Proposition (10.1.2). ■

Proposition (10.2.3). *An object is etalish if and only if it is neatish and flat.*

Proof. An etalish object is obviously neatish and, as it is pre-etale, it is flat. Conversely, let A be a flat neatish object. Then **A** is a filtered colimit of neat objects, say $A = \xrightarrow{lim}_{i \in I} A_i$ (Theorem (8.3.6)). According to Corollary (9.1.4), there exists some $i \in \mathbf{I}$ such that, for any j under $i \in \mathbf{I}$, the object A_j is flat and hence etale. Consequently A is etalish. ■

Definition (10.2.4). *The* category *EtshA* of etalish objects *of A is the full subcategory of A whose objects are the etalish ones.*

The full subcategory **EtshLocA** of **LocA**, and the full subcategory **EtshSimA** of **SimA** are defined similarly.

Theorem (10.2.5). *To any object of A is associated a co-universal etalish object, i.e.* ***EtshA*** *is a coreflective subcategory of A.*

Proof. The proof is the analogue of the proof of Theorem (8.3.4) for neatish objects. For an object A, let $(E(A), \alpha)$ be the colimit of the etale objects above A, i.e.

$$(E(A), \alpha) = \varinjlim_{(B,f) \in (EtA, A)} B$$

and let $u_A : E(A) \to A$ be the morphism defined by $u_A \alpha_f = f$ for any $(B, f) \in (\mathbf{EtA}, A)$. Then $E(A)$ is etalish and we can prove, in the same way as for Theorem (8.3.4), that $u_A : E(A) \to A$ is a co-universal etalish object associated to A. ■

Corollary (10.2.6). *EtshA is closed in A under colimits.*

Theorem (10.2.7). *EtshA is a locally simple category [10], whose category of finitely presentable objects is EtA.*

Proof. The proof is analogous to the proof of Theorem (8.3.6). ■

Proposition (10.2.8). *Finite products, singular quotients, and local objects of etalish objects are etalish.*

Proof. The proof follows from propositions (8.3.7), (9.1.2), (9.1.5), and (9.1.8). ■

Corollary (10.2.9). *The co-unit morphisms $u_A : E(A) \to A$ are prelocal.*

Proposition (10.2.10). *The simple objects in EtshA are precisely the indecomposable etalish objects.*

Proof. The proof follows from Propositions (8.3.9) and (10.2.8). ■

Theorem (10.2.11). *EtshA is a coreflective subcategory* of *NtshA, and the coreflector E : NtshA → EtshA is a morphism of locally simple categories.*

Proof. The first part follows from Theorem (10.2.5) and the fact that **NtshA** is a full subcategory of **A**. Since **EtshA** and **NtshA** are closed in **A** under finite products, the inclusion functor **EtshA** → **NtshA** preserves finite products. Let us prove that the coreflector E : **NtshA** → **EtshA** preserves filtered colimits. Let $(\alpha_i : A_i \to A)_{i \in \mathbf{I}}$ be a filtered colimit in **NtshA**, let $(\beta_i : E(A_i) \to B)_{i \in \mathbf{I}}$ be the colimit in **A** of the diagram $(E(A_i))_{i \in \mathbf{I}}$, and let $f : B \to E(A)$ be the morphism defined by $f\beta_i = E(\alpha_i)$ for any $i \in \mathbf{I}$. The morphism $u_A f$ is the filtered colimit in **A** of the diagram of prelocal morphisms $(u_{A_i})_{i \in \mathbf{I}}$; therefore it is a prelocal morphism, and f is prelocal also. Let us prove that f is an isomorphism. Let C be an etale object and let $(g, h) : C \rightrightarrows B$ be such that $fg = fh$. The coequalizer q of (g, h) is a direct factor (Proposition (8.2.2)) through which f must factor. Therefore it is an isomorphism and $g = h$. Consequently, f is a monomorphism in **EtshA**. Let C be an etale object and let $g : C \to E(A)$ be a morphism. There exist an index $i \in \mathbf{I}$ and a morphism $h : C \to A_i$ such that $u_A g = \alpha_i h$. Then there exists a morphism $g_i : C \to E(A_i)$ such that $u_{A_i} g_i = h$. Then $u_A E(\alpha_i) g_i = \alpha_i u_{A_i} g_i = \alpha_i h = u_A g$, and hence $f \beta_i g_i = E(\alpha_i) g_i = g$. It follows that f is an isomorphism in **EtshA**. Consequently $(E(\alpha_i) : E(A_i) \to E(A))_{i \in \mathbf{I}}$ is a colimit in **EtshA**. As a result, the coreflector E is a morphism of locally simple categories [10]. ∎

Corollary (10.2.12). *The coreflector $E : A \to$ **EtshA** is a morphism of Zariski categories.*

Proposition (10.2.13). ***EtshLocA** is a coreflective subcategory of **LocA**.*

Proof. The proof follows readily from Theorem (10.2.5) and Corollary (10.2.9). ∎

10.3 PURELY UNETALE OBJECTS

Definition (10.3.1). *An object is* purely unetale *if the co-universal etalish object associated to it is the initial object.*

Definition (10.3.2). *The* category ***PunetA*** of purely unetale objects of ***A*** *is the full subcategory of **A** whose objects are the purely unetale ones. It is the inverse image of the singleton subcategory $\{Z\}$ of **A** by the coreflector $E : A \to$ **EtshA**.*

Proposition (10.3.3). ***PunetA** is a multireflective subcategory of **A**, which is closed under filtered colimits and subobjects.*

Proof. Since the coreflector $E : \mathbf{A} \to \mathbf{EtshA}$ preserves limits and filtered colimits, **PunetA** is closed in **A** under connected limits, filtered colimits, and subobjects. For any object A in **A** we obtain a solution set of morphisms from A to **PunetA** by taking the set of regular quotients of A whose codomain is purely unetale. Applying Theorem 3.6.1 of [7] gives a left-multiadjoint to the inclusion functor **PunetA** → **A.** ∎

Proposition (10.3.4). *Purely unneat objects are purely unetale.*

Proof. Let A be a purely unneat object. Since the co-universal etalish object $u_A : E(A) \to A$ associated to A is prelocal, the morphism $N(u_A) : E(A) \to Z$ is prelocal in **NtshA**, i.e. it is monomorphic. It follows that $E(A) \simeq Z$, i.e. A is purely unetale. ∎

10.4 ETALENESS OF MORPHISMS

Definition (10.4.1). *A morphism $f: A \to B$ is* pre-etale (etale, etalish, purely unetale) *if the object (B, f) in the category $A/\mathbf{A}$ is pre-etale (etale, etalish, purely unetale).*

Proposition (10.4.2). *Pre-etale, etale, and etalish morphisms are co-universal.*

Proof. The proof follows from Propositions (1.6.2) and (8.5.2). ∎

Proposition (10.4.3). *For a morphism $f: A \to B$ the following assertions are equivalent.*

(i) *f is pre-etale.*
(ii) *For any $p \in \mathrm{Spec}(A)$, the pushout of f along l_p is pre-etale.*
(iii) *For any $q \in \mathrm{Spec}(B)$, the local morphism of f at q is pre-etale.*

Proof. The proof follows from Proposition (3.6.5) and (8.5.3). ∎

Proposition (10.4.4). *Purely unetale morphisms are prelocal.*

Proof. The proof is similar to the proof of (8.6.3). ∎

Corollary (10.4.5). *Purely unetale morphisms with local codomain are local morphisms.*

Proof. The proof follows from Proposition (2.12.4). ∎

Proposition (10.4.6). *Let $f: A \to B$ be any morphism.*

(i) *If A and f are pre-etale (etale, etalish, purely unetale), then B is also.*
(ii) *If A and B are pre-etale (etale, etalish) then f is also.*
(iii) *If B is purely unetale and f is prelocal, A is purely unetale.*

Proof.

(i) For pre-etale, etale, and etalish, the result follows from Propositions (8.5.7) and (9.1.8). For purely unetale, the proof is similar to the proof for purely unneat in Proposition (8.5.7).

(ii) follows from Propositions (8.5.6) and (9.1.8).

(iii) The morphism $E(f): E(A) \to Z$ is prelocal in **EtshA** (Proposition (7.3.2)) and hence is monomorphic in **EtshA**. Consequently, $E(A) = Z$ and A is purely unetale. ∎

Proposition (10.4.7). *Local pre-etale morphisms are interminable.*

Proof. They are local and flat, and hence are interminable (Proposition (11.3.4) below). ■

Proposition (10.4.8). *The classes of pre-etale, etale, etalish, and purely unetale morphisms are closed under composition.*

Proof. The proof follows from Proposition (10.4.6) by an argument similar to that used in the proof of (8.6.1). ■

Proposition (10.4.9). *Let $f: A \to B$ and $g: B \to C$ be two morphisms.*
(i) *If f and gf are pre-etale (etale, etalish) then g is also.*
(ii) *If gf is purely unetale and g is prelocal, then f is purely unetale.*

Proof. The proof follows readily from Proposition (10.4.6) applied in $A/\mathbf{A}$. ■

Proposition (10.4.10). *Purely unetale morphisms are precisely the morphisms which are right-orthogonal to etale, or etalish, morphisms.*

Proof. The proof is similar to the proof of Proposition (8.6.5). ■

Proposition (10.4.11). *The co-unit morphisms $E(A) \to A$ are purely unetale.*

Proof. The proof is similar to that of Proposition (8.6.6). ■

Theorem (10.4.12). *Any morphism factors in an essentially unique way as the composite of an etalish morphism followed by a purely unetale morphism.*

Proof. The proof is similar to that of Theorem (8.6.7). ■

Proposition (10.4.13). *Etalish morphisms are precisely the morphisms which are left-orthogonal to purely unetale morphisms.*

Proof. The proof is similar to that of Proposition (8.6.8). ■

Corollary (10.4.14). *An object A is etalish if and only if for any purely unetale morphism f, the map $Hom_{\mathbf{A}}(A,f)$ is bijective.*

10.5 HENSELIAN OBJECTS

Definition (10.5.1). *An object H is* henselian *if it is not terminal and any regular epimorphism $q: H \to Q \neq 1$ is purely unetale.*

Proposition (10.5.2). *Henselian objects are local.*

Proof. A henselian object H has a simple quotient $q : H \to K$ which is purely unetale. According to Corollary (10.4.5), H is local. ■

Proposition (10.5.3). *For a local object H, the following assertions are equivalent.*

(i) *H is henselian.*

(ii) *The residue simple quotient $k_H : H \to K_H$ of H is a purely unetale morphism.*

(iii) *Any etalish local morphism $f : H \to L$ whose residue morphism is an isomorphism, is an isomorphism.*

(iv) *Any local morphism $f : H \to L$ whose residue morphism is an isomorphism, is purely unetale.*

Proof.

(i) ⇒ (ii) The proof is obvious.

(ii) ⇒ (iii) If $\bar{f} : K_H \to K_L$ is the residue morphism of f, then $\bar{f}k_H = k_L f$. Since k_H is purely unetale and k_L is local, f is purely unetale (Proposition (10.4.9)) and therefore is an isomorphism.

(iii) ⇒ (iv) Let $f = ue$ be the etalish factorization of f, where $e : H \to E$ and $u : E \to L$. According to Corollary (10.4.5), u is local. Consequently, e is local and its residue morphism is an isomorphism. Therefore e is an isomorphism. Thus f is purely unetale.

(iv) ⇒ (i) A regular epimorphism $q : H \to Q \neq 1$ is a local morphism whose residue morphism is an isomorphism. Therefore it is purely unetale. ■

Proposition (10.5.4). *Let $f : A \to B$ be a purely unetale morphism. If B is henselian, A is henselian.*

Proof. According to Proposition (10.5.2) and Corollary (10.4.5), f is local. The residue morphism $\bar{f} : K_A \to K_B$ of f satisfies the relation $\bar{f}k_A = k_B f$. Since f and k_B are purely unetale, $\bar{f}k_A$ is purely unetale. Since $\bar{f}$ is local, k_A is purely unetale (Proposition (10.4.9)). Then A is henselian (Proposition (10.5.3)). ■

Definition (10.5.5). *The* category ***HenA*** of henselian objects of *A is the full subcategory of **LocA** whose objects are the henselian ones.*

The full subcategories **NtshenA** of **NtshLocA** and **Etshen A** of **EtshLocA** are defined similarly.

Theorem (10.5.6). *To any local object is associated a universal henselian object, i.e. **HenA** is a reflective subcategory of **LocA**.*

Proof. Let L be a local object with residue simple quotient $k_L : L \to K_L$. Let $k_L = kf$ be the etalish factorization of k_L, where $f: L \to H$ is an etalish morphism and $k: H \to K_L$ is a purely unetale morphism. Since k_L is local, f and k are local. Moreover, k factors through the residue quotient $k_H : H \to K_H$ of H in the form $k = gk_H$ and the residue morphism $\bar{f}$ of f satisfies the relation $\bar{f}k_L = k_H f$. Consequently, $g\bar{f}k_L = gk_H f = kf = k_L$. Thus $g\bar{f} = 1_{K_L}$, and hence g is an isomorphism and k is the residue simple quotient of the local object H. Since k is purely unetale, H is henselian (Proposition (10.5.3)). Let $m : L \to M$ be a local morphism with a henselian codomain M. If $k_M : M \to K_M$ is the residue simple quotient of M and $\bar{m}$: $K_L \to K_M$ is the residue morphism of m, we have $\bar{m}kf = k_M m$. Since f is etalish and k_M is purely unetale, there is a unique morphism $g : H \to M$ such that $gf = m$ and $k_M g = \bar{m}k$. The latter relation entails that g is local. Let $g' : H \to M$ be another local morphism such that $g'f = m$ with residue morphism $\bar{g}'$. Then $\overline{g'}k_L = \overline{g'}kf = k_M g'f = k_M m = \bar{m}k_L$. Hence $\overline{g'} = \bar{m}$ and thus $k_M g' = \bar{m}k$. Therefore $g' = g$. As a result, $f: L \to H$ is a universal henselian object associated to L. ■

Corollary (10.5.7). *NtshenA is a reflective subcategory of NtshLocA, Etshen A is a reflective subcategory of EtshLocA, and NtshSimA is a reflective subcategory of NtshenA.*

Proposition (10.5.8). *The simple residue functor EtshenA → NtshSimA is full and faithful.*

Proof. It is faithful because it is the composite of the faithful inclusion functor **EtshenA** → **NtshenA** with the faithful residue functor **NtshenA** → **NtshSimA** (Proposition (8.1.7)). Let H and H' be two objects in **EtshenA** and $g : K_H \to K_{H'}$. Since H is etalish and $k_{H'} : H' \to K_{H'}$ is purely unetale, the map $\mathrm{Hom}_{\mathbf{A}}(H, k_{H'})$ is bijective (Corollary (10.4.14)). Therefore there is a unique morphism $f: H \to H'$ such that $k_{H'}f = gk_H$. This morphism is local; hence it belongs to **EtshenA** and its residue morphism is g. As a result, the functor is full. ■

10.6 STRICTLY LOCAL OBJECTS

Definition (10.6.1). *An object L is* strictly local *if it is local and any local morphism $f: L \to M$ is purely unetale.*

Proposition (10.6.2). *For a local object L, the following assertions are equivalent.*

(i) *L is strictly local.*

(ii) *Any etalish local morphism $f: L \to M$ is an isomorphism.*

Proof.

(i) ⇒ (ii) The proof is obvious.

(ii) ⇒ (i) Let $f = ge$ where $e: L \to A$ is etalish and $g: A \to M$ is purely unetale. According to Corollary (10.4.5), A is local. Thus e is an etalish local morphism, and hence is an isomorphism. Consequently f is purely unetale. ∎

Proposition (10.6.3). *For a local object L in a simply flat simply noetherian Zariski category, the following assertions are equivalent.*

(i) *L is strictly local.*

(ii) *L is injective with respect to interminable etalish morphisms.*

(iii) *L is injective with respect to interminable etale morphisms.*

(iv) *Any interminable etalish morphism $f: L \to A$ has a retraction.*

(v) *Any interminable etale morphism $f: L \to A$ has a retraction.*

Proof.

(i) ⇒ (ii) Let $f: A \to B$ be an interminable etalish morphism, let $g: A \to L$ be a morphism, and let $(\bar{f}: L \to C, \bar{g}: B \to C)$ be the pushout of (g, f). Then $\bar{f}$ is etalish and interminable. According to Proposition (11.6.5) below, the map $\mathrm{Spec}(\bar{f}): \mathrm{Spec}(C) \to \mathrm{Spec}(L)$ is surjective. Therefore there exists some $p \in \mathrm{Spec}(C)$ such that $\mathrm{Spec}(\bar{f})\ (p)$ is the maximal congruence on L, i.e. such that the morphism $l_p\bar{f}: L \to C_p$ is local. Since l_p is etalish, $l_p\bar{f}$ is etalish. Consequently, $l_p\bar{f}$ is an isomorphism. Then the morphism g factors through f. As a result, L is injective with respect to f.

(ii) ⇒ (iii) ⇒ (v) and (ii) ⇒ (iv) ⇒ (v) are obvious.

(v) ⇒ (i)

(a) Let us consider a special case. Let $f: L \to A$ be an etale morphism and let $p \in \mathrm{Spec}(A)$. Then the morphism $l_p f: L \to A_p$ is etalish. Let us assume that this morphism $l_p f$ is local and prove that it is an isomorphism. Let m be the maximal congruence on L and let F be the fibre of the map $\mathrm{Spec}(f)$ at m. Since $l_p f$ is local, $p \in F$. Since f is neat, F is finite (Theorem (9.2.9)). We shall perform the proof by induction on the number of elements of F. According to Proposition (10.4.7), $l_p f$ is interminable and hence f is interminable. Then f has a retraction g. According to Proposition (8.5.5), g is a direct factor, i.e. there is a morphism $g': A \to A'$ such that $(g: A \to L, g': A \to A')$ is a product. Let $f' = g'f$. Since f and g' are etale, f' is etale. The congruence $\mathrm{Spec}(g)\ (m)$ belongs to F. If $p = \mathrm{Spec}(g)\ (m)$, the two localizations l_p and g are isomorphic, so that $l_p f$ is an isomorphism. This case happens whenever $\mathrm{card}(F) = 1$. If $p \neq \mathrm{Spec}(g)\ (m)$, then there is some $p' \in \mathrm{Spec}(A')$ such that $\mathrm{Spec}(g')\ (p') = p$ and $l_p = l_{p'} g'$. As a result, we obtain an etale morphism $f': L \to A'$ and $p' \in \mathrm{Spec}(A')$ such that $l_{p'} f': L \to A_{p'}$ is a local morphism but with card $(\mathrm{Spec}(f')^{-1}(m)) =$ *card*$(F) - 1$. Therefore we obtain the proof by induction on $\mathrm{card}(F)$.

(b) Now let us consider an arbitrary etalish local morphism $f: L \to A$.

In the category $L/\mathbf{A}$ there exists a filtered colimit $(g_i : (A_i, f_i) \to (A, f))_{i \in I}$ with each f_i etale. Each morphism $g_i : A_i \to A$ factors in the form $g_i = h_i l_{pi}$ with $p_i \in \operatorname{Spec}(A_i)$ and h_i local. Then $(h_i : ((A_i)_{pi}, l_{pi} f_i) \to (A, f))_{i \in I}$ is a filtered colimit in $L/\mathbf{A}$. Since f is local, each morphism $l_{pi} f_i$ is local and according to (a) it is an isomorphism. It follows that f is an isomorphism. ∎

Proposition (10.6.4).

(i) *Any strictly local object is henselian.*

(ii) *A henselian object whose residue simple object is strictly local, is strictly local.*

Proof.

(i) Any regular quotient $q : L \to Q \neq 1$ of a strictly local object L is a local morphism and thus is purely unetale.

(ii) Let H be a henselian object with a strictly local residue simple object K_H. Let $f : H \to L$ be a local morphism with residue morphism $\bar{f} : K_H \to K_L$. Both morphisms k_H and $\bar{f}$ are purely unetale. Thus $k_L f = \bar{f} k_H$ is purely unetale. Since k_L is local, f is purely unetale (Proposition (10.4.9)). Therefore H is strictly local. ∎

Proposition (10.6.5). *For a simple object, K, the following assertions are equivalent.*

(i) *K is strictly local.*

(ii) *Any simple extension of K is a purely unetale morphism.*

Proof.

(i) ⇒ (ii) A simple extension is a local morphism.

(ii) ⇒ (i) Let $f : K \to L$ be a local morphism. Then $k_L f : K \to K_L$ is a simple extension of K and thus is a purely unetale morphism. Since k_L is local, f is purely unetale (Proposition (10.4.9)). Consequently, K is strictly local. ∎

Proposition (10.6.6). *Any neatly closed object is strictly local.*

Proof. Any simple extension of a neatly closed object is purely unneat (Proposition (8.7.3)), and thus is purely unetale (Proposition (10.3.4)). The result then follows from Proposition (10.6.5).

Proposition (10.6.7). *For an absolutely flat simple object K, the following assertions are equivalent.*

(i) *K is strictly local.*

(ii) *K is neatly closed.*

Proof. The proof follows from Propositions (8.7.3), (10.6.2), and (10.6.6). ■

Proposition (10.6.8). *In a simply flat simply noetherian Zariski category, let $f: A \to B$ be a purely unetale morphism. If B is strictly local, then A is strictly local.*

Proof. According to Corollary (10.4.5), A is local. Let $g: C \to D$ be an etale interminable morphism and let $h: C \to A$. Since B is injective with respect to g (Proposition (10.6.3)), there is a morphism $u: D \to B$ such that $fh = ug$. Since f is right-orthogonal to g (Proposition (10.4.10)), there is some morphism $v: D \to A$ such that $vg = h$ and $fv = u$. Consequently, A is injective with respect to g. By proposition (10.6.3), A is strictly local. ■

Definition (10.6.9). *The* category ***SLocA*** (***EtshSLocA***) of strictly local (etalish strictly local) objects of A *is the full subcategory of* ***LocA*** *whose objects are the strictly local (etalish strictly local) ones.*

Theorem (10.6.10) *The category* ***EtshSLocA*** *is a groupoid equivalent to the disjoint union of the automorphism groups of a class of simple objects of* **A**.

Proof. Any morphism in **EtshSLocA** is both etalish (Proposition (10.4.6)) and purely unetale, and thus is an isomorphism. Therefore **EtshSLocA** is a groupoid. Since strictly local objects are henselian (Proposition (10.6.4)) **EtshSLocA** is a full subcategory of **EtshenA**. Then the full and faithful residue functor **EtshenA** → **NtshSimA** (Proposition (10.5.8)) sends the groupoid **EtshSLocA** on a full subgroupoid of the category **NtshSimA**. This latter groupoid is equivalent to the disjoint union of its connected components, each of which is equivalent to the automorphism group of a simple object of **A**. ■

Theorem (10.6.11). *In a simply flat simply noetherian Zariski category* A, *to any local object is associated an essentially universal strictly local object, i.e.* ***SLocA*** *is an essentially reflective subcategory of* ***LocA***.

Proof. Let L be a local object, let $q: L \to K$ be its residue simple object, and let $k: K \to \bar{K}$ be the neat closure of K (Theorem (8.7.12)). Let $kq = gf$, where $f: L \to S$ is etalish and $g: S \to \bar{K}$ is purely unetale, be the neatish factorization of kq. The object $\bar{K}$ is neatly closed and thus strictly local (Proposition (10.6.6)). Hence S is strictly local (Proposition (10.6.8)) and the morphism f is local. Let $m: L \to M$ be a local morphism with a strictly local codomain. Since M is injective with respect to the interminable etalish morphism f, there exists some morphism $u: S \to M$ such that $uf = m$. According to Proposition (9.2.10), u is local. Let $v: S \to M$ be another local morphism such that

$vf = m$. Then u and v are purely unetale. According to the essential uniqueness of the etalish factorization, there exists some automorphism $w : S \to S$ such that $wf = f$ and $uw = v$. As a result, $f : L \to S$ is an essentially universal strictly local object associated to L. ■

Notation. The essentially universal strictly local object $f : L \to S$ associated to a local object L is called the *strict localization of* L. It is defined up to isomorphisms. The group of automorphisms $\alpha : S \to S$ such that $\alpha f = f$ is the *Galois group of the strict localization* of L. It is defined up to isomorphisms.

Theorem (10.6.12). *For a local morphism $f : L \to S$ in a simply flat simply noetherian Zariski category, the following assertions are equivalent.*

(i) *f is a strict localization of L.*

(ii) *f is etalish and S is strictly local.*

(iii) *f is an etalish local hull of L, i.e. it factors through any etalish local morphism $g : L \to M$.*

Proof.

(i) ⇒ (ii) See the proof of Theorem (10.6.11).

(ii) ⇒ (iii) Let $g : L \to M$ be an etalish local morphism. Then g is interminable (Proposition (10.4.7)). According to Proposition (10..6.3), S is injective with respect to g and so f factors through g.

(iii) ⇒ (i) Let $m : L \to M$ be a strict localization of L. Then m is etalish, and hence f factors through m in the form $f = gm$. According to Proposition (9.2.10), g is local. Since M is strictly local, g is purely unetale. According to the essential uniqueness of the etalish factorization, g is an isomorphism. Consequently f is a strict localization of L. ■

11

TERMINATORS

The idea of terminators is similar to that of codisjunctors, except that pushouts are used instead of coequalizers. In an arbitrary category **A** with a strict terminal object denoted by 1, let $f: A \to B$ be a given morphism. A morphism $g: A \to C$ is said to terminate f if the pushout of (f, g) is terminal, i.e. the pushout object is 1. In categories of commutative algebras, to any object C is associated a universal von Neumann regular object $n_c: C \to N(C)$ and it turns out that a morphism $g: A \to C$ terminates f if and only if the morphism $n_c g: A \to N(C)$ does. Therefore the study of morphisms terminating f can be restricted, to some extent, to those morphisms $g: A \to C$ which have a von Neumann regular codomain and are called von Neumann morphisms. A terminator of f is then defined as being a universal von Neumann morphism terminating f, i.e. a von Neumann morphism terminating f and such that any von Neumann morphism terminating f factors through it in a unique way.

The existence of a terminator is not assured for any morphism. For example, in the category **CRng** of commutative rings the inclusion morphism $\mathbf{Z} \to \mathbf{Q}$ has no terminator. In any Zariski category **A**, the class of morphisms having a terminator contains the finitely generated regular epimorphisms and the singular epimorphisms, the terminators themselves, and is co-universal and closed under pushouts and finite products. The existence of a terminator can be checked by using a Freyd theorem with a finite terminating solution set condition. In the category **CRng**, any finitely presentable morphism has a terminator. This result is not obvious and is proved by noetherian induction. It is still valid in any category **CAlg**(A) of commutative algebras over some ring A, but no longer holds in an arbitrary Zariski category. In a locally noetherian Zariski category this property is equivalent to the fact that any finitely presentable integral extension has a terminator.

In any Zariski category **A** satisfying the amalgamation property for simple extensions, terminating morphisms and terminators are meaningful from the point of view of prime spectra. A morphism $t: A \to T$ terminates a morphism $f: A \to B$ if and only if the images ImSpec(f) and ImSpec(t) of the maps $\operatorname{Spec}(f): \operatorname{Spec}(B) \to \operatorname{Spec}(A)$ and $\operatorname{Spec}(t): \operatorname{Spec}(T) \to \operatorname{Spec}(A)$ respectively are disjoint subsets in Spec(A). If t is the terminator of f, then ImSpec(f) and ImSpec(t) are complemented subsets in Spec(A). A morphism $f: A \to B$ has a terminator if and only if ImSpec(f) is a constructible subset of Spec(A). Furthermore, any constructible subset of Spec(A) is of the form ImSpec(f) for some morphism $f: A \to B$ having a terminator, and is also of the form ImSpec(t) for some terminator $t: A \to T$ of some morphism $f: A \to B$. Finally, the existence of a terminator for any finitely presentable morphism in **A** implies the following Chevalley theorem: any finitely presentable morphism of schemes on **A** preserves constructible sets.

The non-existence of a terminator for a morphism $f: A \to B$ does not mean that the class of morphisms terminating f is not of interest. In order to assess, for any case, this class of morphisms, we introduce the term of a morphism, i.e. the kernel pair of the class of von Neumann morphisms terminating f and thus a radical congruence on A. For example, in **CRng** the term of the quotient morphism $A \to A/I$ is rad$(A) : I$ and the term of the canonical morphism $A \to A[X][P(X)^{-1}]$ where $P(X) = a_0 + \ldots + a_n X^n \in A[X]$, is rad$(a_0, \ldots, a_n)$. Whenever f has a terminator t, the term of f is naturally the kernel pair of t. If r is the term of f, then the open set $D(r)$ of Spec(A) is precisely the interior of ImSpec(f) in Spec(A). There are two kinds of morphisms with special terms: those whose term is the radical of A, and those whose term is the unit congruence on A.

Morphisms $f: A \to B$ whose term is the radical of A are called preterminal morphisms. For example, in **CRng** the quotient morphism $A \to A/I$ by a finitely generated dense ideal I is preterminal, as is the inclusion morphism $\mathbf{Z} \to \mathbf{Q}$. In any Zariski category, any morphism $f: A \to B$ with an integral domain A must be monomorphic or preterminal. In a locally noetherian Zariski category satisfying the amalgamation property for simple extensions, the fact that finitely presentable morphisms of integral objects cannot be both monomorphic and preterminal is equivalent to the existence of terminators for any finitely presentable morphism. From the point of view of spectra, a morphism $f: A \to B$ is preterminal if and only if the interior of ImSpec(f) in Spec(A) is empty.

Morphisms $f: A \to B$ whose term is the unit congruence on A are called interminable morphisms. They are characterized by one of the following properties: the morphism $A \to 1$ is a terminator of f, any morphism terminating f has a terminal codomain, f is a co-universal prelocal morphism, or f is a co-universal premonomorphism. A morphism $f: A \to B$ is not preterminal if and only if there exists a non-terminal singular epimorphism $d: A \to D$ such that the pushout of f along d is interminable. In **CRng** the canonical morphism $A \to A[X_i]_{i \in I}$ and any integrally dependent ring extension are interminable morphisms. In an arbitrary Zariski category the simple extensions are interminable if and only if the amalgamation property for simple extensions holds. Under this assumption, a morphism f is interminable if and only if the map Spec(f) is surjective.

11.1 THE TERM OF A MORPHISM

Notation. Let us be in a Zariski category **A**. The terminal object is denoted by 1. A morphism of the form $f: A \to 1$ is called a *terminal morphism*. It is the terminal object in the category $A/\mathbf{A}$. A morphism $g: A \to C$ *terminates* a morphism $f: A \to B$ if the pushout of f along g is the terminal morphism. Then f terminates g. A regular object is also called a *von Neumann object*. A morphism $f: A \to B$ whose codomain is a von Neumann object is called a *von Neumann morphism*. After all, it is a von Neumann object in the category $A/\mathbf{A}$. We denote by **RegA** the full subcategory of **A** whose objects are the von Neumann objects and by $J: \mathbf{RegA} \to \mathbf{A}$ the inclusion functor. According to Theorem (2.11.6), **RegA** is a reflective subcategory of **A**. We denote by $N: \mathbf{A} \to \mathbf{RegA}$ the reflector and by $n: 1_{\mathbf{A}} \to JN$ the unit of the reflection. Since **RegA** is closed in **A** under regular subobjects, the morphisms $n_A: A \to N(A)$ are epimorphisms.

Proposition (11.1.1). *A morphism $g: A \to C$ terminates a morphism f: $A \to B$ if and only if the von Neumann morphism $n_c g: A \to N(C)$ does.*

Proof. If an object M is terminal, the object $N(M)$ is terminal. If an object M is not terminal, it has a simple quotient $q: M \to S$ which must factor through $n_M: M \to N(M)$ and thus $N(M)$ is not terminal. Let $(m: B \to M, n: C \to M)$ be the pushout of (f,g) and let $(\bar{n}: N(C) \to \bar{M}, \bar{m}: M \to \bar{M})$ be the pushout of (n, n_c) so that $(\bar{m}m, \bar{n})$ is the pushout of $(f, n_c g)$. If g terminates f, then $M \simeq 1$, so that $\bar{M} \simeq 1$ and $n_c g$ terminates f. If g does not terminate, $M \not\simeq 1$. Then $N(M) \not\simeq 1$, and the existence of a morphism $h: \bar{M} \to N(M)$ satisfying $h\bar{m} = n_M$ and $h\bar{n} = N(n)$ implies $\bar{M} \not\simeq 1$, so that $n_c g$ does not terminate f. ■

It follows from Proposition (11.1.1) that the study of the class of morphisms terminating a given morphism $f: A \to B$ can, to some extent, be restricted to the class of von Neumann morphisms. In order to assess this class, let us introduce the following definition.

Definition (11.1.2). *The* term of a morphism $f: A \to B$ *is the kernel pair of the class of von Neumann morphisms $g: A \to N$ terminating f* i.e. $\operatorname{term}(f) = \wedge \{\ker(g) : g$ is a von Neumann morphism terminating $f\}$.

Proposition (11.1.3). *For any morphism $f: A \to B$*

(i) $\operatorname{term}(f)$ *is a radical congruence on A,*

(ii) $\operatorname{term}(f) = \wedge \{\ker(g) : g$ has a reduced codomain and g terminates $f\}$,

(iii) $\operatorname{term}(f) = \wedge \{\operatorname{rad}(\ker(g)) : g$ terminates $f\}$, *and*

(iv) $\operatorname{term}(f) = \wedge \{\ker(g) : g$ has a simple codomain and g terminates $f\}$.

Proof. The proof follows from Proposition (11.1.1) and the fact that von Neumann objects are reduced (Proposition (2.11.3)), i.e. are subobjects of products of simple objects. ■

In any category of commutative algebras, the term of a morphism is identified with a radical ideal.

Example (11.1.4). *In the category* **CRng**, *the term of the quotient morphism* $q_I : A \to A/I$ *of a ring* A *by an ideal* I *is* $\mathrm{term}(q_I) = \mathrm{rad}(A) : I$.

Proof. Let $T = \mathrm{rad}(A) : I$. Then $T = \mathrm{rad}(A) : \bigcup_{y \in I}(y) = \bigcap_{y \in I}(\mathrm{rad}(A) : (y))$. Let $y \in I$ and $l_y : A \to A[y^{-1}]$ be the canonical localization morphism. Then

$$\begin{aligned} \mathrm{rad}(\ker(l_y)) &= \mathrm{rad}(\{x \in A : \exists p \in \mathbb{N}, xy^p = 0\}) \\ &= \{x \in A : \exists n \in \mathbb{N}, \exists p \in \mathbb{N}, x^n y^p = 0\} = \{x \in A : \exists n \in \mathbb{N}, (xy)^n = 0\} \\ &= \{x \in A : xy \in \mathrm{rad}(A)\} = \{x \in A : x(y) \subset \mathrm{rad}(A)\} \\ &= \mathrm{rad}(A) : (y). \end{aligned}$$

It follows that $T = \bigcap_{y \in I} \mathrm{rad}(\ker(l_y))$. Let $f = q_I$. As the morphism $l_y\,(y \in I)$ terminates f, we have $\mathrm{term}(f) \subset \mathrm{rad}(l_y)$ for any $y \in I$ and so $\mathrm{term}(f) \subset T$. Let $x \in T$ and $g : A \to K$ be a morphism with a simple codomain terminating f. Then g codisjoints I, i.e. $g(I) = K$. Since $xI \subset \mathrm{rad}(A)$, $g(x)g(I) \subset \mathrm{rad}(K)$, and thus $g(x) = 0$ and $x \in \ker(g)$. If follows that $x \in \mathrm{term}(f)$ and thus $T \subset \mathrm{term}(f)$. As a result, $T = \mathrm{term}(f)$. ■

Example (11.1.5). In **CRng**, *for a ring* A *and a polynomial* $P(X) = a_0 + \ldots + a_n X^n \in A[X]$, *the term of the canonical morphism* $f : A \to A[X][P(X)^{-1}]$ *is* $\mathrm{term}(f) = \mathrm{rad}(a_o, \ldots, a_n)$.

Proof. Let $g : A \to K$ be a morphism terminating f whose codomain is a field. Let $Q(X) \in K[X]$ be the polynomial $g(a_0) + \ldots + g(a_n)X^n$. The pushout of f along g is the canonical morphism $K \to K[X][Q(X)^{-1}]$ which must be terminal. Therefore $Q(X) = 0$, i.e. $g(a_0) = \ldots = g(a_n) = 0$, and so g annihilates both the ideal $(a_0, \ldots, a_n)$ and its radical. Hence $\mathrm{rad}(a_0, \ldots, a_n) \subset \mathrm{term}(f)$. Let $p \in \mathrm{Spec}(A)$ such that $(a_0, \ldots, a_n) \subset p$. Let $q : A \to A/p$ be the quotient of A by p, and let $\overline{P(X)} \in A/p[X]$ be the polynomial $q(a_0) + \ldots + q(a_n)X^n$. The pushout of f along q is the canonical morphism $A/p \to A/p[X][(\overline{P(X)})^{-1}]$. But $\overline{P(X)} = 0$ and thus q terminates f. Hence $\mathrm{term}(f) \subset \mathrm{rad}(\ker(q)) = p$. Consequently $\mathrm{term}(f) \subset \mathrm{rad}(a_0, \ldots, a_n)$. As a result $\mathrm{term}(f) = \mathrm{rad}(a_0, \ldots, a_n)$. ■

Proposition (11.1.6). *Let* $f : A \to B$ *be a morphism and let* r *be a congruence on* A.

(i) *If f codisjoints r, then* $\operatorname{term}(f) \leqslant \operatorname{rad}(r)$.
(ii) *If f is the codisjunctor of r, then* $\operatorname{term}(f) = \operatorname{rad}(r)$.

Proof. Let us suppose that f codisjoints $r = (r_1, r_2)$. The quotient morphism $q_r : A \to A/r$ terminates f, and thus $\operatorname{term}(f) \leqslant \operatorname{rad}(\ker(q_r)) = \operatorname{rad}(r)$. Let us suppose that f is the codisjunctor of r and let $g : A \to K$ be a morphism with a simple codomain terminating f. As the pushout of f along g is the codisjunctor of (gr_1, gr_2), the morphism g must coequalize (r_1, r_2). Thence $r \leqslant \operatorname{term}(f)$ and $\operatorname{rad}(r) \leqslant \operatorname{term}(f)$. Consequently $\operatorname{term}(f) = \operatorname{rad}(r)$. ■

Proposition (11.1.7). *For any composable pair of morphisms $f : A \to B$, $g : B \to C$,* $\operatorname{term}(gf) \leqslant \operatorname{term}(f)$.

Proof. The proof follows from the fact that any morphism $h : A \to D$ terminating f terminates gf. ■

11.2 PRETERMINAL MORPHISMS

Definition (11.2.1). *A morphism $f : A \to B$ is* preterminal *if* $\operatorname{term}(f) = \operatorname{rad}(A)$.

Examples.

(i) Any terminal morphism $f : A \to 1$ is preterminal because it is terminated by the unit morphism 1_A, so that $\operatorname{term}(f) = \operatorname{rad}(A)$.

(ii) In **CRng**, for a finitely generated dense ideal I of a ring A, the quotient morphism $q_I : A \to A/I$ is preterminal for $\operatorname{term}(q_I) = \operatorname{rad}(A) : I = \operatorname{rad}(A)$ (cf. Example (11.1.4)).

(iii) In **CRng**, the inclusion morphism $f : \mathbf{Z} \to \mathbf{Q}$ is preterminal. Let p be a prime number. The quotient morphism $q_p : \mathbf{Z} \to \mathbf{Z}/(p)$ terminates f, and thus $\operatorname{term}(f) \subset \ker(q_p) = (p)$. It follows that $\operatorname{term}(f)$ is included in any prime ideal of $\mathbf{Z}$, so that $\operatorname{term}(f) = \{0\}$.

Proposition (11.2.2). *If A is a quasi-primary object, any morphism $f : A \to B$ is premonomorphic or preterminal.*

Proof. Let us suppose that f is not a premonomorphism. According to Proposition (2.5.7), the morphism f terminates a non-terminal singular epimorphism $d : A \to D$. As A is quasi-primary, the pushout along d of any non-terminal singular epimorphism is not terminal. According to Proposition (2.5.7), d is a premonomorphism. Thus $\ker(d) \leqslant \operatorname{rad}(A)$ and hence $\operatorname{term}(f) \leqslant \operatorname{rad}(\ker(d)) \leqslant \operatorname{rad}(A)$. Consequently, f is preterminal. ■

Corollary (11.2.3). *If A is an integral object, any morphism $f : A \to B$ is monomorphic or preterminal.*

Proposition (11.2.4). *If $f: A \to B$ is preterminal and $g: B \to C$ is any morphism, then gf is preterminal.*

Proof. The proof follows from Proposition (11.1.7). ■

11.3 INTERMINABLE MORPHISMS

Definition (11.3.1). *A morphism $f: A \to B$ is* interminable *if its term is the unit congruence on A, i.e.* $\mathrm{term}(f) = 1_{A \times A}$.

Proposition (11.3.2.). *For a morphism f the following assertions are equivalent.*

(i) *f is interminable.*

(ii) *Any morphism terminating f is terminal.*

(iii) *f is a co-universal prelocal morphism.*

(iv) *f is a co-universal premonomorphism.*

Proof. Let $f: A \to B$.

(i) $\Rightarrow$ (ii) Let $g: A \to C$ terminate f. Then $1_{A \times A} = \mathrm{term}(f) \leqslant \mathrm{rad}(\ker(g))$. Thus $\mathrm{rad}(\ker(g)) = 1_{A \times A}$, and hence $\ker(g) = 1_{A \times A}$ (Proposition (2.4.13)) and g is terminal.

(ii) $\Rightarrow$ (iii) Let us assume that f factors in the form $f = gd$, where d is a singular epimorphism codisjunctor of a congruence r on A. The quotient morphism $q_r: A \to A/r$ terminates d and thus terminates f also. Consequently, q_r is terminal, r is codisjointed, and d is an isomorphism. It follows that f is a prelocal morphism. Since assertion (ii) is co-universal, f is indeed a co-universal prelocal morphism.

(iii) $\Rightarrow$ (iv) Let $r = (r_1, r_2) = \ker(f)$. Let $g: A \to K$ be any morphism with a simple codomain and let $(\bar{g}: B \to \bar{C}, \bar{f}: K \to \bar{C})$ be the pushout of (f, g). Then $\bar{f}$ is prelocal and thus is not terminal; therefore it is monomorphic. The equalities $\bar{f}gr_1 = \bar{g}fr_1 = \bar{g}fr_2 = \bar{f}gr_2$ imply $gr_1 = gr_2$, and so g coequalizes r. It follows that r is conjoint (Proposition (2.5.2)) and that f is a premonomorphism (Proposition (2.5.7)). Since assertion (iii) is co-universal, f is indeed a co-universal premonomorphism.

(iv) $\Rightarrow$ (i) Let $g: A \to C$ be a morphism terminating f. The pushout $\bar{f}$: $C \to \bar{C}$ of f along g must be both premonomorphic and terminal, and thus is the terminal isomorphism. Consequently, g is terminal and f is interminable. ■

Proposition (11.3.3).

(i) *Isomorphisms are interminable morphisms.*

(ii) *Interminable morphisms are co-universal.*

(iii) *Composites, finite products, and filtered colimits of interminable morphisms are interminable.*

(iv) *For any composable pair of morphisms* $f: A \to B, g: B \to C$,
(a) *if* gf *is interminable* f *is interminable, and*
(b) *if* g *is interminable,* $\mathrm{term}(gf) = \mathrm{term}(f)$.

Proof.
(i) and (ii) follow from Proposition (11.3.2).
(iii) The composition property follows from Proposition (11.3.2) and Corollary (1.9.3). The finite product property follows from Proposition (11.3.2), and the facts that finite products are co-universal and finite products of prelocal morphisms are prelocal (Proposition (1.9.3)). Let $f: A \to B$ be a filtered colimit of a diagram of interminable morphisms $(f_i : A_i \to B_i)_{i \in I}$ with canonical inductions $(\alpha_i : A_i \to A, \beta_i : B_i \to B)_{i \in I}$. Let $\bar{f}_i : A_i \to \bar{B}_i$ be the pushout of f_i along α_i. Then f_i is interminable and f is the filtered colimit of the diagram of morphisms $(\bar{f}_i)_{i \in I}$. If a morphism $g: A \to C$ terminates f, then there exists $i \in I$ such that g terminates $\bar{f}_i$ and thus g is terminal. Consequently, f is interminable.
(iv) (a) According to Proposition (11.1.7) $\mathrm{term}(gf) \leqslant \mathrm{term}(f)$. Thus if $\mathrm{term}(gf) = 1_{A \times A}$, then $\mathrm{term}(f) = 1_{A \times A}$.
(b) For any morphism $h: A \to H$, let $(\bar{f}: H \to \bar{B},\ \bar{h}: B \to \bar{B})$ be the pushout of (h, f) and let $(\bar{g}: \bar{B} \to \bar{C} = h: C \to \bar{C})$ be the pushout of $(\bar{h}, g)$. Since g is interminable, $\bar{B}$ is terminal if and only if $\bar{C}$ is. Then h terminates $gf \Leftrightarrow \bar{C} = 1 \Leftrightarrow \bar{B} = 1 \Leftrightarrow h$ terminates f. Consequently $\mathrm{term}(gf) = \mathrm{term}(f)$.
■

Proposition (11.3.4).
(i) *Universal von Neumann morphisms* $n_A : A \to N(A)$ *are interminable.*
(ii) *A regular epimorphism* $q_r : A \to A/r$ *is interminable if and only if* r *is conjoint, i.e.* $r \leqslant \mathrm{rad}(A)$.
(iii) *A presingular epimorphism is interminable if and only if it is an isomorphism.*
(iv) *A flat morphism is interminable if and only if it is properly flat.*
(v) *Local flat morphisms are interminable.*
(vi) *An interminable morphism with a reduced domain is monomorphic.*
(vii) *An interminable morphism with a local codomain is local.*
(viii) *An interminable morphism with a pseudo-simple codomain has a pseudo-simple domain.*

Proof.
(i) Let $g: A \to C$ be a non-terminal morphism. Then C has a simple quotient $q: C \to K$. The morphism $qg: A \to K$ factors through n_A. Thus the pushout of (n_A, g) is not terminal, i.e. g does not terminate n_A. As a result, n_A is interminable.
(ii) If q_r is interminable, q_r is premonomorphic (Proposition (11.3.2)). Thus $r = \ker(q_r) \leqslant \mathrm{rad}(A)$ (Proposition (2.5.4)). Conversely, let $r \leqslant$

$\text{rad}(A)$. The pushout of q_r along any morphism $g: A \to C$ is the quotient morphism $q_t: C \to C/t$ with $t = f_{*c}(r) \leqslant f_{*c}(\text{rad}(A)) \leqslant \text{rad}(B)$. Then q_t is premonomorphic (Proposition (2.5.2)), so that q_r is a co-universal premonomorphism and hence an interminable morphism (Proposition (11.3.2)).

(iii) An interminable presingular epimorphism is a prelocal presingular epimorphism and thus is an isomorphism (Theorem (1.9.6)).

(iv) follows from Proposition (11.3.2) and Definition (9.3.1).

(v) follows from (iv) and corollary (9.3.5).

(vi) follows from Propositions (2.5.7) and (11.3.2).

(vii) follows from Propositions (2.12.4) and (11.3.2).

(viii) Let $f: A \to B$ be an interminable morphism with a pseudo-simple codomain B. Let $d: A \to B$ be a non-terminal singular epimorphism. Since the pushout of d along f is not terminal, it is an isomorphism. Then the prelocal morphism f factors through d, so that d is an isomorphism. As a result, A is pseudo-simple. ■

Examples.

(i) In **CRng**, the canonical morphism $A \to A[X_i]_{i \in I}$ from a ring A to the polynomial ring $A[X_i]_{i \in I}$ is interminable because its pushout along any morphism $g: A \to C$ is the monomorphism $C \to C[X_i]_{i \in I}$.

(ii) Any integrally dependent extension of rings is interminable in **CRng**. Let $f: A \to B$ be such an extension and let $g: A \to K$ where K is a field. Let $p = \ker(g)$ and let $\bar{f}: A_p \to \bar{B}$ be the pushout of f along the localization morphism $l_p: A \to A_p$. Since l_p is flat, $\bar{f}$ is monomorphic. Thus $\bar{B}$ is not null. Let $s: \bar{B} \to S$ be a simple quotient and let $s\bar{f} = me$, where $e: A_p \to E$ is a regular epimorphism and $m: E \to S$ is a monomorphism. According to [3], Chapter 5, § 1, Proposition 5, $\bar{f}$ is integrally dependent, and according to [3], Chapter 5, § 1, Corollary 3, m is integrally dependent. It follows that E is a field ([3], Chapter 5, § 2, Lemma 2). It must be the residue field $K(p)$ of A_p so that $e = k_p$. But g factors through $k_p l_p$ in the form $g = hk_p l_p$. By the amalgamation property of field extensions, h does not terminate m. Therefore g does not terminate f. As a result f is interminable.

(iii) In **CRng**, a morphism $f: K \to A$ from a field K to a non-null ring A is interminable. Let m be a maximal ideal in A, let $q_m: A \to A/m = M$ be the quotient field, and let $h = q_m f$. Let $g: K \to C$ be a non-terminal morphism. Let n be a maximal ideal in C, let $q_n: C \to C/n = N$ be the quotient field, and let $k = q_n g$. Then (M, h) and (N, k) are field extensions of K. According to the amalgamation property for extensions of fields, there exist a field extension (L, l) of K and morphisms $\alpha: M \to L$, $\beta: N \to L$ such that $\alpha h = l$ and $\beta k = l$. If follows that g does not terminate f and that f is interminable.

Proposition (11.3.5). *A morphism $f: A \to B$ is not preterminal if and only if there exists a non-terminal singular epimorphism $d: A \to D$ such that the pushout of f along d is interminable.*

Proof. Let $t = \operatorname{term}(f)$. Let us assume that f is non-preterminal, i.e. $\operatorname{rad}(A) < t$. Since t is the join of codisjunctable radical congruences (Proposition (2.4.7)), there exists a codisjunctable radical congruence r on A such that $r \leqslant t$ and $\operatorname{rad}(A) < r$. Let $d: A \to D$ be the codisjunctor of r. Since r is not conjoint (Proposition (2.5.7)), d is not terminal. Let $\bar{f}$ be the pushout of f along d. Let g be a morphism terminating $\bar{f}$. Then gd terminates f. Thus it coequalizes t and also r. But gd also codisjoints r. Therefore gd is terminal and g is also. As a result, $\bar{f}$ is interminable. Conversely, let us assume the existence of a non-terminal singular epimorphism $d: A \to D$ such that the pushout $\bar{f}$ of f along d is interminable. Then d is not preterminal (Proposition (11.1.6)). According to Propositions (11.1.7) and (11.3.3), $\operatorname{term}(f) \geqslant \operatorname{term}(\bar{f}d) = \operatorname{term}(d) > \operatorname{rad}(A)$. Therefore f is not preterminal. ■

Definition (11.3.6). *A morphism is* locally interminable *if its local morphisms are interminable.*

Proposition (11.3.7). *Flat morphisms are locally interminable.*

Proof. Local morphisms of flat morphisms are flat (Proposition (3.6.5)) and thus are interminable by Proposition (11.3.4). ■

11.4 AMALGAMATION PROPERTIES

Notation. A monomorphism $f: A \to B$ is called an *extension* of A. It is called *simple* (*integral, regular*) if both its domain and codomain are simple (integral, regular, i.e. von Neumann). Therefore an extension of a simple (integral, regular) object need not be simple (integral, regular). Let $\mathfrak{M}$ be a class of monomorphisms in **A**, whose members are called $\mathfrak{M}$-extensions. We say that $\mathfrak{M}$-extensions satisfy the *amalgamation property* if, for any pair of $\mathfrak{M}$-extensions $f: A \to B$, $g: A \to C$, there exists a pair of $\mathfrak{M}$-extensions $u: B \to E$, $v: C \to E$ such that $uf = vg$.

Proposition (11.4.1). *For a Zariski category the following assertions are equivalent.*

(i) *The category is amalgamative.*
(ii) *Simple extensions satisfy the amalgamation property.*
(iii) *Simple extensions are interminable.*
(iv) *Extensions of simple objects are interminable.*
(v) *Extensions of regular objects are interminable.*
(vi) *Regular extensions are interminable.*
(vii) *Regular extensions satisfy the amalgamation property.*
(viii) *Integral extensions satisfy the amalgamation property.*
(ix) *Pushouts of pairs of extensions of integral objects are not terminal.*

Proof.

(i) ⇔ (ii) ⇔ (iii) ⇔ (iv) See § 6.1, p. 148, and Proposition (6.1.1).

(iv) ⇒ (v) Let $f: A \to B$ be an extension of a regular object and let $g: A \to K$ be a morphism with a simple codomain K. Let $p = \ker(g)$. Then p is maximal (Proposition (3.2.3)), A_p is simple (Proposition (3.6.1)), and the morphism g factors in the form $g = hl_p$. Since l_p is flat, the pushout $\bar{f}: A_p \to \bar{B}$ of f along l_p is monomorphic and therefore is interminable. It follows that the pushout of f along g, which is identical to the pushout of $\bar{f}$ along h, is not terminal. As a result f is interminable.

(v) ⇒ (vi) is obvious.

(vi) ⇒ (vii) Let $f: A \to B, g: A \to C$ be a pair of regular extensions of A. Since the object C is reduced, there exists a monomorphism $k = (k_i)_{i \in I}: C \to \Pi_{i \in I} K_i$ where K_i are simple. Let $(\bar{f}_i : K_i \to B_i, u_i : B \to B_i)$ be the pushout of $(k_i g, f)$. Since f is interminable, B_i is not terminal and thus it has a simple quotient $q_i : B_i \to L_i$. Let $v = (q_i u_i)_{i \in I}: B \to \Pi_{i \in I} L_i$ and $w = (q_i \bar{f}_i k_i)_{i \in I}: C \to \Pi_{i \in I} L_i$. Then w is a regular extension satisfying the relation $wg = vf$. Let $(\bar{f}: C \to \bar{B}, \bar{g}: B \to \bar{B})$ be the pushout of (g, f) in **RegA**. Since the morphism w factors through $\bar{f}$, $\bar{f}$ is a regular extension. Similarly, $\bar{g}$ is a regular extension. Therefore the amalgamation property holds for regular extensions.

(vii) ⇒ (viii) Let $f: A \to B, g: A \to C$ be a pair of integral extensions. Let $k_A : A \to K(A)$, $k_B : B \to K(B)$, and $k_C : C \to K(C)$ be the universal simple objects associated to A, B, and C respectively, and let $K(f) : K(A) \to K(B)$, $K(g) : K(A) \to K(C)$ be the canonical extensions of f, g. There exist a regular object E and monomorphisms $u : K(B) \to E, v : K(C) \to E$ such that $uK(f) = vK(g)$. The object E has a simple quotient $q : E \to K$ so that qu, qv are simple extensions, and quk_B, qvk_C are integral extensions such that $quk_B f = quK(f)k_A = qvK(g)k_A = qvk_C g$.

(viii) ⇒ (ix) Let $f: A \to B, g: A \to C$ be a pair of extensions of an integral object A. Let $k: A \to K$ be the universal simple object associated to A, let $(\bar{f}: K \to \bar{B}, h: B \to \bar{B})$ be the pushout of (k, f), and let $(\bar{g}: K \to \bar{C}, w: C \to \bar{C})$ be the pushout of (k, g). Since k is flat, $\bar{f}$ and $\bar{g}$ are monomorphic. Then $\bar{B}$ and $\bar{C}$ have simple quotients $m : \bar{B} \to M$ and $n : \bar{C} \to N$ respectively. Then $m\bar{f}$ and $n\bar{g}$ are simple extensions, and so there exist integral extensions $u : M \to E$ and $v : N \to E$ such that $um\bar{f} = vn\bar{g}$. Then $umhf = um\bar{f}k = vn\bar{g}k = vnwg$. Consequently the pushout of (f, g) is not terminal.

(ix) ⇒ (i) Let $f: K \to L, g: K \to M$ be a pair of simple extensions of K. If $(\bar{f}: M \to E, \bar{g}: L \to E)$ is the pushout of (g, f), then E is not terminal. Therefore it has a simple quotient $q : E \to N$ and $q\bar{g}, q\bar{f}$ are simple extensions such that $q\bar{g}f = q\bar{f}g$. ■

11.5 TERMINATORS

Let us be in an amalgamative Zariski category **A**.

Definition (11.5.1). *A* terminator *of a morphism $f: A \to B$ is a von Neumann morphism $t: A \to T$ which terminates f and is such that any von Neumann morphism $g: A \to C$ which terminates f factors uniquely through t.*

A terminator of a morphism is defined up to a unique isomorphism.

Examples.

(i) In **CRng**, let $f: A \to A/(a_1, \ldots, a_p)$ be the quotient of A by the ideal generated by $a_1, \ldots, a_p$. Let $n: A \to N$ be the universal von Neumann ring associated to A, let J be the ideal of N generated by the element $(1 - n(a_1)\overline{n(a_1)}) \ldots (1 - n(a_p)\overline{n(a_p)})$ where $\bar{x}$ denotes the semi-inverse of x, let $q: N \to N/J$ denotes the quotient of N by J, and $t = qn: A \to N/J$. The morphism t terminates f since the pushout of f along t must annihilate the elements $q(n(a_i))$ and thus must also annihilate 1. Let $g: A \to C$ be a von Neumann morphism terminating f. Then $(g(a_1), \ldots, g(a_p)) = C$. For any maximal ideal m of C, we have $i \in [1, p]$ such that $g(a_i) \notin m$. Then the relation $(1 - g(a_i)\overline{g(a_i)})g(a_i) = 0$ entails $1 - g(a_i)\overline{g(a_i)} \in m$. It follows that $(1 - g(a_1)\overline{g(a_1)}) \ldots (1 - g(a_p)\overline{g(a_p)})$ belongs to every maximal ideal of C and thus is null. The morphism g factors through n in the form $g = hn$. Then $h((1 - n(a_1)\overline{n(a_1)}) \ldots (1 - n(a_p)\overline{n(a_p)})) = (1 - g(a_1)\overline{g(a_1)}) \ldots (1 - g(a_p)\overline{g(a_p)}) = 0$. It follows that h factors through q and therefore that g factors through t in a unique way since t is epimorphic. As a result, t is the terminator of f in **CRng**.

(ii) In **CRng**, let $f: A \to A[a^{-1}]$ be the canonical morphism from an object A to its ring of fractions. Let $q: A \to A/(a)$ be the quotient of A by the ideal generated by a, let $n: A/(a) \to N$ be the universal von Neumann ring associated to $A/(a)$, and let $t = nq: A \to N$. The morphism t terminates f since the pushout of f along t both annihilates and inverts $n(q(a))$. Let $g: A \to C$ be a von Neumann morphism terminating t. Then $C[g(a)^{-1}] = \{0\}$, and therefore $g(a)$ must be nilpotent and thus null. If follows that g factors through q in the form $g = hq$ where $h: A/(a) \to C$ factors through n. It follows that g factors uniquely through t so that t is the terminator of f.

(iii) In **CRng**, for a ring A and a polynomial $P(X) = a_0 + \ldots + a_pX^p \in A[X]$, let $f: A \to A[X][P(X)^{-1}]$ be the canonical morphism. Let $q: A \to A/(a_0, \ldots, a_p)$ be the quotient of A by the ideal generated by $a_0, \ldots, a_p$, let $n: A/(a_0, \ldots, a_p) \to N$ be the universal von Neumann ring associated to $A/(a_0, \ldots, a_p)$, and let $t = nq$. The morphism t terminates f since the pushout of f along t both annihilates and inverts a polynomial. Let $g: A \to C$ be a von Neumann morphism terminating f. Let $H(X) = g(a_0) + \ldots + g(a_p)X^p \in C[X]$. Then $C[X][H(X)^{-1}] = \{0\}$, so that $H(X) = 0$ which entails $g(a_0) = \ldots = g(a_p) = 0$. It follows that g factors through q in the form $g = hq$ where h factors through n, so that g factors through t. As a result, t is the terminator of f.

(iv) In **CRng**, for a ring A and a polynomial $P(X) = a_0 + \ldots + a_pX^p \in A[X]$, let $f: A \to A[X]/(P(X))$ be the canonical morphism. Let $p: A \to A/(a_1, \ldots, a_p)[a_0^{-1}]$ be the canonical morphism, let $n: A \to A/(a_1, \ldots, a_p)[a_0^{-1}] \to N$ be the universal von Neumann ring, and let $t = np$. The morphism t terminates f since the pushout of f along t must annihilate an invertible polynomial. Let $g: A \to C$ be a von Neumann morphism terminating f. Let $Q(X) = g(a_0) + \ldots + g(a_p)X^p$. Then $C[X]/(Q(X)) = \{0\}$ which entails that $Q(X)$ is invertible. Since C is a regular, we must have $g(a_0)$ invertible and $g(a_1) = \ldots = g(a_p) = 0$. It follows that g factors through p and through t. As a result, t is the terminator of f.

(v) In **CRng**, the inclusion morphism $f: \mathbf{Z} \to \mathbf{Q}$ has no terminator. Let us assume that t could be a terminator for f. For any prime $p \in \mathbf{Z}$, the quotient $q: \mathbf{Z} \to \mathbf{Z}/(p)$ terminates f and thus factors through t. If follows that $\ker(t) \subset \ker(q) = (p)$ for any prime $p \in \mathbf{Z}$ and therefore that $\ker(t) = \{0\}$, i.e. that t is monomorphic. Since f is flat, the pushout of t along f is monomorphic and consequently t does not terminate f. Therefore we are led to a contradiction.

Proposition (11.5.2). *A morphism $f: A \to B$ is interminable if and only if it has as a terminator the terminal morphism $A \to 1$.*

Proof. The proof follows from the fact that any morphism terminating f is terminal (Proposition (11.3.2)). ∎

Proposition (11.5.3). *A morphism $t: A \to T$ is a terminator of a morphism $f: A \to B$ if and only if it is a von Neumann epimorphism terminating f such that any morphism $g: A \to K$ terminating f and having a simple codomain factors through t.*

Proof. Let t be the terminator of f. Let $N(t) = me$, where $e: N(A) \to E$ is a regular epimorphism and $m: E \to N(T) = T$ is a monomorphism. The object E is von Neumann, as a regular quotient of $N(A)$ (Proposition (2.11.5)). Thence the morphism m is interminable (Proposition (11.4.1)). Since the morphism $men_A = t$ terminates f, the morphism en_A does also. Since en_A is a von Neumann epimorphism, it is indeed a terminator of f, necessarily isomorphic to t. As a result t is epimorphic and satisfies the required property. Conversely, let us assume that t satisfies this property. Let $g: A \to C$ be a von Neumann morphism terminating f. The object C is a regular subobject of the product of simple objects, say $u = (q_i)_{i \in I}: C \to \Pi_{i \in I} K_i$ (Proposition (3.6.1)). For any $i \in I$, the morphism $q_i g: A \to K_i$ terminates f and thus factors through t in the form $q_i g = h_i t$. Then the morphism $h = (h_i)_{i \in I}: T \to \Pi_{i \in I} K_i$ is such that $ht = ug$. Since t is epimorphic, there exists a unique morphism $v: T \to C$ such that $vt = g$. As a result, t is the terminator of f. ∎

Theorem (11.5.4) (Freyd theorem). *A morphism $f: A \to B$ has a terminator if and only if it satisfies the following finite terminating solution set condition: there exists a finite family of morphisms $(g_i : A \to C_i)_{i \in [1,n]}$ terminating f such that any morphism $g : A \to K$ with a simple codomain terminating f factors through some morphism g_i.*

Proof. Let $C = \Pi_{i\in[1,n]} C_i$ and $g = (g_i)_{i \in I} : A \to C$. Since finite products are co-universal, the pushout $\bar{g} : B \to \bar{C}$ of g along f is such that $\bar{C}$ is the finite product of n terminal objects, and thus is terminal. Therefore g terminates f. Moreover, any morphism $h : A \to K$ with a simple codomain terminating f factors through some g_i, and thus through g. Let $N(g) = me$, where $e : N(A) \to T$ is a regular epimorphism and m is a monomorphism, be the regular factorization of the morphism $N(g) : N(A) \to N(C)$. Then T is a regular object (Proposition (2.11.5)) and the morphism m is interminable (Proposition (11.4.1)). The morphism $n_C g = men_A : A \to N(C)$ terminates f and therefore the morphism $t = en_A : A \to T$ does also. Since n_A and e are epimorphic, t is an epimorphism. Let $h : A \to K$ be a morphism with a simple codomain terminating f. Then h factors in the form $h = vg$ with $v : C \to K$. Since K is regular, v factors in the form $v = un_C$ with $u : N(C) \to K$. Then the relations $umt = umen_A = uN(g)n_A = un_C g = h$ imply that h factors through t. According to Proposition (11.5.3), t is the terminator of f. ■

Proposition (11.5.5). *The class of morphisms having a terminator*

(i) *is co-universal,*

(ii) *is closed under finite generalized pushouts,*

(iii) *is closed under finite products and contains the morphism $(f_i)_{i\in[1,n]}$: $A \to \Pi_{i\in[1,n]} B_i$ whenever it contains each f_i, and*

(iv) *contains the finitely generated regular epimorphisms and the singular epimorphisms.*

Proof.

(i) It is easy to check that whenever a morphism $f: A \to B$ has a terminator t, then the pushout of f along some morphism $g : A \to C$ has for terminator the von Neumann morphism associated to the pushout of t along g.

(ii) It is sufficient to prove the property for a family with two members. Let $f: A \to B$ and $g : A \to C$ be morphisms with terminators $t : A \to T$ and $s : A \to S$ respectively. Let $(v : B \to E, w : C \to E)$ be the pushout of (f, g) and let $h = vf = wg$. The morphisms s and t terminate h. Let $k : A \to K$ be a morphism with a simple codomain terminating h. Let $\bar{f} : K \to \bar{B}$ and $\bar{g} : K \to \bar{C}$ be the pushouts of f and g respectively along k. Then the pushout of $(\bar{f}, \bar{g})$ is terminal. According to the amalgamation property (Proposition (11.4.1)), one of the morphisms $\bar{f}$, $\bar{g}$ is not monomorphic and thus is terminal. It follows that k terminates f or g, and thus factors through t or s. According to Theorem (11.5.4), h has a terminator.

(iii) It is easy to check that if $f: A \to B$ and $g: C \to D$ have terminators t and s respectively, then the morphism $f \times g: A \times C \to B \times D$ has for terminator the morphism $t \times s$. It is also easy to see that if $f: A \to B$ and $g: A \to C$ have terminators s and t respectively, then the morphism $(f, g): A \to B \times C$ has for terminator the von Neumann morphism associated to the co-intersection of (s, t).

(iv) Let r be a codisjunctable congruence on A, let $q_r: A \to A/r$ be the quotient by r, and let $d: A \to D$ be the codisjunctor of r. Then d terminates q_r and vice versa. Any morphism $g: A \to K$ with a simple codomain coequalizes or codisjoints r, and thus factors through q_r or d. Thus if g terminates q_r, it factors through d, and if it terminates d, it factors through q_r. It follows that both q_r and d have terminators (Theorem (11.5.4)). Let r be a finitely generated congruence on A. According to Proposition (1.5.3), r is the join of finitely many codisjunctable congruences, say $r = \vee_{i \in [1,n]} r_i$. Then $q_r = \wedge_{i \in [1,n]} q_{ri}$ in $\mathrm{QuotReg}(A)$. According to the preceding discussion each q_{ri} has a terminator, and according to (ii) q_r has a terminator. ■

Proposition (11.5.6). *A terminator t of some morphism f has a terminator t' and is the terminator of t'.*

Proof. Let $f: A \to B$ be a morphism with a terminator $t: A \to T$. The morphism t factors in the form $t = N(t)n_A$ where $N(t): N(A) \to N(T) = T$. Let $N(t) = mq$, where $q: N(A) \to R$ is a regular epimorphism and $m: R \to T$ is a monomorphism, be the regular factorization of $N(t)$. Then R is a von Neumann object (Proposition (2.11.5)). According to the amalgamation property, the morphism m is interminable (Proposition (11.4.1)). Since the morphism $t = mqn_A$ terminates f, the von Neumann qn_A does also and therefore is a terminator of f. Since $N(A)$ is a von Neumann object, the regular epimorphism q is a filtered colimit $q = \xrightarrow{lim}_{i \in I} q_i$ of direct factors $q_i: N(A) \to R_i$. Then the pushout of f along qn_A is the filtered colimit of the pushouts of f along $q_i n_A$. It follows the existence of an $i \in \mathbf{I}$ such that $q_i n_A$ terminates f so that $q_i n_A$ is a terminator of f, necessarily isomorphic to t. It follows that t is of the form $t = qn_A$ where $q: N(A) \to T$ is a direct factor. Let $p: N(A) \to S$ be the direct factor such that $(q: N(A) \to T, p: N(A) \to S)$ is a product. Then pn_A terminates qn_A. Moreover, any morphism $h: A \to K$ with a simple codomain factors through n_A in the form $h = kn_A$, and the morphism $k: N(A) \to K$ factors through q or p. It follows that $t' = pn_A$ and $t = qn_A$ are terminators of each other. ■

Proposition (11.5.7). *If a morphism $f: A \to B$ has a terminator t, then* $\mathrm{term}(f) = \ker(t)$.

Proof. The proof follows readily from the definition of $\mathrm{term}(f)$. ■

11.6 TOPOLOGICAL INTERPRETATIONS

Proposition (11.6.1). *A morphism $g: A \to C$ terminates a morphism f: $A \to B$ if and only if the two sets* $\mathrm{ImSpec}(g)$ *and* $\mathrm{ImSpec}(f)$ *are disjoint in* $\mathrm{Spec}(A)$.

Proof. Let us assume that g terminates f and let $p \in \mathrm{ImSpec}(g) \cap \mathrm{ImSpec}(f)$. According to Proposition (3.3.5), the pushout $\bar{g}$ of g along the quotient $q_p: A \to A/p$ and the pushout $\bar{f}$ of f along q_p are monomorphisms. Then the pushout of $(\bar{f}, \bar{g})$ must be terminal, and this contradicts the amalgamation property (Proposition (11.4.1)). Consequently, $\mathrm{ImSpec}(g)$ and $\mathrm{Imspec}(f)$ are disjoint subsets of $\mathrm{Spec}(A)$. Let us assume that g does not terminate f and let $(u: B \to E, v: C \to E)$ be the pushout of (f, g). If p is a prime congruence on E, then $\mathrm{Spec}(f)((\mathrm{Spec}(u)(p)) = \mathrm{Spec}(g)((\mathrm{Spec}(v)(p)))$ so that $\mathrm{ImSpec}(g)$ and $\mathrm{ImSpec}(f)$ are not disjoint. ■

Proposition (11.6.2). *For any morphism $f: A \to B$,* $\mathrm{ImSpec}(f) = \{p \in \mathrm{Spec}(A) : k_p l_p$ does not terminate $f\}$.

Proof. $\mathrm{ImSpec}(k_p l_p) = \{p\}$, and thus according to Proposition (11.6.1) $p \notin \mathrm{ImSpec}(f) \Leftrightarrow \mathrm{ImSpec}(k_p l_p) \cap \mathrm{ImSpec}(f) = \varnothing \Leftrightarrow k_p l_p$ terminates f. ■

Proposition (11.6.3) *For any morphism $f: A \to B$ the open set* $D(\mathrm{term}(f))$ *is the interior of* $\mathrm{ImSpec}(f)$ *in* $\mathrm{Spec}(A)$.

Proof. If $p \in D(\mathrm{term}(f))$, the morphism $k_p l_p: A \to K(p)$ codisjoints $\mathrm{term}(f)$ (Proposition (3.1.1)). Thus it does not coequalize $\mathrm{term}(f)$ and therefore does not terminate f, so that $p \in \mathrm{ImSpec}(f)$ (Proposition (11.6.2)). Consequently, $D(\mathrm{term}(f)) \subset \mathrm{ImSpec}(f)$. Let $D(r)$ be the interior of $\mathrm{ImSpec}(f)$ in $\mathrm{Spec}(A)$. Let $g: A \to K$ be a morphism terminating f and having a simple codomain. If $p = \mathrm{Ker}(g)$, then g factors in the form $g = hk_p l_p$. The morphism $k_p l_p$ terminates f. Otherwise the pushout of f along $k_p l_p$ would be interminable (Proposition (11.4.1)) so that the pushout of f along g could not be terminal. According to Proposition (11.6.2), $p \notin \mathrm{ImSpec}(f)$ and thus $p \notin D(r)$, i.e. $r \leqslant p$. It follows that $r \leqslant \mathrm{term}(f)$ and $D(r) \subset D(\mathrm{term}(f))$. Finally, $D(\mathrm{term}(f))$ is the interior of $\mathrm{ImSpec}(f)$. ■

Proposition (11.6.4). *A morphism $f: A \to B$ is preterminal if and only the interior of* $\mathrm{ImSpec}(f)$ *in* $\mathrm{Spec}(A)$ *is empty.*

Proof. f is preterminal $\Leftrightarrow \mathrm{term}(f) = \mathrm{rad}(A) \Leftrightarrow D(\mathrm{term}(f)) = D(\mathrm{rad}(A))$ $\Leftrightarrow$ the interior of $\mathrm{ImSpec}(f)$ is $\varnothing$. ■

Proposition (11.6.5). *A morphism $f: A \to B$ is interminable if and only if the map* $\operatorname{Spec}(f)$ *is surjective.*

Proof. f is interminable $\Leftrightarrow$ $\operatorname{term}(f) = 1_{A \times A}$ $\Leftrightarrow$ $D(\operatorname{term}(f)) = \operatorname{Spec}(A)$ $\Leftrightarrow$ $\operatorname{ImSpec}(f) = \operatorname{Spec}(A)$ $\Leftrightarrow$ $\operatorname{Spec}(f)$ is surjective. ■

Proposition (11.6.6). *If $f: A \to B$ is a locally interminable morphism, the map* $\operatorname{Spec}(f) : \operatorname{Spec}(B) \to \operatorname{Spec}(A)$ *is generizing, i.e. has an image closed under generizations.*

Proof. Let $p \in \operatorname{Spec}(B)$ and $q = \operatorname{Spec}(f)(p)$. According to Proposition (3.3.9), $\operatorname{ImSpec}(l_p)$ is the set of generizations of p in $\operatorname{Spec}(B)$ and $\operatorname{ImSpec}(l_q)$ is the set of generizations of q in $\operatorname{Spec}(A)$. Since the local morphism $f_p: A_q \to B_p$ is interminable, the map $\operatorname{Spec}(f_p) : \operatorname{Spec}(B_p) \to \operatorname{Spec}(A_q)$ is surjective (proposition (11.6.5)). Then the relation $\operatorname{Spec}(f) \circ \operatorname{Spec}(l_p) = \operatorname{Spec}(l_q) \circ \operatorname{Spec}(f_p)$ implies $\operatorname{Spec}(f)(\operatorname{ImSpec}(l_p)) = \operatorname{ImSpec}(l_q)$ and the fact that $\operatorname{Spec}(f)$ is generizing. ■

Proposition (11.6.7). *If t is the terminator of $f: A \to B$, then* $\operatorname{ImSpec}(t)$ *is the complement of* $\operatorname{ImSpec}(f)$ *in* $\operatorname{Spec}(A)$.

Proof. Since t terminates f, $\operatorname{ImSpec}(t) \cap \operatorname{ImSpec}(f) = \varnothing$(Proposition (11.6.1)). Let $p \in \operatorname{Spec}(A)$. If $p \notin \operatorname{ImSpec}(f)$, then $k_p l_p$ terminates f (Proposition (11.6.2)) and thus factors through t, so that $p \in \operatorname{ImSpec}(t)$. ■

Proposition (11.6.8). *A morphism $f: A \to B$ has a terminator if and only if there exists a morphism $g: A \to C$ such that* $\operatorname{ImSpec}(g)$ *is the complement of* $\operatorname{ImSpec}(f)$ *in* $\operatorname{Spec}(A)$.

Proof. The condition is necessary by Proposition (11.6.7). Let us assume that it holds. Let $N(g) = me$ where $e: N(A) \to T$ is a regular epimorphism and $m: T \to N(C)$ is a monomorphism. Then T is regular (Proposition (2.11.5)). Let $t = en_A : A \to T$. Then t is a von Neumann epimorphism. The map $\operatorname{Spec}(n_C)$ is bijective (Theorem (3.2.5)) and the map $\operatorname{Spec}(m)$ is surjective (Proposition (3.2.3) and (3.3.13)). Therefore $\operatorname{ImSpec}(g) = \operatorname{ImSpec}(t)$, so that $\operatorname{ImSpec}(t)$ is the complement of $\operatorname{ImSpec}(f)$ in $\operatorname{Spec}(A)$. By Proposition (11.6.1), t terminates f. Furthermore, if $h: A \to K$ is a morphism with a simple codomain terminating f, then $\operatorname{ImSpec}(h) \cap \operatorname{ImSpec}(f) = \varnothing$. Thus $\operatorname{ImSpec}(h) \subset \operatorname{ImSpec}(t)$ and hence h does not terminate t. The morphism h factors in the form $h = \bar{h} n_A$ where $\bar{h}$ does not terminate e. Then the pushout of e along $\bar{h}$ is an isomorphism so that $\bar{h}$ factors through e, and hence h factors through t. As a result, t is the terminator of f. ■

Patches in *Spec(A)*

Let us recall that, following Hochster [25], the patch space of a spectral space X is the topological space $T(X)$ whose elements are those of X and whose topology has the compact open sets of X and their complements as an open sub-basis. It is a boolean space, i.e. a compact Hausdorff space in which the closed open sets form an open basis. In fact, it is the co-universal boolean space associated to it [25]. A closed set in $T(X)$ is called a patch in X([25], § 2). The patch spectrum $\mathrm{Spec}_{Pa}(A)$ of an object A is the *patch space* of $\mathrm{Spec}(A)$ (cf. [25]).

Proposition (11.6.9). *The patches in* $\mathrm{Spec}(A)$ *are precisely the subsets of* $\mathrm{Spec}(A)$ *of the form* $\mathrm{ImSpec}(f)$ *for some morphism* $f: A \to B$.

Proof. Let X be a patch in $\mathrm{Spec}(A)$. According to Theorem (3.2.5), X is the image by the map $\mathrm{Spec}(n_A)$ of a closed set Y in $\mathrm{Spec}(N(A))$. According to Proposition (3.3.11), $Y = \mathrm{ImSpec}(q)$ where $q: N(A) \to Q$ is a regular epimorphism. It follows that $X = \mathrm{ImSpec}(qn_A)$. Conversely, for any morphism $f: A \to B$, the map $\mathrm{Spec}(f): \mathrm{Spec}(B) \to \mathrm{Spec}(A)$ is spectral (cf. § 3.1, p. 86), then $\mathrm{ImSpec}(f)$ is a patch in $\mathrm{Spec}(A)$ ([25], § 2).

Constructible sets in *Spec(A)*

Let X be a spectral space. According to [22], Chapter 0, Definitions 2.3.2 and 2.3.10 and Proposition 2.3.11, a *constructible subset* of X is a subset belonging to the smallest family $\mathfrak{F}$ of subsets of X such that

(i) every compact open set is in $\mathfrak{F}$,
(ii) the complement of a element of $\mathfrak{F}$ is in $\mathfrak{F}$, and
(iii) a finite intersection of elements of $\mathfrak{F}$ is in $\mathfrak{F}$.

Proposition (11.6.10). *The constructible subsets in* $\mathrm{Spec}(A)$ *are precisely the clopen subsets in* $\mathrm{Spec}_{Pa}(A)$.

Proof. According to the definition of $\mathrm{Spec}_{Pa}(A)$, a constructible subset of $\mathrm{Spec}(A)$ is an open set in $\mathrm{Spec}_{Pa}(A)$, and therefore is a clopen set in $\mathrm{Spec}_{Pa}(A)$ since its complement is also an open set in $\mathrm{Spec}_{Pa}(A)$. Conversely, a clopen set in $\mathrm{Spec}_{Pa}(A)$ is compact and thus is a finite union of subsets of the form $U \cap (\mathrm{Spec}(A) \setminus V)$, where U and V are compact open sets of $\mathrm{Spec}(A)$, and therefore is constructible. ■

Theorem (11.6.11). *A morphism* $f: A \to B$ *has a terminator if and only if* $\mathrm{ImSpec}(f)$ *is a constructible subset in* $\mathrm{Spec}(A)$.

Proof. The morphism f has a terminator if and only there exists a morphism $g: A \to C$ such that $\mathrm{ImSpec}(g)$ is the complement of $\mathrm{ImSpec}(f)$ in $\mathrm{Spec}(A)$

(Proposition (11.6.8)). By Proposition (11.6.9), this is equivalent to the fact that the complement of the patch ImSpec(f) in Spec(A) is a patch. By the definition of patches, this is equivalent to the fact that ImSpec(f) is a clopen set in $\text{Spec}_{Pa}(A)$, and by proposition 11.6.10, that ImSpec(f) is a constructible subset in Spec(A). ∎

Theorem (11.6.12). *For a set X of* Spec(A), *the following assertions are equivalent.*

(i) *X is a constructible subset of* Spec(A).

(ii) $X =$ ImSpec(f) *for some morphism $f: A \to B$ having a terminator.*

(iii) $X =$ ImSpec(f) *for some finitely presentable morphism $f: A \to B$ having a terminator.*

(iv) $X =$ ImSpec(t) *for some terminator $t: A \to T$ of some morphism $f: A \to B$.*

Proof.

(i) ⇒ (ii) According to Proposition (11.6.9) there is a morphism $f: A \to B$ such that $X =$ ImSpec(f). According to Theorem (11.6.11), f has a terminator.

(ii) ⇒ (iii) Let $t: A \to T$ be the terminator of f. The morphism f is a filtered colimit $f = \xrightarrow{lim}_{i \in I} f_i$ where $f_i: A \to B_i$ are finitely presentable morphisms. Let $\bar{f}: T \to \bar{B}$ be the pushout of f along t, and let $\bar{f}_i: T \to \bar{B}_i$ be the pushout of f_i along t. Then $\bar{f} = \xrightarrow{lim}_{i \in \mathbf{I}} \bar{f}_i$. But $\bar{f}$ is terminal, and thus there exists $i \in \mathbf{I}$ such that $\bar{f}_i$ is terminal, i.e. such that t terminates f_i. It follows readily that t is the terminator of f_i. Then $X =$ ImSpec$(f) =$ Spec$(A)\backslash$ImSpec$(t) =$ ImSpec(f_i) (Proposition (11.6.7)).

(iii) ⇒ (iv) If t is the terminator of f, then t has a terminator t' (Proposition (11.5.6)). Then $X =$ ImSpec$(f) =$ Spec$(A)\backslash$ImSpec$(t) =$ ImSpec(t').

(iv) ⇒ (i) Since t has a terminator (Proposition (11.5.6)), $X =$ ImSpec(t) is constructible (Theorem (11.6.11)). ∎

11.7 CATEGORIES WITH FINITELY PRESENTABLE TERMINATORS

Let us recall that we are working in an amalgamative Zariski category $\mathbf{A}$ and that a morphism $f: A \to B$ in $\mathbf{A}$ is finitely presentable if it is a finitely presentable object in the category $A/\mathbf{A}$.

Definition (11.7.1) *The category* $\mathbf{A}$ *is said* to have finitely presentable terminators *if any morphism $f: A \to B$ with a finitely presentable domain and codomain has a terminator.*

Proposition (11.7.2) *The following assertions are equivalent.*

(i) $\mathbf{A}$ *has finitely presentable terminators.*

(ii) *Any finitely presentable morphism in* $\mathbf{A}$ *has a terminator.*

Proof.

(i) ⇒ (ii) According to the proof of Proposition (6.1.2), any finitely presentable morphism is the pushout along some morphism of a morphism with a finitely presentable domain and codomain, and therefore it has a terminator (Proposition (11.5.5)).

(ii) ⇒ (i) Any morphism with a finitely presentable domain and codomain is finitely presentable. ■

Theorem (11.7.3). *If* **A** *is locally noetherian, the following assertions are equivalent.*

(i) *A has finitely presentable terminators.*

(ii) *Finitely presentable integral extensions have terminators in A.*

(iii) *Finitely presentable integral extensions are not preterminal in A.*

Proof.

(i) ⇒ (ii) follows from Proposition (11.7.2).

(ii) ⇒ (iii) Let $f: A \to B$ be a finitely presentable integral extension with a terminator $t: A \to T$. Then $\mathrm{term}(f) = \ker(t)$. According to the amalgamation property, no monomorphism can terminate f (Proposition (11.4.1)) and thus t is not a monomorphism, i.e. $\ker(t) \neq \Delta_A$. It follows that $\mathrm{term}(f) \neq \mathrm{rad}(A)$, i.e. f is not preterminal.

(iii) ⇒ (i) Let $f: A \to B$ be a morphism with a finitely presentable domain and codomain. Then the object A is noetherian, i.e. the ordered set $\mathrm{QuotReg}(A)$ satisfies the descending-chain condition. We are going to prove by noetherian induction that any $q \in \mathrm{QuotReg}(A)$ is such that the pushout of f along q has a terminator.

First let us assume that, for any proper regular quotient $q: A \to Q$ of A, the pushout of f along q has a terminator, and let us prove that f has a terminator.

(a) Let us assume that A is not reduced. Then the universal reduced object $q_{\mathrm{rad}(A)}: A \to A/\mathrm{rad}(A)$ is a proper regular quotient of A. Thus the pushout $\bar{f}$ of f along $q_{\mathrm{rad}(A)}$ has a terminator t and it can readily be seen that the morphism $tq_{\mathrm{rad}(A)}$ is a terminator of f.

(b) Let us assume that A is not irreductible. There are proper regular quotients $q_1: A \to Q_1, q_2: A \to Q_2$ of A such that $q_1 \vee q_2 = 1_A$ in $\mathrm{QuotReg}(A)$. Then the pushout of f along q_1 (or q_2) has a terminator t_1 (or t_2). The morphisms $t_1 q_1$ and $t_2 q_2$ terminate f. Let us prove that this pair of morphisms provide a terminating solution for f (cf. Theorem (11.5.4)). Let $g: A \to K$ be a morphism with a simple codomain terminating f. Let $g = mp$, where $p: A \to E$ is a regular epimorphism and $m: E \to K$ is a monomorphism. Then E is integral so that p is the quotient of A by a prime congruence. According to Proposition (2.2.13), the relation $p \leqslant q_1 \vee q_2$ entails ($p \leqslant q_1$ or $p \leqslant q_2$). Thus p and g factor through q_1 or q_2. It follows that g factors through $t_1 q_1$ or $t_2 q_2$. According to Theorem (11.5.4), f has a terminator.

(c) Let us assume that A is reduced and irreducible, and that f is not a monomorphism. Then A is integral (Proposition (2.10.3)) and f is not a premonomorphism (Proposition (2.5.7)) so that there exists a non-terminal singular epimorphism $d: A \to D$ terminating f. If d is the codisjunctor of $r \in \mathrm{Cong}(A)$, then r is not unit, so that the quotient $q_r: A \to A/r$ is proper. It follows that the pushout of f along q_r has a terminator t. Then tq_r terminates f. Let us prove that the pair of morphisms (d, tq_r) provides a terminating solution set for f. Let $g: A \to K$ be a morphism with a simple codomain terminating f. Then g coequalizes r or codisjoints r. In the first case, g factors through q_r and thus through tq_r. In the second case, g factors through d. According to Theorem (11.5.4), f has a terminator.

(d) Let us assume that A is integral and f is monomorphic. Let $k: A \to K$ be the universal simple object associated to A and let $(\bar{k}: B \to \bar{B}, \bar{f}: K \to \bar{B})$ be the pushout of (f, k). Since k is flat, $\bar{f}$ is monomorphic and thus $\bar{B}$ is not terminal. Let $s: \bar{B} \to S$ be a simple quotient of $\bar{B}$, and $s\bar{k} = me$ where $e: B \to E$ is a regular epimorphism and $m: E \to S$ is a monomorphism. Then E is integral. Since the category is locally noetherian, E is a finitely presentable object, and consequently $h = ef: A \to E$ is a finitely presentable integral extension. According to the hypothesis, h is not preterminal and thus neither is f (Proposition (11.2.4)), i.e. $\mathrm{term}(f) \neq \Delta_A$. Let $q: A \to A/\mathrm{term}(f)$ be the quotient of A by $\mathrm{term}(f)$. Then q is proper, so that the pushout of f along q has a terminator t. The morphism tq is a terminator of f because it terminates f, and any morphism $g: A \to L$ with a simple codomain terminating f coequalizes $\mathrm{term}(f)$ and thus factors through q and therefore through tq.

Now let us consider a regular quotient $q: A \to Q$ of A and let $\bar{f}: Q \to \bar{B}$ be the pushout of f along q. Since A, B are finitely presentable and the category is locally noetherian, the objects Q and $\bar{B}$ are finitely presentable. Let us assume that any regular quotient $q_1: A \to Q_1$ of A such that $q_1 < q$ in $\mathrm{QuotReg}(A)$ is such that the pushout of f along q_1 has a terminator. According to the previous result, the morphism $\bar{f}$ has a terminator. By noetherian induction, it follows that f has a terminator. ■

Example (11.7.4). *The category* **CRng** *has finitely presentable terminators.*

Proof. Let $f: A \to B$ be a finitely presentable monomorphism in **CRng** with integral domain and codomain. We are going to prove that f is not preterminal. The ring B is an A-algebra generated by finitely many elements, say $x_1, \ldots, x_n$. Let $S = A - \{0\}$ and $K = A[S^{-1}]$ be the field of fractions of A with the canonical morphism $g: A \to K$. Let $(\bar{g}: B \to B[S^{-1}], \bar{f}: K \to B[S^{-1}])$ be the pushout of (f, g). Since g is flat, $\bar{f}$ is monomorphic. Moreover, $\bar{f}$ is a finitely presentable morphism, i.e. $B[S^{-1}]$ is a finitely presentable K-algebra. According to the Noether normalization lemma ([3] Chapter V, § 3, Theorem 1) there are algebraically independent elements $y_1, \ldots,$

$y_p \in B[S^{-1}]$ such that $B[S^{-1}]$ is integrally dependent over $K[y_1, \ldots, y_p]$. Indeed, there exists a finite subset S_0 in S such that $y_1, \ldots, y_p \in B[S_0^{-1}]$ and each element $\bar{g}(x_1), \ldots, \bar{g}(x_n)$ is integrally dependent over $A[S_0^{-1}][y_1, \ldots, y_p]$ (cf. [3], Chapter V, § 3, 1). Let $s = \Pi_{t \in S_0} t$. Then $s \in S, A[s^{-1}] = A[S_0^{-1}]$, and $B[s^{-1}] = B[S_0^{-1}]$. Let $g_0 : A \to A[s^{-1}], \overline{g_0} : B \to B[s^{-1}]$ be the canonical morphisms, let $f_0 : A[s^{-1}] \to B[s^{-1}]$ be the canonical extension of f, let $m : A[s^{-1}][y_1, \ldots, y_p] \to B[s^{-1}]$ be the inclusion morphism, and let $h : A[s^{-1}] \to A[s^{-1}][y_1, \ldots, y_p]$ be the morphism induced by f_0. Since $y_1, \ldots, y_p \in B[s^{-1}]$ are algebraically independent over $A[s^{-1}]$, the morphism h is interminable (Example (i) on p. 238). Since $\bar{g}(x_1), \ldots, \bar{g}(x_n)$ generates the $A[s^{-1}]$-algebra $B[s^{-1}], B[s^{-1}]$ is integrally dependent over $A[s^{-1}][y_1, \ldots, y_p]$. Thus the morphism m is interminable (Example (ii) on p. 238). Then the morphism $f_0 = mh$ is interminable. According to Propositions (11.1.6), (11.1.7), and (11.3.3), $\mathrm{term}(f) \geqslant \mathrm{term}(\overline{g_0}f) = \mathrm{term}(f_0 g_0) = \mathit{term}(g_0) = (s) \neq (0)$. Thus $\mathrm{term}(f) \neq (0)$, i.e. f is not preterminal. ■

Example (11.7.5). *The category **AlgCAlg**(k) of algebraic algebras over a field k has finitely presentable terminators.*

Proof. Let $f : A \to B$ be a morphism in **AlgCAlg**(k) with a finitely presentable domain and codomain. Then A, B are finitely presentable objects in the category **CAlg**(k) ([10], Proposition 1.12.1. and Example 1.12.3). Therefore $f : A \to B$ is a finitely presentable morphism in **CAlg**(k). According to Example (11.7.4), f has a terminator $t : A \to T$ in **CAlg**(k). If $L(T)$ is the algebraic closure of k in T, then $L(T)$ is a von Neumann object and the inclusion morphism $L(T) \to T$ is interminable in **CAlg**(k) (Proposition (11.4.1)), so that the morphism $t' : A \to L(T)$ induced by t terminates f, is isomorphic to t, and is the terminator of f in **AlgCAlg**(k). ■

Example (11.7.6). *The category **Mod** has finitely presentable terminators.*

Proof. If $(\varphi, f) : (A, E) \to (B, F)$ is a finitely presentable morphism in **Mod**, then $\varphi : A \to B$ is a finitely presentable morphism in **CRng** so that it has a terminator $t : A \to T$ in **CRng**, and the morphism $(t, 0) : (A, E) \to (T, \{0\})$ is a terminator of the morphism (φ, f) in **Mod**. Similarly, the categories **Mod(Red)** of modules on variable reduced commutative rings, **CAlg** of commutative algebras on variable commutative rings, and **CAlg(Red)** of commutative algebras on variable commutative reduced rings have finitely presentable terminators. ■

Example (11.7.7). *Any locally noetherian locally simple category [10] satisfying the amalgamation property has finitely presentable terminators.*

Proof. Since any object is von Neumann, monomorphisms are interminable (Proposition (11.4.1)). According to Theorem (11.7.3), the category has finitely presentable terminators.

For example, the categories **RegAlgCAlg**(k) and **AlgCAlgSep**(k), and more generally, any locally pre-Galois category ([10], § 2.16.2), have finitely presentable terminators.

Proposition (11.7.8). *If the category A has finitely presentable terminators then for any object A in A, the category A/A has finitely presentable terminators.*

Proof. The projection functor $P: A/\mathbf{A} \to \mathbf{A}$ preserves and reflects limits, pushouts, and coequalizers, and thus terminating pairs of morphisms. It follows that P preserves and reflects von Neumann objects, von Neumann morphisms, and terminators. Moreover, for any object $(B, b) \in A/\mathbf{A}$, the category $(B, b)/(A/\mathbf{A})$ is canonically equivalent to the category $B/\mathbf{A}$. Then for any morphism f: $(B, b) \to (C, c)$ in $A/\mathbf{A}$, the following assertions are equivalent: f: $(B, b) \to (C, c)$ is a finitely presentable morphism in $A/\mathbf{A}$; $((C, c), f)$ is a finitely presentable object in $(B, b)/(A/\mathbf{A})$; (C, f) is a finitely presentable object in $B/\mathbf{A}$; $f: B \to C$ is a finitely presentable morphism in $\mathbf{A}$. Hence the functor P preserves and reflects finitely presentable morphisms. Consequently, the category $A/\mathbf{A}$ has finitely presentable terminators.

■

Example. Since **CRng** has finitely presentable terminators, it follows that, for any $A \in$ **CRng**, the category **CAlg**$(A) \simeq A/$**CRng** has finitely presentable terminators.

Proposition (11.7.9). *If the category A is locally noetherian and has finitely presentable terminators, the category* **RedA** *of reduced objects in A also has finitely presentable terminators.*

Proof. Let $R: A \to$ **RedA** be the reflector. Let A be a finitely presentable object in **RedA**. Then in **A**, A is a filtered colimit of finitely presentable objects, say $A = \xrightarrow{lim}_{i \in \mathbf{I}} A_i$. Since R preserves colimits, $A \simeq R(A) = \xrightarrow{lim}_{i \in \mathbf{I}} R(A_i)$. Then there exists $i \in \mathbf{I}$ such that the unit morphism $1_A : A \to A$ factors through the induction $R(A_i) \to A$. Therefore A is a split quotient of RA_i. Since the category **A** is locally noetherian, the object RA_i is finitely presentable as a regular quotient of A_i, and the object A is finitely presentable as a regular quotient of RA_i. As a result, any finitely presentable object in **RedA** is finitely presentable in **A**. It follows that if $f: A \to B$ is a morphism in **RedA** with a finitely presentable domain and codomain, then f has a terminator t in **A**, which is easily seen to be a terminator of f in **RedA** also.

■

Examples. Since the category **CRng** has finitely presentable terminators, the category **RedCRng** has finitely presentable terminators and, according to Proposition (11.7.8), the category **RedCAlg**(A) has finitely presentable terminators for any reduced ring A. Since **AlgCAlg**(k) has finitely presentable terminators **RedAlgCAlg**(k) has finitely presentable terminators.

Proposition (11.7.10). *If the category A is locally noetherian and has finitely presentable terminators, then, for a simple object K, the following assertions are equivalent.*

(i) *K is algebraically closed.*

(ii) *K is injective with respect to the class of interminable monomorphisms with a finitely presentable domain and codomain.*

(iii) *K is injective with respect to the class of interminable integral extensions with a finitely presentable domain and codomain.*

Proof.

(i) $\Rightarrow$ (ii) follows from Proposition (6.1.3) since morphisms with a finitely presentable domain and codomain are finitely presentable.

(ii) $\Rightarrow$ (iii) is immediate.

(iii) $\Rightarrow$ (i) Since the category is locally noetherian, finitely generated objects are finitely presentable. Therefore let us consider a monomorphism $m: K \to M$, a monomorphism $f: A \to B$ with a finitely presentable domain and codomain, and morphisms $g: A \to K$, $n: B \to M$ such that $mg = nf$. Since M is not terminal, it has a simple quotient $s: M \to S$. Let $g = hq$ and $sn = kp$ be the regular factorizations of the morphisms g and sn respectively. There exists a unique morphism $e: C \to E$ such that $ke = smh$ and $eq = pf$. Since sm and h are monomorphic, e is monomorphic. Since A and B are finitely presentable objects, C and E are finitely presentable objects. Therefore e is a finitely presentable integral extension. According to Theorem (11.7.3), e is not preterminal. According to Proposition (11.3.5), there exists a non-terminal singular epimorphism $d: C \to D$ such that, in the pushout $(r: D \to \Delta, \delta: E \to \Delta)$ of (d, e), the morphism r is interminable. Since d is flat, r is monomorphic. According to Corollary (1.8.6), D and Δ are finitely presentable objects. According to Proposition (2.2.9), D and Δ are integral. Therefore r is an interminable integral extension with a finitely presentable domain and codomain. Let $(\bar{h}: D \to \bar{D}, \bar{d}: K \to \bar{D})$ be the pushout of (d, h). Then $\bar{h}$ is monomorphic, so that $\bar{d}$ is not terminal and hence is an isomorphism. Consequently, the morphism h factors through d in the form $h = dt$ with $t: D \to K$. According to the hypothesis, there exists a morphism $v: \Delta \to K$ such that $vr = t$. Then the morphism $v\delta p: B \to K$ is such that $v\delta pf = v\delta eq = vrdq = tdq = hq = g$. As a result, the object K is algebraically closed.

■

11.8 CHEVALLEY'S THEOREM IN ZARISKI CATEGORIES

We are still working in an amalgamative Zariski category **A**.

Constructible subsets of a scheme on A

Following Hochster ([25], § 16), a *locally spectral space* is a topological space, each point of which has a spectral open neighbourhood. Let X be such a space. Following [22], Chapter 0, Definition 2.3.10, a *constructible subset of* X is a subset Y of X, each point of which has a compact open neighbourhood U in X such that $U \cap Y$ is a constructible subset in the spectral space U. These notions apply to schemes on **A** (Definition (4.3.4)). If X is a spectral space, the two notions of constructible sets of X coincide ([22], Chapter 0, Corollary 2.3.9).

Finitely presentable morphisms of schemes

A morphism of schemes $f: X \to Y$ on **A** is *locally finitely presentable* if, for each point $x \in X$, there exist an affine open neighbourhood U of x and an affine open neighbourhood V of $f(x)$ such that $f(U) \subset V$ and the morphism $\varrho_U^{f^{-1}(V)} f_V^{\#}: O_Y(V) \to O_X(U)$ is finitely presentable in **A** (cf. [22], Chapter 1, Definition 6.2.1). The morphism $f: X \to Y$ is *compact* if, for any compact open set V of $Y, f^{-1}(V)$ is compact in X (cf. [22], Chapter 1, Definition 6.1.1). The morphism $f: X \to Y$ is *quasi-separated* if its diagonal $X \to \ker(f)$ is compact (cf. [22], Chapter 1, Definition 6.1.3). The morphism $f: X \to Y$ is *finitely presentable* if it is locally finitely presentable, compact, and quasi-separated (cf. [22], Chapter 1, Definition 6.3.7). It can easily be proved that the notion of locally finitely presentable (compact, quasi-separated, finitely presentable) morphism is universal, i.e. stable under pullback along any morphism. Also, if $f: X \to Y$ is a quasi-separated morphism of schemes and Y is a separated scheme, then X is a separated scheme.

Theorem (11.8.1). *If the category A has finitely presentable terminators and $f: X \to Y$ is a compact locally finitely presentable morphism of schemes on A, then* $\operatorname{Im}(f)$ *is a constructible subset of* Y.

Proof. Let V be an affine open set in Y and let $U = f^{-1}(V)$. Then U is a compact open set in X such that $f(U) = \operatorname{Im}(f) \cap V$. Let $x \in U$ and $y = f(x)$. There exists an affine open neighbourhood U_x (or V_y) of x (or y) such that $f(U_x) \subset V_x$ and the morphism $\varrho_{Ux}^{f^{-1}(V_y)} f_{V_y}^{\#}: O_Y(V_y) \to O_X(U_x)$ is finitely presentable in A. Let U'_x be an affine open neighbourhood of x included in $U_x \cap U$. According to Proposition (4.4.11), the morphism $\varrho_{U'_x}^{U_x}: O_X(U_x) \to O_X(U'_x)$ is a singular epimorphism, and thus is finitely presentable (Proposition (1.8.5)). Then the morphism $\varphi = \varrho_{U'_x}^{f^{-1}(V_y)} f_{V_y}^{\#}: O_Y(V_y) \to O_X(U'_x)$ is finitely presentable. According to the data φ has a terminator, and according to Theorem (11.6.12), ImSpec(φ) is a constructible subset of Spec$(O_Y(V_y))$. Up to a canonical isomorphism, the image of Spec(φ) is the image of the morphism $U'_x \to V_y$ induced by f. If follows that $f(U'_x)$ is constructible in V_y and thus in V also ([22], Chapter 0, Proposition 2.3.8). The family $(U'_x)_{x\in U}$ make up an open covering of U from which a finite covering $(U_i)_{i\in I}$ can be extracted. Then $\mathrm{Im}(f) \cap V = f(U) = \bigcup_{i\in I} f(U_i)$ is a constructible set in V. Consequently, $\mathrm{Im}(f)$ is a constructible set in Y. ■

Theorem (11.8.2). *If the category A has finitely presentable terminators, then for a subset Y of a compact separated scheme X* on A, *the following assertions are equivalent.*

(i) *Y is a constructible subset of X.*

(ii) $Y = \mathrm{Im}(f)$ *for some finitely presentable morphism of schemes* $f: \mathrm{Spec}(A) \to X$ *with an affine domain.*

(iii) $Y = \mathrm{Im}(f)$ *for some finitely presentable morphism of schemes* $f: E \to X$.

Proof.

(i) ⇒ (ii) The space X is a finite union of affine open sets, say $X = \bigcup_{i\in \mathbf{I}} X_i$. For each $i \in \mathbf{I}$, the open immersion $g_i : X_i \to X$ is obviously locally finitely presentable and quasi-separated. It is also compact since X is separated. Therefore, each g_i is finitely presentable. The subset $Y_i = X_i \cap Y$ is constructible in X_i. According to Theorem (11.6.12), there exists a finitely presentable morphism of affine schemes $f_i : \mathrm{Spec}(A_i) \to X_i$ whose image $\mathrm{Im}(f_i)$ is precisely Y_i. Then the scheme $\coprod_{i\in\mathbf{I}} \mathrm{Spec}(A_i) \simeq \mathrm{Spec}(\prod_{i\in\mathbf{I}} A_i)$ is affine and the morphism $\langle g_i f_i \rangle_{i\in\mathbf{I}} : \coprod_{i\in\mathbf{I}} \mathrm{Spec}(A_i) \to X$ is finitely presentable and has $\bigcup_{i\in\mathbf{I}} Y_i = Y$ as its image.

(ii) ⇒ (iii) is immediate.

(iii) ⇒ (i) follows from Theorem (11.8.1). ■

Theorem (11.8.3) (Chevalley theorem). *If the category A has finitely presentable terminators, then finitely presentable morphisms of schemes on A preserve constructible subsets.*

Proof. Let $f: X \to Y$ be a finitely presentable morphism of schemes on $\mathbf{A}$ and let E be a constructible subset of X. Let V be an affine open set in Y and let $U = f^{-1}(V)$. Then U and V are compact separated schemes. The morphism of schemes $g: U \to V$ induced by f is finitely presentable and $E \cap U$ is a constructible subset of U. According to Theorem (11.8.2), there is a finitely presentable morphism of schemes $h: W \to U$ such that $h(W) = E \cap U$. Then $gh: W \to V$ is a finitely presentable morphism of schemes whose image is $g(h(W)) = f(E \cap U) = f(E) \cap V$. According to Theorem (11.8.2), $f(E) \cap V$ is a constructible subset of V. It follows that $f(E)$ is a constructible subset of Y. ■

12

SOME CONSTRUCTIONS OF ZARISKI CATEGORIES

Some general constructions of categories can be used to build up Zariski categories starting with a Zariski category **A**. The coslice construction is extensively used: for any fixed object A in **A**, the coslice category $A/\mathbf{A}$ of objects of **A** under A is a Zariski category. The functor category construction can also be used: if **C** is a small category with a strict initial object, the functor category $\mathbf{A}^{\mathbf{C}}$ is a Zariski category. In particular, the category $\mathbf{A}^{\rightarrow}$ of morphisms of **A** is a Zariski category. Finite products of Zariski categories are Zariski categories. Some categories of sheaves are Zariski categories: if X is a boolean topological space the category $Sh[X, \mathbf{A}]$ of sheaves on X with values in **A** is a Zariski category. Some full subcategories of **A** are Zariski categories, as the subcategories of reduced, regular, neatish, or etalish objects of **A**.

12.1 COLISCE ZARISKI CATEGORIES

Theorem (12.1.1). *If A is a Zariski category and A is an object in A, the coslice category A/A of objects of A under A is a Zariski category and the projection functor A/A → A is a morphism of Zariski categories.*

Proof. It is well known that the category $A/\mathbf{A}$ is complete and cocomplete, and that the projection functor $U: A/\mathbf{A} \to \mathbf{A}$ creates limits and non-empty connected colimits, in particular coequalizers, pushouts, and filtered colimits. It follows that the functor U preserves and reflects codisjointed pairs of morphisms, and therefore that it preserves and reflects codisjunctors. Moreover, U has as a left-adjoint the functor $F: \mathbf{A} \to A/\mathbf{A}$ defined by $FB = (A \amalg B, i_B)$, where $i_B: A \to A \amalg B$ is the canonical induction, and by $Fg = 1_A \amalg g$. The functor F preserves codisjunctable objects, for if B is a codisjunctable object in **A** and $(\bar{g}, \bar{h}): FB \rightrightarrows (C, c)$ is a pair of morphisms in $A/\mathbf{A}$, then $\bar{g}$ and $\bar{h}$ are of the form $\bar{g} = \langle c, g\rangle$ and $\bar{h} = \langle c, h\rangle$, where $(g, h): B \rightrightarrows C$ and the codisjunctor $d: C \to D$ of (g, h) in **A** is such that the morphism $d: (C, c) \to (D, dc)$ is the codisjunctor of $(\bar{g}, \bar{h})$ in $A/\mathbf{A}$. It follows that, if $\mathcal{G}$ is a proper generating set in **A** whose objects are finitely presentable and flatly codisjunctable, then $\{FG : G \in \mathcal{G}\}$ is a proper generating set of finitely presentable flatly codisjunctable objects in $A/\mathbf{A}$. Because all the axioms of Zariski categories are described in terms of limits, coequalizers, pushouts, and filtered colimits, it follows easily that $A/\mathbf{A}$ is a Zariski categories and that U is a morphism of Zariski categories. ■

Example. For any ring $A \in \mathbf{CRng}$, the category $A/\mathbf{CRng}$ is isomorphic to the category $\mathbf{CAlg}(A)$ of commutative algebras on A and the projection functor $A/\mathbf{CRng} \to \mathbf{CRng}$ is isomorphic to the forgetful functor $\mathbf{CAlg}(A) \to \mathbf{CRng}$. Therefore this forgetful functor is a morphism of Zariski categories. This result extends naturally to algebras equipped with a given extra structure.

Corollary (12.1.2). *If* **A** *is a Zariski category and* $f: A \to B$ *is a morphism in* A, *the functor* $f^*: B/A \to A/A$ *defined by* $f^*(C, g) = (C, gf)$ *and* $f^*(h) = h$ *is a morphism of Zariski categories.*

Proof. The category B/A is isomorphic to the category $(B, f)/(A/\mathbf{A})$, and up to this isomorphism the functor f^* is the projection functor: $(B, f)/(A/\mathbf{A}) \to A/\mathbf{A}$. ■

Corollary (12.1.3). *If A is a Zariski category and X is a boolean topological space, the category $Sh[X, A]$ of sheaves on X with values in A is a Zariski category.*

Proof. Let B denote the boolean algebra of clopen sets of X. As the category **A** is locally indecomposable (Theorem (3.10.1)), the direct factor functor $\Delta_{\mathbf{A}}: \mathbf{A} \to \mathbf{Bool}$ is defined ([10], § 1.3.5). Let us prove that the category $\mathbf{Sh}[X, \mathbf{A}]$ is equivalent to the comma category $(B, \Delta_{\mathbf{A}})$. Let us define the functor $G: \mathbf{Sh}[X, \mathbf{A}] \to (B, \Delta_{\mathbf{A}})$ on an object F by $G(F) = (F(1), \gamma_F)$, where $\gamma_F: B \to \Delta(F(1))$ is the map which assigns to $b \in B$ the direct factor $F(b \to 1): F(1) \to F(b)$ of $F(1)$. The map γ_F is indeed a homomorphism of boolean algebras because it is order preserving and it preserves the zero element, the unit element, and finite disjoint unions. For a morpism $\alpha: F \to H$, the morphism $G(\alpha): G(F) \to G(H)$ is defined by $G(\alpha) = \alpha_1: (F(1), \gamma_F) \to (H(1), \gamma_H)$. This is justified by the fact that the relation $\alpha_1 \gamma_F = \gamma_H$ holds. On the other hand, let us define the functor $C: (B, \Delta_{\mathbf{A}}) \to \mathbf{Sh}[X, \mathbf{A}]$ as follows. For any object (A, γ) in $(B, \Delta_{\mathbf{A}})$, let $C(A, \gamma): B^{\mathrm{op}} \to \mathbf{A}$ be the functor defined by $C(A, \gamma)(b) = \mathrm{Codom}(\gamma(b))$. It satisfies $C(A, \gamma)(\mathrm{O}) = \mathrm{Codom}(\gamma(\mathrm{O})) = 1$ and, for any pair of elements $b_1, b_2 \in B$ such that $b_1 \wedge b_2 = 0$,

$$\begin{aligned} C(A, \gamma)(b_1 \vee b_2) &= \mathrm{Codom}(\gamma(b_1 \vee b_2)) \\ &= \mathrm{Codom}(\gamma(b_1) \vee \gamma(b_2)) \simeq \mathrm{Codom}(\gamma(b_1)) \\ &\times \mathrm{Codom}(\gamma(b_2)) = C(A, \gamma)(b_1) \times C(A, \gamma)(b_2) \end{aligned}$$

It follows that $C(A, \gamma)$ is a sheaf on X with values in **A** (cf. [9], Propositions 14.0 and 14.1). For a morphism $f: (A, \gamma) \to (E, \varphi)$ in $(B, \Delta_{\mathbf{A}})$, let $C(f)$:

$C(A,\gamma) \to C(E,\varphi)$ be the morphism whose value at $b \in B$ is the unique morphism $C(f)_b$ which satisfies the relation $C(f)_b\gamma(b) = \varphi(b)f$. In this way we obtain a functor $C: (B,\Delta_{\mathbf{A}}) \to \mathbf{Sh}[X,\mathbf{A}]$ which is easily shown to be quasi-inverse to the functor $G: \mathbf{Sh}[X,\mathbf{A}] \to (B,\Delta_{\mathbf{A}})$. As a result, the categories $\mathbf{Sh}[X,\mathbf{A}]$ and $(B,\Delta_{\mathbf{A}})$ are equivalent. On the other hand, the functor $\Delta_{\mathbf{A}}: \mathbf{A} \to \mathbf{Bool}$ has a left-adjoint $K_{\mathbf{A}}: \mathbf{Bool} \to \mathbf{A}$ ([10], Proposition 1.8.2). The adjunction isomorphism associated to it, $\varphi: \mathrm{Hom}_{\mathbf{A}}(K_{\mathbf{A}}(.),—) \xrightarrow{\sim} \mathrm{Hom}_{\mathbf{Bool}}(.,\Delta_{\mathbf{A}}(—))$, provides an isomorphism of comma categories $\varphi_B: (K_{\mathbf{A}}(B),\mathbf{A}) \xrightarrow{\sim} (B,\Delta_{\mathbf{A}}(—))$. Consequently, the category $\mathbf{Sh}[X,\mathbf{A}]$ is equivalent to the category $(K_{\mathbf{A}}(B),\mathbf{A}) = K_{\mathbf{A}}(B)/\mathbf{A}$. According to Theorem (12.1.1), $K_{\mathbf{A}}(B)/\mathbf{A}$ is a Zariski category. Therefore $\mathbf{Sh}[X,\mathbf{A}]$ is a Zariski category. ■

12.2 FUNCTOR ZARISKI CATEGORIES

Proposition (12.2.1). *Finite products of Zariski categories are Zariski categories.*

Proof. The singleton category **1** is a Zariski category. Let **A**, **B** be a pair of Zariski categories. The limits and colimits in $\mathbf{A} \times \mathbf{B}$ are computed argument by argument. It follows that the category $\mathbf{A} \times \mathbf{B}$ satisfies Axioms (1.2.1), (1.2.3), and (1.2.5) of Zariski categories. Let A be a finitely presentable object in **A** and let B be a finitely presentable object in **B**. For any filtered colimit $\varinjlim_{i\in\mathbf{I}}(A_i,B_i)$ in $\mathbf{A} \times \mathbf{B}$,

$$\begin{aligned}
&\mathrm{Hom}_{\mathbf{A}\times\mathbf{B}}((A,B), \varinjlim_{i\in\mathbf{I}}(A_i,B_i)) \\
&= \mathrm{Hom}_{\mathbf{A}\times\mathbf{B}}((A,B), (\varinjlim_{i\in\mathbf{I}}A_i, \varinjlim_{i\in\mathbf{I}}B_i)) \\
&\simeq \mathrm{Hom}_{\mathbf{A}}(A, \varinjlim_{i\in\mathbf{I}}A_i) \times \mathrm{Hom}_{\mathbf{B}}(B, \varinjlim_{i\in\mathbf{I}}B_i) \\
&\simeq (\varinjlim_{i\in\mathbf{I}}\mathrm{Hom}_{\mathbf{A}}(A,A_i)) \times (\varinjlim_{i\in\mathbf{I}}\mathrm{Hom}_{\mathbf{B}}(B,B_i)) \\
&\simeq \varinjlim_{i\in\mathbf{I}}(\mathrm{Hom}_{\mathbf{A}}(A,A_i) \times \mathrm{Hom}_{\mathbf{B}}(B,B_i)) \\
&\simeq \varinjlim_{i\in\mathbf{I}}\mathrm{Hom}_{\mathbf{A}\times\mathbf{B}}((A,B),(A_i,B_i)).
\end{aligned}$$

It follows that the object (A,B) is finitely presentable in $\mathbf{A} \times \mathbf{B}$. Consequently, the terminal object $(1,1)$ in $\mathbf{A} \times \mathbf{B}$ is finitely presentable, so that Axiom (1.2.4) is valid, and a proper generating set of finitely presentable objects in $\mathbf{A} \times \mathbf{B}$ is obtained by taking the set of objects of the form (A,B) where A and B run over the proper generating sets of objects in **A** and **B** respectively mentioned in Axiom (1.2.2) of Zariski categories. Axioms (1.2.2) and (1.2.6) are valid in $\mathbf{A} \times \mathbf{B}$ because the notions of codisjointness, codisjunctors, flatness, and congruences are computed argument by argument. As a result, $\mathbf{A} \times \mathbf{B}$ is a Zariski category. ■

Theorem (12.2.2). *If A is a Zariski category and $\mathbf{C}$ is a small category with a strict initial object, then the category $A^{\mathbf{C}}$ of functors from $\mathbf{C}$ to $\mathbf{A}$ is a Zariski category.*

Proof.

(i) The category $\mathbf{A}^{\mathbf{C}}$ is complete and cocomplete, where the limits and colimits are computed argument by argument, the terminal object is the constant functor with value 1 denoted by T, and the regular epimorphisms are the pointwise regular epimorphisms. It follows that Axioms (1.2.1), (1.2.3), and (1.2.5) are satisfied and that T has no proper subobject. Let O denote the initial object in $\mathbf{C}$. Let $(\alpha_k : F_k \to F)_{k \in \mathbf{K}}$ be a filtered colimit in $\mathbf{A}^{\mathbf{C}}$ and let $\beta : T \to F$ be a morphism. Then $(\alpha_{k\mathrm{o}} : F_k(O) \to F(O))_{k \in \mathbf{K}}$ is a filtered colimit in $\mathbf{A}$. Because the terminal object $T(O) = 1$ is finitely presentable in $\mathbf{A}$, there exist $k \in \mathbf{K}$ and $\gamma : T(O) \to F_k(O)$ such that $\alpha_{k0}\gamma = \beta_0$. Then $F_k(O) = 1$. Hence $F_k = T = F$. Thus β factors through α_k and T is a finitely presentable object in $\mathbf{A}^{\mathbf{C}}$. Therefore Axiom (1.2.4) is satisfied.

(ii) Let $\mathcal{G}$ be a proper generating set in $\mathbf{A}$ whose objects are finitely presentable and flatly codisjunctable. For any pair (U, G) of an object U in $\mathbf{C}$ and an object $G \in \mathcal{G}$, let us define the generalized representable functor [37] $H_{U,G} : \mathbf{C} \to \mathbf{A}$ by $H_{U,G}(V) = \coprod_{\mathrm{Hom}_{\mathbf{C}}(U,V)} G$ for $V \in \mathbf{C}$, and for $y : V \to W$ in $\mathbf{C}$ by $H_{U,G}(y) : H_{U,G}(V) \to H_{U,G}(W)$ defined by $H_{U,G}(y)\iota_x = \iota_{yx}$ where ι_x, ι_{yx} are the canonical inductions into the coproducts. According to [37], for any F in $\mathbf{A}^{\mathbf{C}}$, we obtain a one-to-one correspondence $\mathrm{Hom}_{\mathbf{A}^{\mathbf{C}}}(H_{U,G}, F) \to \mathrm{Hom}_{\mathbf{A}}(G, F(U))$ by assigning to $\alpha : H_{U,G} \to F$ the morphism $\alpha_U \iota_{1_U} : G \to F(U)$, and this correspondence is natural with respect to F in $\mathbf{A}^{\mathbf{C}}$. Let us prove that the set of functors $\{H_{U,G}\}$, where U runs over $\mathrm{Obj}(\mathbf{C})$ and G over $\mathcal{G}$, is an adequate proper generating set in $\mathbf{A}^{\mathbf{C}}$. It is a proper generating set, for, if $\alpha : F \to H$ is a morphism in $\mathbf{A}^{\mathbf{C}}$ such that the map $\mathrm{Hom}_{\mathbf{A}^{\mathbf{C}}}(H_{U,G}, \alpha)$ is bijective for any $(U, G) \in \mathrm{Obj}(\mathbf{C}) \times \mathcal{G}$, then the maps $\mathrm{Hom}_{\mathbf{A}}(G, \alpha_U)$ are bijective. Thus the morphisms α_U are isomorphisms so that α is an isomorphism. The objects $H_{U,G}$ are finitely presentable, for if $F = \lim_{k \in \mathbf{K}} F_k$ is a filtered colimit in $\mathbf{A}^{\mathbf{C}}$, then

$$\mathrm{Hom}_{\mathbf{A}^{\mathbf{C}}}(H_{U,G}, F) \simeq \mathrm{Hom}_{\mathbf{A}}(G, F(U)) \simeq \mathrm{Hom}_{\mathbf{A}}(G, \varinjlim_{k \in \mathbf{K}} F_k(U))$$
$$\simeq \varinjlim_{k \in \mathbf{K}} \mathrm{Hom}_{\mathbf{A}}(G, F_k(U)) \simeq \varinjlim_{k \in \mathbf{K}} \mathrm{Hom}_{\mathbf{A}^{\mathbf{C}}}(H_{U,G}, F_k).$$

(iii) Let us compute the codisjunctors in $\mathbf{A}^{\mathbf{C}}$. Let us prove that a pair of morphisms $(\alpha, \beta) : H \rightrightarrows F$ is codisjunctable in $\mathbf{A}^{\mathbf{C}}$ if and only if the pair $(\alpha_0, \beta_0) : H(O) \rightrightarrows F(O)$ is codisjunctable in $\mathbf{A}$ and that the codisjunctor $\delta : F \to D$ of (α, β) is such that $\delta_0 : F(O) \to D(O)$ is the codisjunctor of (α_0, β_0) and, for any $V \in \mathbf{C}$ and $x : O \to V$, $(\delta_V : F(V) \to D(V), D(x) : D(O) \to D(V)$ is the pushout of $(F(x), \delta_0)$. Let us assume that (α, β) is codisjunctable with codisjunctor $\delta : F \to D$. Then the coequalizer of $(\delta\alpha, \delta\beta)$ is T, and hence the coequalizer of $(\delta_0\alpha_0, \delta_0\beta_0)$ is 1. Thus δ_0 codisjoints

(α_0, β_0). Let $f: F(O) \to A$ be a morphism which codisjoints (α_0, β_0). There is a pair (P, φ) of a functor $P: \mathbf{C} \to \mathbf{A}$ and a natural transformation $\varphi: F \to P$ such that, for any object V in $\mathbf{C}$ with the unique morphism $x: O \to V$, $(P(x): A \to P(V), \varphi_V: F(V) \to P(V))$ is the pushout of $(f, F(x))$. The coequalizer of $(\varphi\alpha, \varphi\beta)$ is T because the coequalizer of $(\varphi_0\alpha_0, \varphi_0\beta_0) = (f\alpha_0, f\beta_0)$ is 1. Thence φ codisjoints (α, β). Therefore φ factors uniquely through δ, and hence $\varphi_0 = f$ factors through δ_0. Moreover, δ_0 is epimorphic because δ is. Thus δ_0 is the codisjunctor of (α_0, β_0). Conversely, let us assume that (α_0, β_0) is codisjunctable with codisjunctor $d: F(O) \to D_0$. There is a unique pair (D, δ) of a functor $D: \mathbf{C} \to \mathbf{A}$ and a natural transformation $\delta: F \to D$ such that, for any object V in $\mathbf{C}$ with the unique morphism $x: O \to V$, $(D(x): D_0 \to D(V), \delta_V: F(V) \to D(V))$ is the pushout of $(d, F(x))$. The morphism δ codisjoints (α, β) because $\delta_0 = d$ codisjoints (α_0, β_0). Moreover, any morphism $\varphi: F \to P$ which codisjoints (α, β) is such that φ_0 codisjoints (α_0, β_0). Therefore φ_0 factors uniquely through δ_0, and, by the universal property of pushouts, any morphism φ_V factors uniquely through δ_V in a morphism $\psi_V: D(V) \to P(V)$ which defines a natural transformation $\psi: D \to P$. It follows that δ is the codisjunctor of (α, β).

(iv) It follows immediately from (iii) that the objects $H_{U,G}$ with $(U, G) \in \mathrm{Obj}(\mathbf{C}) \times \mathcal{G}$ are codisjunctable in $\mathbf{A}^{\mathbf{C}}$, because $H_{U,G}(O) = G$ for $U \simeq O$ and $H_{U,G}(O) = Z$ for $U \not\simeq O$ are codisjunctable objects in $\mathbf{A}$, and thus any pair $(\alpha_0, \beta_0): H_{U,G}(O) \rightrightarrows F(O)$ is codisjunctable in $\mathbf{A}$ so that any pair $(\alpha, \beta): H_{U,G} \rightrightarrows F$ is codisjunctable in $\mathbf{A}^{\mathbf{C}}$. Moreover, these objects are flatly codisjunctable because the value at O of the associated codisjunctors are flat morphisms, and flat morphisms are co-universal. Thus Axiom (1.2.2) is satisfied.

(v) Let r, s be a pair of codisjunctable congruences on an object F in $\mathbf{A}^{\mathbf{C}}$ such that $r \overset{c}{\vee} s = 1_{F \times F}$, with codisjunctors δ and φ respectively. Then r_0, s_0 is a pair of codisjunctable congruences on $F(O)$ such that $r_0 \overset{c}{\vee} s_0 = 1_{F(O) \times F(O)}$ with codisjunctors δ_0 and φ_0 respectively. Therefore we have a co-universal co-union $\delta_0 \vee \varphi_0 = 1_{F(O)}$, which implies $\delta_V \vee \varphi_V = 1_{F(V)}$ by pushing out along the morphism $F(x): F(O) \to F(V)$. As a result, $\delta \vee \varphi = 1_F$. This result extends immediately to a finite sequence of codisjunctable congruences. Thus Axiom (1.2.6) is satisfied. ■

Corollary (12.2.3). *If A is a Zariski category, the category $A^{\to}$ of morphisms of A is a Zariski category.*

Proof. The category $\mathbf{A}^{\to}$ is the category $\mathbf{A}^{\mathbf{C}}$, where $\mathbf{C}$ is the category with two objects 0, 1 and one non-identity morphism $0 \to 1$. ■

Example. The category $\mathbf{CRng}^{\to}$ is isomorphic to the category $\mathbf{CAlg}$ of commutative algebras on variable rings. Since $\mathbf{CRng}$ is a Zariski category, it follows that **CAlg** is a Zariski category. This result extends naturally to rings which have a given extra structure.

12.3 ZARISKI FULL SUBCATEGORIES

Proposition (12.3.1). *If A is a Zariski category, any full reflective subcategory of A closed under singular quotients and filtered colimits, such that the inclusion functor preserves regular epimorphisms and the associated adjunction preserves codisjointed pairs of morphisms, is a Zariski subcategory of A.*

Proof. Let **B** be such a subcategory, let $U : \mathbf{B} \to \mathbf{A}$ be the inclusion functor, let $R : \mathbf{A} \to \mathbf{B}$ be the reflector, and let $\eta : 1_{\mathbf{A}} \to UR$ be the unit of reflection. We assume that $RU = 1_{\mathbf{B}}$.

(i) The category **B** is complete and cocomplete ([35], Theorem 16.6.1).

(ii) First let us show that U reflects regular epimorphisms. Let $f : B \to C$ be a morphism in **B** such that $U(f)$ is a regular epimorphism in **A**. The kernel pair (r_1, r_2) of f in **B** is also the kernel pair of $U(f)$. Then $U(f)$ is the coequalizer of $(U(r_1), U(r_2))$ and consequently f is the coequalizer of (r_1, r_2) in **B**. Thus f is a regular epimorphism in **B**. Let us prove now that regular epimorphisms are universal in **B**. Let $q : A \to Q$ be a regular epimorphism in **B**, let $h : P \to Q$ be some morphism in **B**, and let $(g : B \to A$, $p : B \to P)$ be the pullback of (q, h). Then $(U(g), U(p))$ is the pullback of $(U(q), U(h))$. Because U preserves regular epimorphisms, $U(q)$ is a regular epimorphism and thus $U(p)$ is a regular epimorphism. As U reflects regular epimorphisms, p is a regular epimorphism. As a result, **B** satisfies Axiom (1.2.3).

(iii) The existence of the morphism $\eta_1 : 1 \to UR(1)$ implies $UR(1) \simeq 1 \simeq U(1)$. Then $R(1) \simeq 1$. Because the object 1 is finitely presentable in **A** and the functor U preserves filtered colimits, the object $1 \simeq R(1)$ is finitely presentable in **B**. Moreover, 1 has no proper subobject in **B** for its has none in **A**. Therefore Axiom (1.2.4) is satisfied.

(iv) Let $(p : A \to P, q : A \to Q)$ be a product in **B** and let $f : A \to B$ be some morphism in **B**. Then $(U(p), U(q))$ is a product in **A**, and so $U(p)$, $U(q)$ are singular epimorphisms (Propositiön (1.8.4)). Let $(p_1 : U(B) \to P_1$, $v : U(P) \to P_1)$ be the pushout of $(U(f), U(p))$, and let $(q_1 : U(B) \to Q_1, w : U(Q) \to Q_1)$ be the pushout of $(U(f), U(q))$. Then p_1, q_1 are singular epimorphisms (Proposition (1.8.4)) in **A**. Because **B** is closed in **A** under singular quotients, the morphisms p_1, q_1 belong to **B**. Then v, w also belong to **B**, and (p_1, v) is the pushout of (f, p), and (q_1, w) is the pushout of (f, q). Because finite products are co-universal in **A**, (p_1, q_1) is a product in **A** and thence it is a product in **B**. As a result, products of pairs of objects are co-universal in **B**. Therefore Axiom (1.2.5) is satisfied.

(v) First let us prove that the adjunction φ associated to U reflects codisjointed pairs. Let us suppose that $(g, h) : R(A) \rightrightarrows B$ is a pair of morphisms in **B** such that $(\varphi_{A,B}(g), \varphi_{A,B}(h))$ is codisjointed. Let $p : B \to P$ be the coequalizer of (g, h) in **B**. Then $U(p)\varphi_{A,B}(g) = \varphi_{A,P}(pg) = \varphi_{A,P}(ph) = U(p)\varphi_{A,B}(h)$. Therefore $U(P) \simeq 1$; hence $P \simeq 1$ and (g, h) is codisjointed. Now let us consider a proper generating set $\mathcal{G}$ in **A** whose objects are finitely presentable and flatly codisjunctable, and let us prove that the set $\{R(G) : G \in \mathcal{G}\}$ is an adequate generating set in **B**. It is well known and easy to check that it is a proper generating set of finitely presentable objects in **B**. Let $(g, h) : R(G) \rightrightarrows A$, where $G \in \mathcal{G}$, be a pair of morphisms in **B**. Let $d : U(A) \to D$ be the codisjunctor of $(\varphi_{G,A}(g), \varphi_{G,A}(h)) : G \rightrightarrows U(A)$. Because **B** is closed in **A** under singular quotients, the morphism d belongs to **B**. The pair (dg, dh) is codisjointed because the pair $(\varphi_{G,D}(dg), \varphi_{G,D}(dh)) = (U(d)\varphi_{G,A}(g), U(d)\varphi_{G,A}(h))$ is codisjointed. Thus d codisjoints (g, h). Moreover, if $f : A \to B$ is a morphism in **B** which codisjoints (g, h), the pair $(U(f)\varphi_{G,A}(g), U(f)\varphi_{G,A}(h)) = (\varphi_{G,B}(fg), \varphi_{G,B}(fh))$ is codisjointed. Thus $U(f)$ factors through $U(d)$, i.e. f factors through d. As a result, d is the codisjunctor of (g, h). Therefore the objects of the form $R(G)$ with $G \in \mathcal{G}$ are codisjunctable in **B**. Let $m : A \to M$ be a monomorphism in **B**. Then $U(m)$ is a monomorphism in **A**. Let $(e : U(M) \to N, n : U(D) \to N)$ be the pushout of $(U(m), U(d))$ in **A**. Because G is flatly codisjunctable, the codisjunctor $U(d)$ is flat. Thus n is monomorphic. Moreover, e is a singular epimorphism in **A**. As **B** is closed under singular quotients, the morphism e belongs to **B**. Then n belongs to **B**, (e, n) is the pushout of (m, d) in **B**, and n is monomorphic in **B**. As a result, the objects $R(G)$ with $G \in \mathcal{G}$ are flatly codisjunctable in **B**.

(vi) Let A be an object in **B** and let $(s = (s_1, s_2) : S \rightrightarrows A, t = (t_1, t_2) : T \rightrightarrows A)$ be a pair of codisjunctable congruences on A such that $s \overset{c}{\vee} t = 1_{A \times A}$ in $\mathrm{Cong}(A)$. Because the set $\{R(G) : G \in \mathcal{G}\}$ is a proper generating set in **B**, there exist epimorphic families of morphisms $(f_i : R(G_i) \to S)_{i \in I}$ and $(f_j : R(G_j) \to T)_{j \in J}$ where $G_i, G_j \in \mathcal{G}$ and $I \cap J = \phi$. Let $K = I \cup J$ and, for any $k \in K$, let $(g_k, h_k) : R(G_k) \rightrightarrows A$ be the pair $(s_1 f_k, s_2 f_k)$ or $(t_1 f_k, t_2 f_k)$. Let $q_k : A \to Q_k$ be the coequalizer of (g_k, h_k) for $k \in K$. Then $\wedge_{k \in K} q_k = O_A$ in $\mathrm{QuotReg}(A)$. Because $\mathrm{QuotReg}(A)$ is a cocompact lattice there exists a finite subset K_0 of K such that $\wedge_{k \in K_0} q_k = O_A$. Let $p_k : U(A) \to P_k$ be the coequalizer of the pair $(\varphi_{G_k, A}(g_k), \varphi_{G_k, A}(h_k)) : G_k \rightrightarrows U(A)$ for $k \in K_0$, and let $p = \wedge_{k \in K_0} p_k$ in $\mathrm{QuotReg}(U(A))$. Then $R(p_k)$ is the coequalizer of $(R(\varphi_{G_k, A}(g_k)), R(\varphi_{G_k, A}(g_k))) = (g_k, h_k)$, i.e. $R(p_k) = q_k$. Thus $R(p) = \wedge_{k \in K_0} R(p_k) = \wedge_{k \in K_0} q_k = O_A$. The regular epimorphism $p : U(A) \to P$ is the coequalizer of some pair of morphisms $(g, h) : B \rightrightarrows UA$ in **A**. Then $R(p)$ is the coequalizer of the pair $(R(g), R(h)) = (\varphi^{-1}_{B,A}(g), \varphi^{-1}_{B,A}(h))$. Therefore this pair is codisjointed; hence (g, h) is codisjointed and $p = O_{U(A)}$, i.e. $\wedge_{k \in K_0} p_k = O_{U(A)}$. Let $d_k : U(A) \to D_k$ be the codisjunctor of $(\varphi_{G_k,A}(g_k), \varphi_{G_k,A}(h_k))$. Because **A** is a Zariski category, $(d_k : U(A) \to D_k)_{k \in K}$ is a co-universal effective monomorphic family of morphisms in **A**. Because **B** is closed in **A** under singular quotients, this family is also a co-universal effective monomorphic family in **B**. According to (v), $(d_k : A \to D_k)_{k \in K_0}$ is the family of codisjunctors of the family of pairs $((g_k, h_k) : R(G_k) \rightrightarrows A)_{k \in K_0}$. Let $\delta_1 : A \to \Delta_1$ and $\delta_2 : A \to \Delta_2$ be the codisjunctors of s and t respectively. Then any morphism $d_k : A \to D_k$ for $k \in K_0$ factors through δ_1 or δ_2. It follows that the pair $(\delta_1 : A \to \Delta_1, \delta_2 : A \to \Delta_2)$ is a co-universal effective monomorphic pair and thus it is a co-union, i.e. $\delta_1 \vee \delta_2 = 1_A$. As a result, the category **B** satisfies Axiom (1.2.6) of Zariski categories for a pair of codisjunctable congruences. Indeed, the proof holds for any finite sequence of codisjunctable congruences. ■

Theorem (12.3.2) *If A is a Zariski category, the category **RedA** of reduced objects of A is a Zariski subcategory of A.*

Proof. According to Theorem (2.3.2), **RedA** is a full reflective subcategory of **A**. By Proposition (2.3.3), **RedA** is closed in **A** under singular quotients and filtered colimits. Because the reflector $R : \mathbf{A} \to \mathbf{RedA}$ is such that the unit morphisms $A \to R(A) = A/\mathrm{rad}(A)$ are regular epimorphisms in **A**, the inclusion functor **RedA** → **A** preserves regular epimorphisms. Let A be an object in **A** such that $R(A) = 1$. Then $A/\mathrm{rad}(A) = 1$,and thus $\mathrm{rad}(\Delta_A) = \mathrm{rad}(A) = 1_{A \times A}$. According to Proposition (2.4.13), $\Delta_A = 1_{A \times A}$ and hence $1 = A$. It follows that the reflector R reflects the terminal object. Let $(g, h) : \mathbf{R}(A) \rightrightarrows B$ be a codisjointed pair of morphisms in **RedA**. Let $q : B \to Q$ be the coequalizer of (g, h) in **A**. Then $R(q)$ is the coequalizer of (g, h) in **RedA**. Thus $R(q) = O_B$ and hence $q = O_B$, i.e. the pair (g, h) is codisjointed in **A**. As a result, the associated adjunction preserves codisjointed pairs. According to Proposition (12.3.1), **RedA** is a Zariski subcategory of **A**. ■

Theorem (12.3.3). *If A is a Zariski category, the category **RegA** of regular objects of A is a Zariski subcategory of A. It is the co-universal locally simple category associated to A.*

Proof. According to Theorem (2.11.6) **RegA** is a full reflective subcategory of **A**. According to Proposition (2.11.5), **RegA** is closed in **A** under filtered colimits, regular quotients, and flat quotients, and therefore it is closed under singular quotients (Proposition (1.8.3)). As the inclusion functor **RegA** → **A** lifts coequalizers, it preserves them and consequently it preserves regular epimorphisms. Let A be a non-terminal object in **A**. According to Corollary (2.1.9), A has a simple quotient $k : A \to K$. If $f : A \to B$ is the universal regular object asociated to A, then k factors through f. Consequently, B is not terminal. It follows that the reflector **RegA** → **A** reflects the terminal object. It follows that the associated adjunction preserves codisjointed pairs (cf. proof of Theorem (12.3.2)). According to Proposition (12.3.1), **RegA** is a Zariski subcategory of **A** . Since any object in **RegA** is regular, the category **RegA** is indeed locally simple ([10], Proposition 2.1.2). It is the co-universal locally simple category associated to **A**, since any morphism of Zariski categories $U: \mathbf{B} \to \mathbf{A}$ with a locally simple domain **B** preserves regular objects (Proposition (7.3.2)) and thus induces a morphism of locally simple categories $\mathbf{B} \to$ **RegA.** ■

REFERENCES

1 Alonso, M. E. and Roy, M. F. Real strict localizations, *Math. Z.*, **194** (1987), 429–41.

2 Barr, M., Exact categories and categories of sheaves. Lecture Notes in Mathematics Vol. 236, 1–120. Springer-Verlag, Berlin, 1971.

3 Bourbaki, N., *Algèbre commutative*. Hermann, Paris, 1964.

4 Brumfiel, G. W., *Partially ordered rings and semi algebraic geometry*. London Mathematical Society Lecture Note Series Vol. 37. Cambridge University Press, Cambridge, 1979.

5 Coste, M., *Localisation dans les catégories de modèles*. Thesis, Université de Paris Nord, 1977.

6 Coste, M., Localisation, spectra and sheaf representation. Applications of sheaves. Lecture Notes in Mathematics Vol. 753, 212–38, Springer-Verlag, Berlin, 1979.

7 Diers, Y., Familles universelles de morphismes, *Ann. Soc. Sci. Bruxelles*, **93**(III) (1979), 175–95.

8 Diers, Y., Sur les familles monomorphiques régulières de morphismes, *Cah. Topol. Géom. Différ.*, **21**(4) (1980), 411–25.

9 Diers, Y., Une description axiomatique des catégories de faisceaux de structures algébriques sur les espaces topologiques booléens, *Adv. Math.*, **47** (1983), 258–99.

10 Diers, Y., *Categories of Boolean sheaves of simple algebras*. Lecture Notes in Mathematics Vol. 1187, Springer-Verlag, Berlin, 1986.

11 Diers, Y., Codisjoncteurs dans les catégories d'algèbres commutatives. *Cah. Topol. Géom. Différ.*, **28**(1) (1987), 5–28.

12 Diers, Y., Codisjunctors and singular epimorphisms in the category of commutative rings, *J. Pure. Appl. Alg.*, **53** (1988), 39–57.

13 Diers, Y., Locally Hilbert categories. *Categorical algebra and its applications*. Lecture Notes in Mathematics Vol. 1348, 87–101. Springer-Verlag, Berlin, 1988.

14 Fakir, S., *Objets algébriquement clos et injectifs dans les catégories localement présentables, Bull. Soc. Math. Fr. Mém.*, **42** (1975).

15 Freyd, P. J., and Kelly, G. M., Categories of continuous functors 1, *J. Pure. Appl. Alg.*, **2** (1972), 169–91.

16 Gabriel, P. and Ulmer, F., *Lokal präsentierbare Kategorien*. Lecture Notes in Mathematics Vol. 221. Springer-Verlag, Berlin, 1971.

17 Grätzer, G., *Lattice theory*, W. H Freeman, San Francisco, CA, 1971.

18 Gray, J. W., Fibred and cofibred categories. Proceedings of the Conference on Categorical Algebra, La Jolla, 1965, 21–83. Springer-Verlag, Berlin, 1966.

19 Grillet, P. A., Regular categories. Lecture Notes in Mathematics Vol. 236, 121–222. Springer-Verlag, Berlin, 1971.

20 Grothendieck, A., Séminaire de géométrie algébrique, *Pub. Math. I.H.E.S.*, **32**(IX) (1965).

21 Grothendieck, A., Artin, M., and Verdier, J. L., *Théorie des topos et cohomologie étale des schémas*. Lecture Notes in Mathematics Vol. 269. Springer-Verlag, Berlin, 1972.

22 Grothendieck, A., and Dieudonné, J. A., *Eléments de géométrie algébrique I*, Springer-Verlag, Berlin, 1971.

23 Hakim, M., *Topos annelés et schémas relatifs*, Springer-Verlag, Berlin, 1972.

24 Hartshorne, R., *Algebraic geometry*. Springer-Verlag, Berlin, 1977.

25 Hochster, M., Prime ideal structure in commutative rings, *Trans. Am. Math. Soc.*, **142** (1969), 43–60.

26 Johnstone, P., *Topos Theory*, Academic Press, New York, 1977.

27 Keigher, W. K., Prime differential ideals in differential rings. *Contributions to algebra. A collection of papers dedicated to Ellis Kolchin*. Academic Press, New York, 1977.

28 Keimel, K., The representation of lattice ordered groups and rings by sections in sheaves. *Lectures on the applications of sheaves to ring theory*. Lecture Notes in Mathematics Vol. 248, 1–98. Springer-Verlag, Berlin, 1971.

29 Kock, A., *Synthetic differential geometry*. London Mathematical Society Lecture Notes Series Vol. 51. Cambridge University Press, Cambridge, 1981.

30 Lazard, D., Epimorphismes plats d'anneaux, *C.R. Acad. Sci. Paris*, **266** (1968), 314–16.

31 Mac Lane, S., *Categories for the working mathematician*. Springer-Verlag, Berlin, 1971.

32 Moerdijk, I. and Reyes, G. E., Rings of smooth functions and their localizations I, J. Alg., **99** (1986), 324–36.

33 von Neumann, J., Regular rings, *Proc. Natl Acad. Sci. U.S.A.*, **22** (1936), 707–13.

34 Raynaud, M., *Anneaux locaux henséliens*. Lecture Notes in Mathematics Vol. 169. Springer-Verlag, Berlin, 1970.

35 Schubert, H., *Categories*. Springer-Verlag, Berlin, 1972.

36 Stenström, B., *Rings of Quotients*. Springer-Verlag, Berlin, 1975.

37 Ulmer, F., Representable functors with values in arbitrary categories, *J. Alg.*, **8** (1967), 96–129.

38 Van Osdol, D. H., Sheaves in Regular Categories. Lecture Notes in Mathematics Vol. 236, 223–39, Springer-Verlag, Berlin, 1971.

INDEX OF SYMBOLS

INDEX OF TERMS